Biomechanics of Sport and Exercise

Peter McGinnis, PhD
State University of New York, Cortland

Human Kinetics

Library of Congress Cataloging-in-Publication Data

McGinnis, Peter Merton, 1954-
    Biomechanics of sport and exercise / Peter M. McGinnis.
       p.   cm.
    Includes bibliographical references and index.
    ISBN 0-87322-955-X
    1. Human mechanics.   I. Title.
    QP303.M36   1999
    612.7'6--DC20

ISBN: 0-87322-955-X   96-45731
                                     CIP

Acquisitions Editors: Rick Frey and Loarn Robertson; **Developmental Editors**: Julia Anderson and Kent Reel; **Assistant Editors**: Jacqueline Eaton Blakley and Laurie Stokoe; **Editorial Assistant**: Coree Schutter; **Copyeditor**: Barbara Field; **Proofreader**: Jim Burns; **Indexer**: Nancy Ball; **Graphic Designer**: Judy Henderson; **Graphic Artists**: Francine Hamerski and Sandra Meier; **Photo Editor**: Boyd LaFoon; **Cover Designer**: Keith Blomberg; **Photographer (cover)**: Will Zehr; **Line Drawings Illustrator**: Paul To; **Mac Illustrators**: Joe Bellis and Denise Lowry; **Medical Illustrator**: Kristin Mount; **Cartoons**: Tom Roberts; **Printer**: Edwards Brothers, Inc.

Printed in the United States of America

10 9 8 7 6 5 4 3 2 1

Human Kinetics
Web site: http://www.humankinetics.com/

United States: Human Kinetics
P.O. Box 5076, Champaign, IL 61825-5076
1-800-747-4457
e-mail: humank@hkusa.com

Canada: Human Kinetics
475 Devonshire Road Unit 100, Windsor, ON N8Y 2L5
1-800-465-7301 (in Canada only)
e-mail: humank@hkcanada.com

Europe: Human Kinetics
P.O. Box IW14, Leeds LS16 6TR, United Kingdom
+44 (0)113-278 1708
e-mail: humank@hkeurope.com

Australia: Human Kinetics
57A Price Avenue, Lower Mitcham, South Australia 5062
(08) 82771555
e-mail: humank@hkaustralia.com

New Zealand: Human Kinetics
P.O. Box 105-231, Auckland Central
09-523-3462
e-mail: humank@hknewz.com

Contents

Preface   vii

Acknowledgments   ix

Part I   Fundamental Biomechanical Concepts, Principles, and Terminology   1

Chapter 1   Why Study Biomechanics?   3

What Is Biomechanics?   5

What Are the Goals of Exercise and Sport Biomechanics?   6

The History of Sport Biomechanics   12

The Future of Sport Biomechanics   14

Summary   14

Chapter 2   Anatomical Movement Terminology   17

Body Segment, Bone, Joint, and Muscle Terminology   18

Spatial and Directional Terminology   24

Planes and Axes of Motion   25

Joint Actions   29

Summary   33

Chapter 3   Fundamental Concepts and Principles of Mechanics   39

The Organization of Mechanics   40

Basic Dimensions Used in Mechanics   41

Newton's Laws   44

Summary   46

Part II   External Biomechanics: External Forces and Their Effects on the Body and Its Movement   47

Chapter 4   Forces: Maintaining Equilibrium or Changing Motion   49

What Are Forces?   50

Classifying Forces   51

Friction   52

Addition of Forces: Force Composition   57

Resolution of Forces   62

Static Equilibrium   67

Summary   72

Chapter 5   Linear Kinematics: Describing Objects in Linear Motion   75

Motion   76

Linear Kinematics   78

Summary   93

Chapter 6   Linear Kinetics: Explaining the Causes of Linear Motion   95

Law of Inertia   96

Law of Acceleration   100

Impulse and Momentum   105

Law of Action-Reaction   109

Summary   110

**Chapter 7   Work, Power, and Energy: Explaining the Causes
of Motion Without Newton   113**

Work   114

Energy   117

The Work-Energy Relationship   119

Power   123

Summary   125

**Chapter 8   Torques and Moments of Force: Maintaining Equilibrium
or Changing Angular Motion   127**

What Are Torques?   128

Forces and Torques in Equilibrium   136

What Is Center of Gravity?   140

Summary   153

Chapter 9   Angular Kinematics: Describing Objects in Angular Motion   157

Angular Position and Displacement   158

Angular and Linear Displacement   161

Angular Velocity   163

Angular and Linear Velocity   163

Angular Acceleration   166

Angular and Linear Acceleration   166

Summary   167

Chapter 10   Angular Kinetics: Explaining the Causes of Angular Motion   171

Angular Inertia   172

Angular Momentum   178

Angular Interpretation of Newton's First Law of Motion   180

Angular Interpretation of Newton's Second Law of Motion   183

Angular Impulse and Angular Momentum   184

Angular Interpretation of Newton's Third Law of Motion   185

Summary   186

Chapter 11   Fluid Mechanics: The Effects of Water and Air   191

Buoyant Force: Force Due to Immersion   192

Dynamic Fluid Force: Force Due to Relative Motion   196

Summary   207

Part III   Internal Biomechanics: Internal Forces and Their Effects on the Body and Its Movement   211

Chapter 12   Mechanics of Biological Materials: Stresses and Strains on the Body   213

Stress   214

Strain   225

Mechanical Properties of Materials: The Stress-Strain Relationship   227

Mechanical Properties of the Musculoskeletal System   230

Summary   235

Chapter 13   The Skeletal System: The Rigid Framework of the Body   239

Bones   241

Joints   244

Summary   251

Chapter 14   The Muscular System: The Motors of the Body   255

The Structure of Skeletal Muscle   256

Muscle Action   260

Muscle Contraction Force   265

Summary   276

Chapter 15   The Nervous System: Control of the Musculoskeletal System   279

The Nervous System and the Neuron   280

The Motor Unit   282

Receptors and Reflexes   284

Summary   288

Part IV   Applying Biomechanical Principles   291

Chapter 16   Qualitative Biomechanical Analysis to Improve Technique   293

Types of Biomechanical Analysis   294

Qualitative Biomechanical Analysis to Improve Technique   295

Example Analyses   306

Summary   321

Chapter 17   Qualitative Biomechanical Analysis to Improve Training   323

Biomechanics and Training   324

Qualitative Anatomical Analysis Method   325

Example Analyses   329

Summary   342

Chapter 18   Qualitative Biomechanical Analysis to Understand Injury Development   347

Steven T. McCaw

Mechanical Stress and Injury   348

Tissue Response to Stress   350

Mechanism of Overuse Injury   353

Individual Differences in Tissue Threshold   354

Intrinsic and Extrinsic Factors Affecting Injury   354

Running   358

Summary   367

Appendix A   Trigonometric Functions   369

Appendix B   Answers to Selected Review Questions   379

Glossary   385

References and Suggested Readings   395

Index   399

About the Author   406

Preface

The purpose of this text is to introduce undergraduate students to the field of exercise and sport biomechanics. Today, students of human movement, whether majoring in kinesiology, exercise science, or physical education, have a variety of career options available to them besides teaching. All of these careers involve some observation, evaluation, or prescription of human movement. A knowledge of mechanics is thus valuable and necessary for success in these careers. Many human movement professionals' only formal instruction in the mechanics of human movement occurs during a single undergraduate course in kinesiology or biomechanics. This book has been developed with this constraint in mind. My goal was to write a concise, clear, and effective textbook for a one-semester undergraduate course in biomechanics for students of human movement.

The book is divided into four parts. Part I consists of three chapters. Chapter 1 provides an introduction to biomechanics, which includes justifications for the study of biomechanics. Chapters 2 and 3 review fundamental concepts and principles of anatomy and mechanics, respectively. The material included in these two chapters might be considered the prerequisite material for a course in biomechanics. This material is the foundation for the material presented later in the book.

Part II is concerned with external biomechanics, or external forces and their effects on the body and its movement. Rigid-body mechanics with applications to human movement is the primary topic of this part. Mechanics is one of the most difficult topics for undergraduate students of human movement to understand, so this is the most important part of the book. Thus, part II makes up the largest part of the book. The order of presentation of topics in this part differs from most other kinesiology and biomechanics texts. It begins with the presentation of the concepts of force and statics in chapter 4, before discussing movement in chapter 5 and the causes of linear motion in chapter 6. Mechanical work and energy principles are discussed in chapter 7. Torque, moment of force, and center of gravity are introduced in chapter 8, prior to discussing angular kinematics in chapter 9. The causes of angular motion are presented in chapter 10. Logically, it makes more sense to understand the mechanics of nonmoving bodies and the forces and torques that maintain the equilibrium of these bodies before trying to understand the mechanics of moving bodies. Part II concludes with a discussion of fluid mechanics in chapter 11.

Part III is concerned with internal biomechanics, or internal forces and their effects on the body and its movement. This part begins with a discussion of the mechanics of biological materials in chapter 12. Stress and strain are introduced in this chapter, along with various concepts of material strength. Overviews of the skeletal system, muscular system, and nervous system control are then presented in chapters 13, 14, and 15.

Part IV is concerned with the application of biomechanics. Commonsense methods of applying biomechanics to the analysis of sport or human movement skills are presented in the three chapters of this part. The first of these chapters, chapter 16, presents procedures for completing qualitative biomechanical analyses to improve technique. Chapter 17 presents a method of qualitative biomechanical analysis to improve training. A qualitative procedure for identifying active muscle groups in phases or parts of movements is emphasized in this chapter. Chapter 18 concludes the book with an examination of how qualitative biomechanical analysis can be used to help understand the causes of injury. This chapter was contributed by Steven McCaw.

Throughout the book, and especially in part II, the objective has been to allow students to discover the principles of mechanics for themselves, rather than presenting the principles directly. Common activities are observed, and explanations for these activities are then developed. The resulting explanations reveal the underlying mechanical concepts. This discovery process requires more active participation by the reader, but it results in a better understanding of the subject matter.

What makes this book unique among biomechanics textbooks is its order of presentation. In most undergraduate biomechanics textbooks, functional anatomy is presented before mechanics. This textbook presents mechanics first. Bones and ligaments are the structural elements that support the human body. Muscles are the motors that move this structure. Understanding how the forces that bones and ligaments exert support the body and how the forces and torques that muscles produce do work to move the body's limbs requires a knowledge of forces and their effects. Mechanics is the study of forces and their effects. Thus mechanics should precede the study of the musculoskeletal system.

This book is also unique in its order of presentation of mechanical topics. The mechanics section of most biomechanics textbooks begins with linear kinematics, followed by linear kinetics, angular kinematics, and finally angular kinetics. In this book, forces are introduced first, prior to linear kinematics. Because forces are the causes of changes in motion and forces are in equilibrium if no changes in motion occur, it makes sense to define and understand forces before discussing motion. This is especially true if projectile motion and the equations for projectile motion are discussed under linear kinematics. Since projectile motion is influenced by the force of gravity, an understanding of this force should precede discussion of the effects of the force. Similarly, torques are introduced prior to angular kinematics.

Since mechanics utilizes equations to describe relationships or to define quantities, some knowledge of mathematics (primarily algebra) is necessary. I have tried to write the book in such a way that even those students who are insecure about their math skills can succeed in learning biomechanics. However, success in learning the material will come more easily to those who are better prepared mathematically. The trigonometry information in appendix A will help as students work through some material in the book.

Certain elements contained in each chapter of this book may help the reader learn the material. Each chapter begins with an opening scenario leading to questions that can be answered by reading and understanding the material presented in the chapter. A list of objectives is also presented at the beginning of each chapter. Practical examples of concepts are integrated into the text throughout each chapter. Review questions appear at the end of each chapter to test the reader's understanding of the material presented in the chapter. Answers to some of these questions appear in appendix B.

I have tried to explain the concepts in this book as clearly as possible and in such a way that you, the reader, are actively involved in discovering them. Still, you might find that some of the material is challenging. Occasionally, you might find yourself distracted or confused while reading. Don't give up! Your effort will be worthwhile. A knowledge of biomechanics will be valuable to you as you pursue a career in a human movement profession.

Acknowledgments

Numerous individuals are responsible for guiding my learning and teaching of biomechanics, for inspiring me to write this book, for motivating me to finish it, or for contributing directly to its development. I thank the following: the engineering faculty at Swarthmore College who taught me mechanics; the graduate students and faculty at the University of Illinois who taught me biomechanics; and the graduate and undergraduate students in my biomechanics courses at the University of Oregon, the University of Northern Colorado, and SUNY College at Cortland who taught me about teaching.

Thanks to Steve McCaw of Illinois State University, who wrote chapter 18 and by doing so helped me finally get this book completed. Thanks to the two anonymous reviewers whose comments improved the book, and to the extremely patient current and former staff at Human Kinetics who assisted and encouraged me, especially Loarn Robertson, Kent Reel, Jackie Blakely, Richard Partin, Rick Frey, Cindy Yates, and Julia Anderson. Finally, I am especially grateful to my wife, Boodie, for her patience and understanding.

Fundamental Biomechanical Concepts, Principles, and Terminology

Biomechanics is a multidisciplinary field. Knowledge from a variety of fields, including human anatomy and physiology, mechanics, engineering, and mathematics, is utilized in biomechanics. Before we proceed with our study of biomechanics, we should all have a common background knowledge of certain fundamental terminology, principles, and concepts of biomechanics that are derived from these other fields of study. This background material is presented in part I, which includes chapters 1, 2, and 3. Chapter 1 is an introduction, which addresses the need for biomechanics. Chapter 2 is a review of anatomical terminology and joint motion description. Chapter 3 is a review of fundamental terms and principles of mechanics. Some of the material may be new to some of you, others may already have an understanding of all of the material, and some of you may not have an understanding of any of this material. In any case, a review of the material presented in the next three chapters will ensure that all of us are working with the same background knowledge.

Why Study Biomechanics?

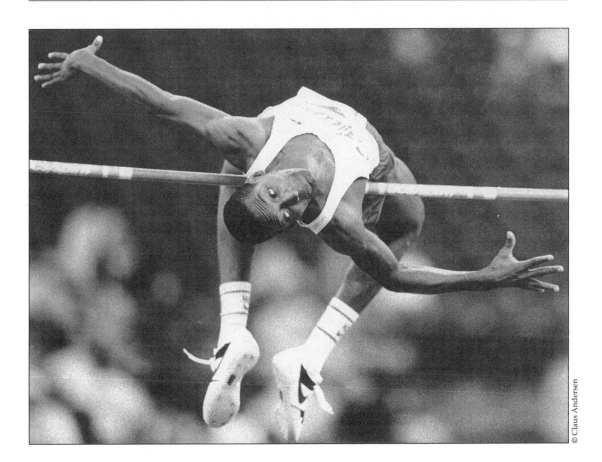

© Claus Andersen

You are watching the Olympic Games on television when you see a high jumper successfully jump over a crossbar set more than a foot above his head. The technique he uses looks very awkward. He approaches the bar from the side, and when he jumps off the ground, he turns his back toward the bar. His head and arms clear the bar first; he then arches his back and finally kicks his legs out and up to get them over the bar. He lands in the pit in an ungainly position—on his shoulders and back with his legs extended in the air above him. You think to yourself, *How can he jump so high using such an odd-looking technique?* Certainly there must be another technique that is as effective but more graceful looking. Biomechanics might provide you with some insights to answer this and other questions you have about human movement.

OBJECTIVES

When you finish this chapter, you should be able to do the following:

▌ Define biomechanics

▌ Define exercise and sport biomechanics

▌ Identify the goals of exercise and sport biomechanics

▌ Describe the methods used to achieve the goals of exercise and sport biomechanics

▌ Be somewhat familiar with the history and development of exercise and sport biomechanics

The title of this chapter poses a question about your motivation for studying biomechanics. What can you gain from learning about biomechanics? How will a working knowledge of biomechanics assist you in your future endeavors? Will the time you spend learning about biomechanics be worthwhile? These questions should be considered before investing a lot of time learning about biomechanics.

If you are like most readers of this book, you are probably an undergraduate physical education student. If so, your answer to the question "Why study sport biomechanics?" may be that you are enrolled in a required course in biomechanics for which this is the required text. You are studying biomechanics to earn a required credit so that you can graduate with a degree in physical education. If this is your answer, and for the majority of readers this may be true, you are probably unable to answer the other questions because you do not have enough prior knowledge of biomechanics to know how it can benefit you. So let me give you some reasons for studying biomechanics. Hopefully, I can provide you with some intrinsic motivation to get started on the task of learning about biomechanics.

Most of you are probably planning a career as a physical education teacher, coach, or some other physical activity specialist; and you probably are or have been active as a participant in one or more sports or fitness activities. Suppose a student or athlete asks you, "Why do I have to do this skill this way?" or "Why isn't this technique better?" (see figure 1.1). Perhaps you even asked such questions when you were a student or athlete. Was the coach or teacher able to answer your questions? Were you asked these questions? Could *you* answer them? Traditional teaching and coaching methods tell you *what* techniques to teach or coach, whereas biomechanics tells you *why* those techniques are best to teach or coach.

A good knowledge of biomechanics will enable you to evaluate techniques used in unfamiliar sports skills as well as to better evaluate new techniques in sports with which you are familiar.

> Perhaps the best outcome of studying and using biomechanics will be improved performances by your athletes or the accelerated learning of new skills by your students.

Athletic training and physical therapy students as well as other students of sports medicine will also benefit from a knowledge of biomechanics. A good knowledge of biomechanics will help in diagnosing the causes of an injury. It can provide the mechanical basis for taping techniques, braces, and orthotic devices. A knowledge of biomechanics may also guide therapists in their prescriptions for rehabilitation and indicate to exercise specialists what exercises may be dangerous to certain individuals.

BUT COACH,
I LIKE TO
HOLD THE BAT
THIS WAY!

| Figure 1.1 | Studying biomechanics will help you understand why some sport techniques work and some don't. |

What Is Biomechanics?

What is this science that promises so much? Before going any further, we should agree on a definition of the word *biomechanics*. What is biomechanics? How have you heard this word used by others? How have you used this word? Your immediate response may be that biomechanics has something to do with determining the best techniques used by athletes in various sports skills. Indeed, some biomechanists may be involved in such work, and we just highlighted technique evaluation as a primary reason for studying biomechanics. But biomechanics encompasses much more than that.

Let's look in the library for clues to the definition. There are several journals in the library in which the word *biomechanics* or a derivation of it appears in the title. These include the *Journal of Biomechanics*, the *Journal of Biomechanical Engineering*, and the *Journal of Applied Biomechanics*. Looking through the tables of contents of these journals, we find that the *Journal of Applied Biomechanics* contains articles about sport biomechanics, such as: "Timing of the lower limb drive and throwing limb movement in baseball pitching"; "The snatch technique of world-class weight lifters at the 1985 World Championships"; "Approach run strategies in the long jump"; "Biomechanical model of the press handstand in gymnastics"; and so forth. These articles support our sport-related definition of biomechanics. However, a look through the other journals reveals a broad range of topics that at first may appear unrelated, such as: "Knee joint motion and ligament forces before and after ACL reconstruction"; "A theoretical description of blood flow through the mitral orifice"; "Heat generation in laser-irradiated tissue"; "A model of oxygen exchange between an arteriole or venule and the surrounding tissue"; and so forth. There's even an article entitled "Biomechanics of fruits and vegetables"! From these titles we can deduce that biomechanics is not limited to sports; it is not even limited to human activities; indeed, it is not even limited to animal activities! The range of titles indicates that biomechanics could include not only the study of the movement of an athlete, but also the study of the flow of the athlete's blood and the strength of the athlete's tissues.

Let's return to the word itself and examine it directly for clues about its definition. The word *biomechanics* can be divided into two parts: the prefix *bio-* and the root word *mechanics*. The prefix *bio-* indicates that biomechanics has something to do with living or biological systems. The root word *mechanics* indicates that biomechanics has something to do with the analysis of forces and their effects. So it appears that biomechanics is the study of forces and their effects on living systems. This comes very close to the definition of biomechanics presented by Herbert Hatze in 1974: "Biomechanics is the study of the structure and function of biological systems by means of the methods of mechanics" (p. 189). This is a much broader field of study than we may have at first thought! The study of the structure and function of plants as well as animals is encompassed by the definition of biomechanics.

Biomechanics is the study of forces and their effects on living systems.

What Are the Goals of Exercise and Sport Biomechanics?

Now let's focus on our specific topic of interest in biomechanics. Biomechanics includes the study of all living things, plant and animal; animal biomechanics includes only animals as subjects of study; human biomechanics includes only humans; and exercise and sport biomechanics includes only humans involved in exercise and sport. We might define exercise and sport biomechanics as the study of forces and their effects on humans in exercise and sport.

Performance Improvement

The ultimate goal of exercise and sport biomechanics is performance improvement in exercise or sport. A secondary goal is injury prevention and rehabilitation. This secondary goal is closely related to the first and could almost be considered part of the primary goal, because an uninjured athlete will perform better than an injured athlete. Well, how do biomechanists work toward achieving these goals?

The ultimate goal of exercise and sport biomechanics is performance improvement in exercise or sport.

Technique Improvement

The most common method for improving performance in many sports is to improve an athlete's technique. This was highlighted as one motivation for studying biomechanics, and it is probably what you thought of when asked how a biomechanist went about trying to improve an athlete's performance.

The application of biomechanics to improve technique may occur in two ways: Teachers and coaches may use their knowledge of mechanics to correct actions of a student or athlete in order to improve the execution of a skill, or a biomechanics researcher may discover a new and more effective technique for performing a sports skill. In the first instance, teachers and coaches use qualitative biomechanical analysis methods in their everyday teaching and coaching to affect changes in technique. In the second instance, a biomechanics researcher uses quantitative biomechanical analysis methods to discover new techniques, which then must be communicated to the teachers and coaches who will implement them.

Let's look at a simple example of the first case. As a coach, suppose you observe that your gymnast is having difficulty completing a double somersault in the floor exercise. You might suggest three things to the gymnast to help her successfully complete the stunt: (1) jump higher, (2) tuck tighter, and (3) swing her arms more vigorously before takeoff. These suggestions may all result in improved performance and are based on biomechanical principles. Jumping higher will give the gymnast more time in the air to complete the somersault. Tucking tighter will cause the gymnast to rotate faster due to conservation of angular momentum. Swinging the arms more vigorously prior to takeoff will generate more angular momentum, thus also causing the gymnast to rotate faster. This general procedure is the most common type of situation where biomechanics has an effect on the outcome of a skill. Coaches and teachers use their knowledge of biomechanics to determine what actions may improve performance.

Coaches and teachers use their knowledge of biomechanics to determine what actions may improve performance.

The second general situation in which biomechanics contributes to improved performance through improved technique occurs when biomechanics researchers develop new and more effective techniques. Despite the common belief that new and revolutionary techniques are regularly developed by biomechanists, such developments are rare. Perhaps this is because biomechanics as a discipline is a relatively new science. The much more common outcome of biomechanics research is the discovery of small refinements in technique. One example of biomechanics research that did greatly affect the technique and performances in a sport occurred in swimming in the late sixties and early seventies. Research done by Ronald Brown and James "Doc" Councilman (1971) indicated that the lift forces acting on the hand as it moved through the water were much more important in propelling a swimmer through the water than previously thought. This research indicated that rather than pulling the hand in a straight line backward through the water to produce a propulsive drag force, the swimmer should move the hand back and forth in a sweeping action as it was pulled backward through the water to produce propulsive lift forces as well as propulsive drag forces (see figure 1.2). This technique is now taught by swimming teachers and coaches throughout the world.

Other examples of sports in which dramatic changes in technique produced dramatic improvement in performance include javelin throwing, high jumping, and cross-country skiing. In 1956, prior to the Summer Olympic Games in Melbourne, Felix Erasquin, a javelin thrower from the Basque region of Spain, experimented with an unconventional way of throwing the javelin. Rather than throwing it over the shoulder with one hand from a run, Erasquin spun around like a hammer thrower and slung the javelin from both hands, which guided the implement. To reduce the frictional forces acting on the javelin as it slid through his hands, the javelin was dunked in soapy water to make it slippery. The outstanding results achieved by Erasquin with this technique attracted international attention (he had several throws that were beyond the existing world record). The International Amateur Athletic Federation (IAAF), the governing body for track and field, became so alarmed that they altered the rules for the event, and this unconventional technique became illegal (see figure 1.3).

In 1968, most world-class high jumpers used the straddle technique. But at the Olympics in Mexico City, the gold medalist in the high jump used a technique few had ever seen. Dick Fosbury, an American from Oregon State University, used a back layout technique to jump 7 ft 4-1/4 in. (2.24 m). The technique became known as the Fosbury Flop (see figure 1.4). Its advantages over the straddle technique were its faster approach run and its ease of learning. No biomechanics researcher developed this technique. Fosbury achieved success with it in high school and continued using and jumping higher with it, despite its dramatic differences from the conventional straddle technique. His successes led others to

Figure 1.2 Swimming techniques have been influenced by biomechanics.

© Claus Andersen

UPI/Corbis-Bettmann

Figure 1.3 Current IAAF rules require athletes to throw the javelin over the shoulder with one hand.

With the exception of the swimming example, the above examples of new and dramatically different techniques leading to improved performances all happened without the apparent assistance of biomechanics. Maybe this situation is evidence of the skill of teachers, coaches, and athletes. Through repeated observation, trial and error, and possibly some application of mechanical principles, they have successfully developed excellent techniques for performing skills in most sports without the assistance of biomechanical researchers. But perhaps these improved techniques would have been developed sooner if more teachers and coaches had a working knowledge of biomechanics.

Equipment Improvement

How else can biomechanics contribute to performance improvement? What about improved designs for the equipment used in various sports? Shoes and apparel constitute the equipment used in almost every sport. The equipment worn may have an effect on the performance, either directly or through injury prevention. Besides shoes and apparel, many sports require the use of some sort of implement. Think of sports in which an implement is used. How have changes in sports implements changed performances in these sports? What about bicycling, skiing, tennis, golf, pole vaulting, javelin throwing . . . ? Lighter and better designed implements have not only contributed to improved performances by elite athletes in these sports, they have contributed to improved performances by recreational participants as well.

Let's examine javelin throwing as an example of a sport where a basic application of mechanics to the equipment design changed the event dramatically. In 1952, Frank "Bud" Held made the United States Olympic team in the javelin. At the 1952 Olympics in Helsinki, he placed ninth behind his American teammates, who won the gold and silver medals. Upon returning to the United States, Bud met with his brother, Dick Held, who had some engineering expertise, and together they designed and built a more aerodynamic javelin. The increased surface area of their new javelin gave it more lift, causing it to "fly" farther. In 1953, Bud Held used one of his javelins to break the existing world record for the javelin throw. The Held brothers were not biomechanists, but their knowledge of mechanics enabled them to improve the design of the javelin. The records continued to be broken as others began using the Held javelin. In 1955, the IAAF implemented rules that limited the size of the javelin so

adopt it, and now all world-class high jumpers use the Fosbury Flop.

In the late seventies, Bill Koch, an American cross-country skier, began experimenting with a new technique he had observed marathon skiers using in Europe. The skating technique he experimented with was much different from the traditional diagonal stride skiing technique in which cross-country skiers moved their skis parallel to each other in set tracks. In the 1976 Olympic Games in Innsbruck, Austria, Koch surprised the world by winning a silver medal in the 30-km cross-country skiing event. More surprising were his performances in the 1982-83 season, when he became the first American ever to win the World Cup. Koch used the skating technique in achieving this title. By the mid-1980s, the skating technique was used by virtually all elite Nordic ski racers (see figure 1.5). Beginning with the 1992 Winter Olympics, there were separate competitions for traditional (diagonal stride) and freestyle (skating) cross-country skiing.

a

UPI/Corbis-Bettmann

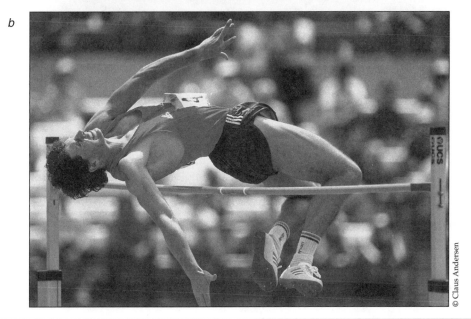

b

© Claus Andersen

Figure 1.4   Most high jumpers used the straddle technique (*a*) until Dick Fosbury won the Olympic gold medal using the Fosbury Flop (*b*) in 1968.

that further increases in its surface area and lift were constrained. Prior to 1953, the world record in the javelin was 258 ft 2-3/8 in. (78.70 m), set in 1938. Using modern aerodynamic javelins based on the Held design, the world record in the event eventually progressed to 343 ft 10 in. (104.80 m) in 1984. In 1986, the IAAF effectively reduced the distance of the javelin throw by again changing the rules governing construction of the javelin. The new specifications prevented the javelin from "sailing" so far. Despite this attempt to limit performances,

by 1990 the world record with the new javelin rules exceeded 300 ft (91.44 m). This is a rare example of mechanics being applied to limit performance in a sport.

Training Improvement

How else can biomechanics contribute to improved performance in sports and physical activities? What about training? Biomechanics has the potential of leading to modifications in training and thus

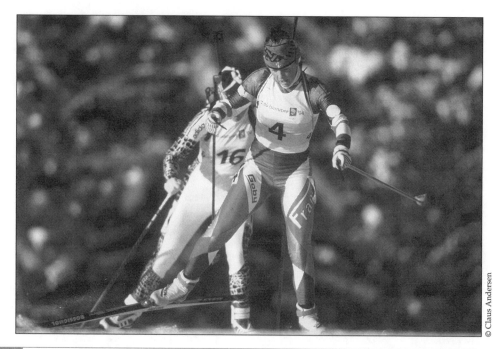

© Claus Andersen

Figure 1.5   Cross-country skiers now employ a skating stride rather than the traditional diagonal stride.

improvements in performance. This application of biomechanics can occur in several ways. An analysis of the technique deficiencies of an athlete can assist the coach or teacher in identifying the type of training the athlete requires to improve. The athlete may be limited by the strength or endurance of certain muscle groups, by speed of movement, or by one specific aspect of his or her technique. Sometimes, the limitation may be obvious. For example, a gymnast attempting an iron cross maneuver requires tremendous strength in the adductor muscles of the shoulder (see figure 1.6). A mechanical analysis of the maneuver would reveal this, but it is already obvious to gymnastics coaches and observers. In other sport skills, the strength requirements may not be so obvious.

> **An analysis of the technique deficiencies of an athlete can assist the coach or teacher in identifying the type of training the athlete requires to improve.**

Consider the sport of pole vaulting. High school-age boys learning this event will often reach a plateau in their performances. A common technique deficiency among beginning pole vaulters concerns the athlete's movement on the pole. The vaulter does not get his hips above his head or hands during the latter stages of the vault. This prevents the vaulter from achieving an inverted position, and the heights the vaulter clears are modest. This flaw is easily identified by the coach or teacher, but despite repeated instructions to the vaulter about getting his hips up, he is still unable to correct this flaw in technique. Why? Most young vaulters do not have the shoulder extensor strength they require to swing upside down into an inverted position. A biomechanical analysis of this flaw in the vaulter's technique would reveal that greater shoulder extensor strength is required. The coach or teacher could then devise a training program that would sufficiently strengthen the vaulter's shoulder extensor muscles so that he could complete this aspect of the vault successfully.

Figure skating provides another example of how biomechanical analysis can lead to changes in training and ultimately performance improvement. Training camps for junior female skaters held at the U.S. Olympic Training Center in Colorado Springs in the mid-1980s included biomechanical analyses of the skaters attempting double and some triple jumps. Many of the skaters who attempted triple twisting jumps were unsuccessful. An initial analysis revealed that some of the skaters were unsuccessful in the triples because they were not bringing their arms in tight enough to cause them to spin faster while they were in the air (see figure 1.7). Further biomechanical analysis revealed that they were unable to bring their arms in tight enough or

| Figure 1.6 | A male gymnast's ability to perform an iron cross maneuver may be limited by the strength of his shoulder adductor muscles. |

© Claus Andersen

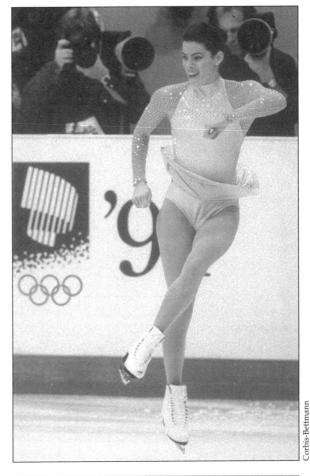

| Figure 1.7 | Successfully completing triple jumps in figure skating requires strong arms and shoulders. |

Corbis-Bettmann

quickly enough due to inadequate strength in their arm and shoulder musculature. After their training programs were modified to include upper body strength training to increase arm and shoulder strength, several of the skaters were able to complete triple jumps successfully in subsequent training camps.

One final example of a simple biomechanical analysis that revealed deficiencies in training occurred in cross-country skiing in the late 1970s. An analysis of an international competition included timing skiers over specific sections of the course. The results of this analysis indicated that the American skiers were just as good as the race leaders on the flat and downhill sections of the course, but they were losing on the uphills. This result encouraged the coaches to allocate more training time to uphills and place more emphasis on perfecting the uphill skiing techniques of the skiers.

Injury Prevention and Rehabilitation

Some believe that the secondary goal of exercise and sport biomechanics, injury prevention and rehabilitation, should be the primary goal. A knowledge of biomechanics is useful to sports medicine professionals in identifying what forces may have caused an injury, how to prevent the injury from recurring (or occurring in the first place), and what exercises may assist with rehabilitation from the injury. Biomechanics can be used to provide the basis for alterations in technique, equipment, or training to prevent or rehabilitate injuries.

> Some believe that the secondary goal of exercise and sport biomechanics, injury prevention and rehabilitation, should be the primary goal.

Techniques to Reduce Injury

An example of how biomechanics may aid in reducing injury is in gymnastics. Research funded in part by the United States Olympic Committee and the United States Gymnastics Association is concerned with the impact forces gymnasts experience when landing from stunts and what strategies they use to reduce these forces (McNitt-Gray, 1991; McNitt-Gray, Yokoi, and Millward, 1993; McNitt-Gray, Yokoi, and Millward, 1994). Judges award higher points to gymnasts who "stick" their landings. But such landings may involve greater and more dangerous impact forces. These impact forces are the cause of overuse injuries in many gymnasts. A landing in which the gymnast flexes at the knees, hips, and ankles may reduce these impact forces, but it also results in a lower score. One outcome of this research may be new rules allowing for landing strategies that reduce these impact forces without penalty to the gymnast's score.

Equipment Designs to Reduce Injury

An example of biomechanics affecting the design of sports equipment to reduce injury involves the running shoe industry. Following Frank Shorter's marathon gold medal in the 1972 Olympics, the United States experienced a running boom. Unfortunately, this boom in participation was also accompanied by a boom in running-related injuries. The increase in injuries led runners to be more sophisticated about their selection of running shoes. Thus a boom in biomechanics research into running and running shoes began in the 1970s. An annual shoe ranking published in *Runner's World* magazine included results of biomechanical tests conducted on the shoes at a university biomechanics laboratory. Some of the shoe companies hired biomechanists as consultants and some funded biomechanics research in other university biomechanics laboratories. In 1980, Nike established the Nike Sport Research Laboratory to further the development of athletics and athletic shoes by means of studies in biomechanics, exercise physiology, and functional anatomy.

Running shoes available in the early 1970s were too stiff for many inexperienced runners, and impact injuries such as shin splints and stress fractures became common. Shoe manufacturers responded to this by producing softer shoes. However, the softer shoes did not provide as much stability or control as the harder shoes, and as a result, ankle, knee, and hip joint injuries increased among runners. Biomechanics research sponsored by various shoe companies has led to many of the features offered in modern running shoes, which provide stability as well as cushioning. These improvements have resulted in fewer running injuries.

Exercise and sport biomechanics can lead to performance improvement and may aid in injury prevention and rehabilitation through improvements in technique, equipment design, and training. Most of the examples in the previous sections illustrate how biomechanics can play a role in performance improvement. A few of the examples, such as the technique changes in the javelin throw, high jump, and cross-country skiing, demonstrate that radically different and improved techniques occur in sport without the apparent assistance of biomechanics researchers. In fact, there are very few examples of biomechanics contributing to the development of new techniques or equipment to improve performance. Why? The answer may be that the people who can most affect sports techniques—the teachers, coaches, and athletes—for the most part are not knowledgeable in biomechanics, but their constant trial-and-error motivation allows them to stumble upon improved techniques. As more teachers, coaches, and athletes become exposed to biomechanics, improvements in techniques may occur more rapidly. Exercise and sport biomechanics is still a relatively young field, though. The population of practicing sport biomechanists is too small to affect changes in many sports. When the technique changes described above for the javelin and high jump occurred, the number of people with a knowledge of biomechanics was very small. In fact, prior to 1960, the word *biomechanics* was used by only a handful of people. A brief review of the history of sport biomechanics can give us more insight into why biomechanics has not had the impact on sports that it seems capable of.

The History of Sport Biomechanics

The history of sport biomechanics is partly the history of kinesiology. The word *kinesiology* was first used in the late 19th century and became popular during the 20th century, whereas the word *biomechanics* did not become popular until the 1960s.

The roots of the word *kinesiology* give its definition as the study of movement, but in its present-day usage, kinesiology is defined as the study of human movement.

An individual course in kinesiology has been a required part of the undergraduate curriculum in physical education at many American schools for most of the 20th century. The major content of such a course was usually applied anatomy with some mechanics and possibly physiology thrown in. Usually this was the only course in which a future coach or physical education teacher received any exposure to mechanics. But in many instances, with the emphasis on applied anatomy, the amount of mechanics the future practitioner was exposed to was not enough to be of much practical use.

Researchers concerned with the biomechanics of human movement have been active throughout the 20th century, although the mechanics of human and animal motion have intrigued scientists since at least the time of Aristotle (1912). In the last decades of the 19th century, Etienne Jules Marey wrote *Le Mouvement* (1895/1972), in which he described the use of a variety of devices, including cameras and pressure-sensitive instruments, to measure and record forces and motions produced by man (and animals) in a variety of activities. His well-instrumented "biomechanics" laboratory was the precursor to modern biomechanics and exercise physiology laboratories.

> **The mechanics of human and animal motion have intrigued scientists since at least the time of Aristotle.**

An early example of exercise and sport biomechanics research appeared in *The Baseball Magazine* in 1912. The publishers commissioned a study to determine the speed of a baseball pitched by Walter Johnson. At that time, Johnson was the "king of speed pitchers." Although biomechanics is now thought of as a relatively new field of study, the prelude to this article in *The Baseball Magazine* states, "*The Baseball Magazine* makes an excursion into an absolutely new field of scientific investigation in its study of the National Game" (F.C. Lane, 1912, p. 25).

Archibald V. Hill conducted studies of the mechanics and energetics of sprinting in the 1920s (cited in Braun, 1941). This work was continued by Wallace Fenn in the 1930s (cited in Cureton, 1939). Although he was more well known as an exercise physiologist, Thomas Cureton also wrote about the mechanics of swimming (1930) and various track and field skills in the 1930s (Cureton, 1939). He also described techniques for analyzing movements in sport using motion picture cameras (Cureton, 1939). During this time, Arthur Steindler wrote one of the first "biomechanics" textbooks (1935). The 1940s saw World War II, and research in the biomechanics of sport was not a priority. In 1955, the book *Scientific Principles of Coaching* by John Bunn was published. This was one of the first texts to emphasize the mechanics rather than the anatomical aspects of human movement in sports.

By the 1960s, use of the term *biomechanics* became popular, and more people were involved in exercise and sport biomechanics research. In 1967, the First International Seminar on Biomechanics was held in Zurich, Switzerland. The majority of papers presented at this conference dealt with the mechanics of human movement. This seminar was a success, and international conferences on biomechanics have been held every two years since then. In 1968, the *Journal of Biomechanics* was first published. Several papers in the first volume of this journal dealt with sport biomechanics. During the 1960s, several graduate programs in biomechanics were established within physical education departments, and a few of these programs offered doctoral degrees.

In 1973, the International Society of Biomechanics was formed, followed closely by formation of the American Society of Biomechanics in 1977. Exercise and sport biomechanists were involved in the formation of each of these organizations, although the membership of the societies included scientists with a variety of interests. In the early 1980s, the International Society of Biomechanics in Sport was formed to represent the interests of sport biomechanists. In 1985, the *International Journal of Sports Biomechanics* began publication, and in 1992 it changed its name to the *Journal of Applied Biomechanics*.

Research in exercise and sport biomechanics has increased steadily throughout the 1970s, 1980s, and 1990s. The number of people involved in exercise and sport biomechanics has also increased tremendously during this time. One reason for this boom has been the advent of the modern digital computer, which allows for easier data collection and analysis from the high-speed film or video cameras and electronic force measuring platforms used in biomechanics research. Without a computer, the time required to compute measurements accurately from film data and do quantitative biomechanical research was grossly excessive, accounting for the dearth of research in exercise and sport biomechanics prior to the 1960s. Anatomical research was not as difficult to complete, and thus courses in kinesiology were weighted toward applied anatomy. With the increase in exercise and sport biomechanics research in the last two decades, the content of many kinesiology courses has been reexamined, and mechanics is covered more thoroughly now. Many kinesiology courses have been renamed as biomechanics courses.

We began this section on the history of sport biomechanics by asking why biomechanics has not had the impact on sports that it seems capable of. The answer seems clearer now. Biomechanics and education in biomechanics have not been in existence long enough to have had a great impact yet. But will biomechanics ever have a significant impact on sports? Let's look toward the future.

The Future of Sport Biomechanics

The proclaimed goal of exercise and sport biomechanics is performance improvement. The impact of exercise and sport biomechanics was held back prior to the 1960s due to lack of technology. The impact of exercise and sport biomechanics may be much greater in the future due to rapid advances in technology. This technology may include improved techniques for measuring biomechanical and anthropometric parameters, improved computer simulation techniques, and other unimagined applications of technology to improve biomechanics methodology. All these advances will increase the productivity of sport biomechanists. But the real future of sport biomechanics lies with you, the practitioners, teachers, coaches, and exercise science professionals who must implement the discoveries made by researchers and whose questions will provoke future research in exercise and sport biomechanics. Your knowledge of biomechanics and your insight and experience in the activities you teach and coach may also lead to your own discoveries of new and better techniques. At the very least, the future of biomechanics will include many more teachers and coaches educated in biomechanics. These people will be more effective because of their knowledge of biomechanics. Students will be more enthusiastic about physical activities because they will learn skills more quickly due to their teachers' knowledge and application of biomechanics. Improved performances by athletes will occur more readily due to more effective instruction and training of the athlete by a coach knowledgeable in biomechanics.

> **The impact of exercise and sport biomechanics may be much greater in the future due to rapid advances in technology.**

The first step toward this future is for you and other future teachers and coaches to learn biomechanics. Now that we know what biomechanics is and what it can offer us, let's take the first step and begin our formal study of biomechanics!

Summary

Biomechanics is the study of forces and their effects on living systems, whereas exercise and sport biomechanics is the study of forces and their effects on humans in exercise and sport. Biomechanics may be a useful tool for physical educators, coaches, exercise scientists, athletic trainers, physical therapists, and others involved in human movement. Application of biomechanics may lead to performance improvement or the reduction and rehabilitation of injury through improved techniques, equipment, or training.

Biomechanics is a relatively young term that only became popular in the 1960s. Biomechanical studies of human movements have been reported throughout the 20th century. Professional societies devoted to biomechanics began meeting and journals devoted to biomechanics appeared in the late 1960s and early 1970s. The advent and widespread use of electronic digital computers made biomechanical research more feasible throughout the 1970s and 1980s.

Key Terms

biomechanics   *(p. 6)*

exercise and sport biomechanics   *(p. 6)*

kinesiology   *(p. 12)*

REVIEW QUESTIONS

1. Explain the difference between kinesiology and biomechanics.

2. In what ways will biomechanics be useful to you in your career?

3. Give an example of how a physical education teacher might utilize biomechanics.

4. Give an example of how a coach might utilize biomechanics.

5. Give an example of how an athletic trainer might utilize biomechanics.

6. Give an example of how a physical therapist might utilize biomechanics.

7. Give an example of how an aerobics teacher might utilize biomechanics.

8. What are the goals of exercise and sport biomechanics?

9. What are the methods for achieving the goals of exercise and sport biomechanics?

10. What professional societies are specifically concerned with biomechanics?

11. What technology contributed to the growth of biomechanics in the 1970s?

2

Anatomical Movement Terminology

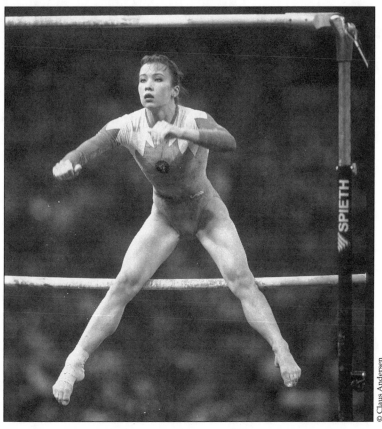

© Claus Andersen

Imagine watching an athletic competition and then trying to describe to your friend a particular skill you saw performed. Would your description be accurate enough for your friend to visualize what you just saw? For example, suppose you've just finished watching a gymnast competing on the uneven parallel bars. One particular stunt the gymnast performed was especially impressive. Your friend missed seeing the routine, but you want to tell her about the stunt that impressed you. Is there some sort of terminology that is suited to this task of human movement description? The terminology developed by anatomists may be what we're looking for.

OBJECTIVES

When you finish this chapter, you should be able to do the following:

▪ Name and identify the segments of the body

▪ Name and identify most of the large bones of the body

▪ Name and identify most of the joints of the body

▪ Name and identify the large superficial muscles of the body

▪ Use the following terms to locate the relative position of any part of the body: anterior/posterior, superior/inferior, medial/lateral, proximal/distal, and superficial/deep

▪ Identify the sagittal, frontal, and transverse planes and the corresponding axes: transverse, anteroposterior, and longitudinal

▪ Identify the plane and axis of motion for any human movement

▪ Use the following terms to describe joint actions: flexion/extension, abduction/adduction, internal rotation/external rotation, pronation/supination, plantar flexion/dorsiflexion, and circumduction

Many terms used in biomechanics are borrowed from anatomy. Biomechanics also makes use of a system for describing body positions and relative motions of limbs at joints that was developed by anatomists. The purpose of this chapter is to review the anatomical terminology used in biomechanics and to explain and define the anatomical planes and axes of motion used to describe relative motions of body segments. A secondary purpose of this chapter is to review the major segments, joints, bones, and superficial muscles of the human body.

Body Segment, Bone, Joint, and Muscle Terminology

Before learning anything further about anatomical terminology, we should know the anatomical names given to the segments of the body. A review of the major bones and joints and the superficial skeletal muscles is also appropriate at this point.

Segments of the Body

The human body can be viewed as a group of links or segments that are connected to each other at joints (see figure 2.1). The segments of the upper extremity are the arm (the segment between the shoulder and elbow), the forearm (the segment between the elbow and wrist), and the hand. The segments of the lower extremity are the thigh (the segment between the hip and knee), the leg (the segment between the knee and ankle), and the foot. The trunk is often divided into two segments: the thorax (the upper trunk segment or chest) and the abdomen (the lower trunk segment or belly). The head and neck segments complete the body.

Bones and Joints of the Body

The bones and joints of the body define the segments of the body. The humerus is the bone of the arm, whereas the radius and ulna define the forearm. The bones of the wrist are the carpals, and the bones of the hand are the metacarpals. The finger bones are the phalanges. The scapula (shoulder

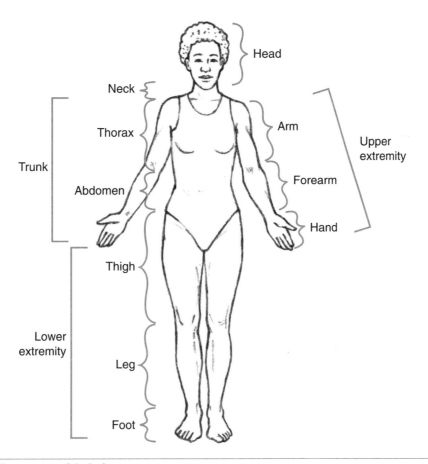

| Figure 2.1 | The segments of the body. |

blade) and clavicle (collarbone) are also considered bones of the upper extremity.

> **The bones and joints of the body define the segments of the body.**

The bones of the lower extremity are larger than those of the upper extremity. The thighbone is called the femur. The bones of the leg are the tibia and fibula. The bones of the foot include the tarsals (anklebones), the metatarsals, and the phalanges (the bones of the toes). The pelvis is also considered part of the lower extremity, although it forms the lower end of the abdomen.

Individual vertebrae linked together form the vertebral column, the supporting column of the trunk. Most ribs attach to the vertebrae in the back and to the sternum in front. The bones of the neck are vertebrae also, and the skull forms the head.

The major joints of the body are the shoulder, elbow, proximal radioulnar joint, distal radioulnar joint, wrist, hip, knee, and ankle. Figure 2.2 shows the important bones and joints of the body.

Major Muscles of the Body

The body's shape is defined by the superficial muscles. The major muscles of the body are shown in figure 2.3. The deltoid muscle shrouds the shoulder. The biceps brachii defines the bulge on the front of the arm. You can easily feel its tendon at the front of the elbow. The triceps brachii gives bulk to the back of the arm. The wrist and finger flexor muscles give bulk to the front of the forearm, and the wrist and finger extensor muscles lie on the back of the forearm. Look at your forearm as you squeeze your hand into a fist and then extend it. You should be able to see some indication of the movement of these wrist and finger flexor and extensor muscles beneath the skin.

> **The body's shape is defined by the superficial muscles.**

The gluteus maximus and gluteus medius cover the hip to the rear and side and shape the buttocks.

©K GalasynWright '94

Figure 2.2a   Anterior view of the bones and joints of the body.

Clavicle

Humerus

Ribs

Elbow joint

Pelvis

Hip joint

Scapula

Vertebral column

Radius

Ulna

Wrist joint

Femur

Knee joint

Tibia

Fibula

©K GalasynWright '94

Figure 2.2b Posterior view of the bones and joints of the body.

Deltoid

Pectoralis major

Biceps brachii

Rectus abdominis

External oblique

Wrist and finger flexors

Pectineus

Adductor longus

Gracilis

Sartorius

Vastus lateralis

Rectus femoris

Vastus medialis

Tibialis anterior

©K GalasynWright '94

Figure 2.3a   Anterior view of the major muscles of the body.

Trapezius

Triceps brachii

Latissimus dorsi

Gluteus medius

Wrist and
finger extensors

Gluteus maximus

Biceps femoris

Semitendinosus

Semimembranosus

Gastrocnemius

Soleus

©K GalasynWright '94

Figure 2.3b   Posterior view of the major muscles of the body.

The four quadriceps femoris muscles (the rectus femoris, vastus lateralis, vastus intermedius, and vastus medialis) lie on the front of the thigh. The sartorius, the longest muscle of the body, lies on the front and medial side of the thigh. The three hamstring muscles (the biceps femoris, semimembranosus, and semitendinosus) lie on the back of the thigh. You can feel the tendons of these muscles behind and to the inside and outside of the knee. The groin muscles or adductor group (the adductor magnus, adductor longus, adductor brevis, and gracilis) form the inside of the thigh. The gastrocnemius and soleus muscles (the triceps surae group) give shape to the back of the leg. The tibialis anterior is the major muscle on the front of the leg.

The major muscles of the front of the trunk include the pectoralis major on the chest and the abdominal muscles (the rectus abdominis and external oblique) on the belly. The trapezius covers most of the upper back, whereas the latissimus dorsi covers most of the lower back. The erector spinae muscle group runs up and down the back along either side of the vertebral column (not shown in figure 2.3).

Spatial and Directional Terminology

Try to describe where a specific freckle or mole or hair is on your body. It's a difficult task. You probably identified the location of the freckle or mole or hair by describing where it was in relation to some other body part. A similar situation occurs if you try to describe the movement of a limb. To describe the location or movement of a body part, the other parts of the body are used as a reference. But the human body can adopt many different positions, and the orientation of the limbs may change as well, so a common reference position of the body must be used. The most commonly used reference position of the human body is called anatomical position. Early anatomists suspended cadavers in this position so as to study them more easily. The body is in anatomical position when it is standing erect, facing forward, both feet aligned parallel to each other, toes forward, arms and hands hanging straight below the shoulders at the sides, fingers extended, and palms facing forward. The body is shown in anatomical position in figure 2.1.

Anatomical position is the standard reference position for the body when describing locations, positions, or movements of limbs or other anatomical structures. What are the anatomical terms used to describe locations, positions, or movements of body parts? Definitions of the most commonly used terms are presented below and shown graphically in figure 2.4. Alternative or synonymous terms are presented in parentheses. Remember, when these terms are used, no matter what position the body may actually be in, the terms refer to the position or location as if the body were in anatomical position.

> **Anatomical position is the standard reference position for the body when describing locations, positions, or movements of limbs or other anatomical structures.**

- Anterior (ventral) and posterior (dorsal) mean toward the front and toward the rear of the body, respectively. Ventral and dorsal are synonymous with anterior and posterior. The sternum is anterior to the vertebral column. The nose is on the anterior surface of the head. The fingernails are on the posterior or dorsal side of the fingers.
- Superior (cranial or cephalo) indicates toward or closer to the head, whereas inferior (caudal) indicates toward or closer to the feet. The feet are inferior to the knees. The eyes are superior to the mouth and nose.
- Medial and lateral refer to directions or positions relative to the midline of the body. Medial refers to a location or direction closer to the midline of the body, whereas lateral refers to a position away from or farther from the midline of the body. The sternum is medial to the shoulder. The medial side of the right knee is the left side of that knee.
- Proximal and distal refer to directions or relative positions along a limb. Proximal refers to a location or direction toward or closer to the attachment of the limb to the body, whereas distal refers to a location or direction toward or closer to the end of the limb farthest from its attachment to the body. The fingers are distal to the elbow. The hip is proximal to the knee.
- Superficial and deep refer to position relative to the exterior surface of the body. Superficial muscles are those that lie just beneath the skin, whereas deep muscles are those that are farther from the skin.

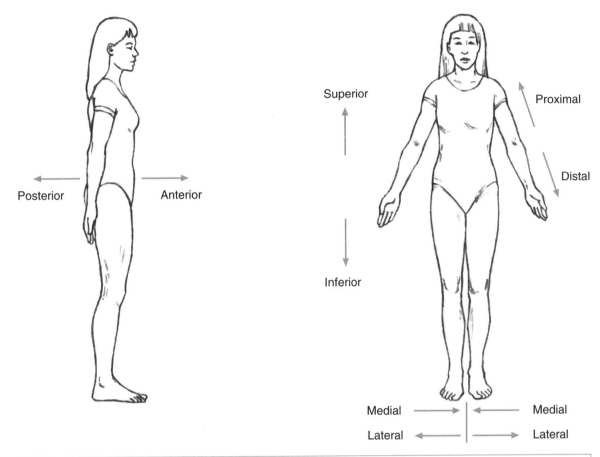

Figure 2.4   Terms used in anatomy to describe relative locations or directions.

Planes and Axes of Motion

Anatomists have developed names to identify specific planes that pass through the body. Each plane has a corresponding axis that passes perpendicularly through the plane. These planes are useful to anatomists in describing planes of dissections or imaginary dissections. The planes may also be useful in describing relative movements of body parts, with the axes used to describe the lines around which these motions occur.

Anatomical Planes

A plane is a flat two-dimensional surface. A sagittal plane, also called an anteroposterior plane, is an imaginary plane running anterior (front) to posterior (back) and superior (top) to inferior (bottom), dividing the body into right and left parts. A frontal plane, also called a coronal or lateral plane, runs side to side and superior to inferior, dividing the body into anterior and posterior parts. A transverse plane, or horizontal plane, runs from side to side and anterior to posterior, dividing the body into superior and inferior parts. All sagittal planes are perpendicular to all frontal planes, which are perpendicular to all transverse planes.

Many sagittal planes can be imagined to pass through the body, but they are all parallel to each other. Likewise, many frontal or transverse planes can be imagined to pass through the body. A cardinal plane is a plane that passes through the midpoint or center of gravity of the body. The center of gravity is the point on which the body would balance if it were supported at only one point. The cardinal sagittal plane (midsagittal or median plane) is the plane that divides the body into equal right and left halves. The cardinal sagittal, frontal, and transverse planes of the body are shown in figure 2.5.

Biomechanically, the anatomical planes may be useful for locating anatomical structures, but their greatest worth is for describing limb movements. How can a plane be useful for describing movement? Movements of most limbs occur as rotations of the limbs. Rotations occur around specific axes

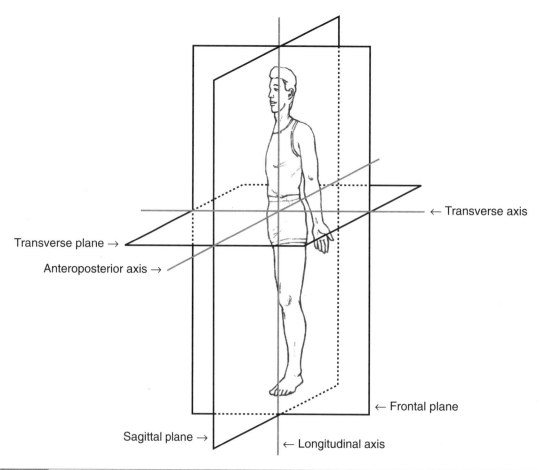

Transverse axis ←

Transverse plane →

Anteroposterior axis →

Frontal plane ←

Sagittal plane →   Longitudinal axis ←

Figure 2.5   Cardinal anatomical planes and axes of the body.

of rotation and within specific planes of movement. Descriptions of the limb movements relative to each other are thus facilitated by identification of the axis of rotation around which the limb moves and the plane within which the limb moves.

> **Biomechanically, the anatomical planes may be useful for locating anatomical structures, but their greatest worth is for describing limb movements.**

Anatomical Axes

We've defined several specific anatomical planes. What are the specific anatomical axes? The anatomical axes correspond to lines that are perpendicular to the previously defined anatomical planes. An **anteroposterior axis** (sagittal, sagittal-transverse, or cartwheel axis) is an imaginary line running from anterior to posterior and perpendicular to the frontal planes. The anteroposterior axis is often abbreviated as AP axis. An AP axis is defined by the intersection

of a transverse plane with a sagittal plane, so it may also be called a sagittal-transverse axis. A **transverse axis** (lateral, frontal, mediolateral, frontal-transverse, or somersault axis) is an imaginary line running from left to right and perpendicular to the sagittal planes. A transverse axis is defined by the intersection of a transverse plane with a frontal plane, so it may also be called a frontal-transverse axis. A **longitudinal axis** (vertical, frontal-sagittal, or twist axis) is an imaginary line running from top to bottom and perpendicular to the transverse planes. A longitudinal axis is defined by the intersection of a frontal plane with a sagittal plane, so it may also be called a frontal-sagittal axis. All AP axes are perpendicular to all transverse axes, which are perpendicular to all longitudinal axes. An infinite number of these axes pass through the body. Examples of AP, transverse, and longitudinal axes are shown in figure 2.5.

Identifying Planes and Axes of Motion

Now let's see how axes and planes are used to describe human motion. Imagine a bicycle wheel. The wheel turns about an axle. The line along and through the axle of the wheel defines the axis of rotation of the wheel. This is the axis around which the wheel rotates. The spokes of the wheel are perpendicular to the axle or axis of rotation. Thus the spokes must lie in the plane of motion of the wheel. Now let's look at an example from the human body. Stand in anatomical position and flex (bend) your right arm at the elbow without moving your upper arm, as shown in figure 2.6. Think of your forearm as a spoke of the bicycle wheel and your elbow as the axle. What plane does your forearm lie in throughout the movement? It is moving within a sagittal plane. What axis is perpendicular to a sagittal plane? A transverse axis is perpendicular to a sagittal plane. The movement of your forearm and hand is in a sagittal plane and around a transverse axis.

Now let's describe the method we use to identify the plane or axis of motion for any limb movement. First, one principle should be noted. If you can identify either the plane or the axis of motion, the other is easily identified. If the plane within which the motion occurs is known, there is only one axis around which the motion can occur, and that is the axis perpendicular to the plane of motion. Likewise, if the axis around which the motion occurs is known, there is only one plane within which the motion can occur, and that is the plane perpendicular to the axis of motion. Table 2.1 lists each of the three planes of motion along with the corresponding axis that is perpendicular to that plane of motion.

If you can identify either the plane or the axis of motion, the other is easily identified.

Prior to describing techniques for determining the plane of motion, a more precise and accurate definition of a plane may be helpful. Just as two points geometrically define a line in space, three noncolinear points or two intersecting lines define a plane in space. If a table or chair has only three legs, each of those legs will always touch a plane, such as the floor, as it only takes three points to define a plane. A table with four legs will always have three legs on one plane but the fourth leg will only touch that plane if its end lies on the plane defined by the other three legs.

Most of our limbs are longer in one dimension than in the others, so they can be thought of as long cylinders or even line segments. These line segments move by swinging or rotating around joints. If you can imagine the line segment defined by the limb at the start of motion and the line segment defined by the limb at any other instant during its motion, these line segments will intersect at the joint (if it's a single-joint motion). The plane of motion is defined by these two line segments. For example, stand in anatomical position. Imagine the line segment defined by a line drawn from your right shoulder to your right wrist. Now abduct at the shoulder (lift your right arm up and to the side, laterally away from your body, until it is shoulder height) as shown in figure 2.7. Imagine a line segment drawn through your shoulder and wrist now. What planes do both line segments fall within? When you were in anatomical position, your right arm was in a sagittal plane and a frontal plane.

Elbow flexion

| Figure 2.6 | Imagining a bicycle wheel can help you identify the plane and axis of motion. |

TABLE 2.1 Anatomical Planes of Motion and Their Corresponding Axes of Motion	
Plane	Axis
Sagittal	Transverse
Frontal	Anteropost
Transverse	Longitudin

Shoulder
abduction

Figure 2.7   Frontal view of shoulder abduction.

right arm at the shoulder) from a position in front of (anterior to, as shown in figure 2.7) or behind (posterior to) the person moving, you would see the full length of the arm throughout the movement, and the arm wouldn't move away from or toward you. If you viewed the motion from the side (sagittally, as shown in figure 2.8a), the full length of the arm would be in view at the start, but as the arm moved upward, it would appear to shorten until you would only see the hand when the arm reached shoulder height. The arm would move toward you. If you viewed the motion from above (transversely, as shown in figure 2.8b), only the shoulder would be in view at the start, but the full length of the arm would be in view when it reached shoulder height. The arm would move toward you. The best view was from the front, so the view was a frontal view and the movement was in a frontal plane. The axis perpendicular to a frontal plane is an anteroposterior axis.

When you completed the movement, your arm was in a transverse plane and a frontal plane. Your arm was in the frontal plane at the start and end of the movement (and throughout the movement), so the movement occurred within a frontal plane. Because frontal planes are perpendicular to AP axes, the movement occurred around an anteroposterior axis.

Here's another procedure for determining the plane of motion. If you could view the movement from any vantage point, what vantage point would give you the best view so that you could always see the entire length of the moving limb? What vantage point would you view from so that the moving limb was not moving toward or away from you, but across your field of view? If the best viewpoint is from in front or behind, the view of the body is a frontal view and the movement is in a frontal plane. If the best viewpoint is from the left or right side, the view of the body is a sagittal view and the movement is in a sagittal plane. If the best viewpoint is from above or below (admittedly difficult vantage points to view from, but imagine the view if you could get into a position above or below a person), the view of the body is a transverse view and the movement is in a transverse plane.

For example, if you viewed the movement described in the previous example (abduction of the

a

b

Figure 2.8   Sagittal (*a*) and transverse (*b*) views of shoulder abduction.

In some limb movements, the length of the limb doesn't swing around an axis, but the limb rotates or twists about its length. In this case, it is easier to determine the axis of rotation of the limb first and then determine its plane of motion. The axis of rotation of a limb twisting about its length is defined by the direction of the line running the length of the limb from the proximal to the distal end. If this line is parallel to a longitudinal axis, it is a longitudinal axis. If it is parallel to an anteroposterior axis, it is an anteroposterior axis. If it is parallel to a transverse axis, it is a transverse axis.

For example, stand in anatomical position and turn the palm of your right hand toward your side and then to the rear as shown in figure 2.9. Your arm twisted around an axis through its length. A line drawn from the proximal to the distal end of your arm is a vertical line and parallel to a longitudinal axis. The axis of rotation for this movement was a longitudinal axis. A longitudinal axis is perpendicular to a transverse plane, so the twisting motion of your right arm occurred in the transverse plane.

So far, the examples we've used have all been movements that occurred within one of the three planes of motion we defined. Sagittal, transverse, and frontal planes are primary planes of motion. Other planes exist that are not primary planes of motion, but movements can occur within these planes. For example, what plane do your arms move in when you swing a golf club as shown in figure 2.10? The best viewpoint wouldn't be directly in front or directly above but in front and above. The plane of motion is a diagonal plane between the transverse plane and the frontal plane, and the axis of motion is a diagonal axis between a longitudinal and an AP axis. An infinite number of diagonal planes and axes exist within or around which our limbs can move. The principal planes and axes give us standard planes and axes from which other, diagonal planes and axes can be described.

Joint Actions

We can identify the plane and action of limb motions, but what are the terms that describe the limb motions? Human movement description uses terminology that describes the relative movements of two limbs on either side of a joint rather than terminology that describes individual limb movements. The terms thus describe joint actions, the

Figure 2.9 Movement around a longitudinal axis.

Figure 2.10 Movement in a diagonal plane.

relative angular movements of the limbs on the distal and proximal sides of a joint. From anatomical position, the joint actions that occur when limbs move around transverse joint axes and within sagittal planes are flexion, extension, hyperextension, plantar flexion, and dorsiflexion. The joint actions that occur when limbs move around anteroposterior joint axes and within frontal planes are abduction, adduction, radial deviation (radial flexion), ulnar deviation (ulnar flexion), inversion, eversion, elevation, depression, lateral flexion to the right, and lateral flexion to the left. The joint actions that occur when limbs move around longitudinal axes and within transverse planes are internal (inward or medial) rotation, external (outward or lateral) rotation, pronation, supination, horizontal abduction (horizontal extension), horizontal adduction (horizontal flexion), rotation left, and rotation right.

> **Human movement description uses terminology that describes the relative movements of two limbs on either side of a joint rather than terminology that describes individual limb movements.**

Movements Around Transverse Axes

Flexion, extension, and hyperextension are joint actions occurring at the wrist, elbow, shoulder, hip, knee, and intervertebral joints. Starting from anatomical position, flexion is the joint action that occurs around the transverse axes through these joints and causes limb movements in sagittal planes through the largest range of motion. Extension is the joint action that occurs around the transverse axes through these joints and causes the opposite limb movements in sagittal planes that return the limbs to anatomical position. Hyperextension is the joint action that occurs around the transverse axes and is a continuation of extension past anatomical position. Elbow flexion thus occurs when the forearm is moved forward and upward and the angle between the forearm and the upper arm at the anterior side of the elbow joint gets smaller. Elbow extension occurs when the forearm is returned to anatomical position. Elbow hyperextension would occur if the forearm could continue extending past anatomical position. Figure 2.11a-g shows the flexion, extension, and hyperextension joint actions that occur at the wrist, elbow, shoulder, hip, knee, trunk, and neck joints, respectively.

Dorsiflexion and plantar flexion are joint actions that occur at the ankle. Starting from anatomical position, dorsiflexion is the joint action that occurs around the transverse axis through the ankle joint and causes the foot to move in a sagittal plane such that it moves forward and upward toward the leg. When you lift your toes off the ground and put your weight on your heels, you are dorsiflexing at your ankles. Plantar flexion is the joint action that occurs around the transverse axis through the ankle joint and causes the opposite movement of the foot in a sagittal plane so that the foot moves downward away from the leg. When you stand on your toes you are plantar flexing at your ankles. Dorsiflexion and plantar flexion are also shown in figure 2.11g.

Movements Around Anteroposterior Axes

Abduction and adduction are joint actions occurring at the shoulder and hip joints. Starting from anatomical position, abduction is the joint action that occurs around the AP axes through these joints and causes limb movements in frontal planes through the largest range of motion. Adduction is the joint action that occurs around the AP axes through these joints and causes limb movement in a frontal plane back toward anatomical position. Shoulder abduction thus occurs when the arm is moved upward and laterally away from the body. Shoulder adduction occurs when the arm is returned to anatomical position. Figure 2.12a and b shows the abduction and adduction joint actions that occur at the shoulder and hip joints.

Ulnar deviation (adduction or ulnar flexion) and radial deviation (abduction or radial flexion) are joint actions occurring at the wrist joint (see figure 2.12c). Starting from anatomical position, ulnar deviation is the joint action that occurs around the AP axis through the wrist and causes hand movement in a frontal plane toward the little finger. Radial deviation is the joint action that occurs around the AP axis through the wrist joint and causes the opposite movement of the hand in a frontal plane, moving it laterally toward the thumb.

Inversion and eversion are frontal plane movements that occur at the ankle joint (see figure 2.12c). These joint actions occur around an AP axis through the foot. Starting from anatomical position, inversion occurs when the medial side of the sole of the foot is lifted. The return to anatomical position and the movement of the foot beyond anatomical position where the lateral side of the sole of the foot is lifted is eversion.

Figure 2.11 Sagittal plane joint actions at the wrist, elbow, shoulder, hip, knee, trunk, and neck and ankle.

a

b

c

d

e

Figure 2.12 Frontal plane joint actions at the shoulder, hip, wrist and ankle, trunk, and neck.

The movements of the scapula (the shoulder blade) and shoulder girdle occur primarily in the frontal plane as well. These movements include abduction (movement of the scapula away from the midline) and adduction (movement of the scapula toward the midline), elevation (superior movement of the scapula) and depression (inferior movement of the scapula), and upward rotation (such that the medial border moves inferiorly and the shoulder joint moves superiorly) and downward rotation (such that the medial border moves superiorly and the shoulder joint moves inferiorly).

Lateral flexion to the left or right also occurs in the frontal plane. Lateral flexion of the trunk to the left occurs when you lean to the left, and lateral flexion to the right occurs when you lean to the right. Likewise, lateral flexion of the neck to the left occurs when you lean your head toward your left shoulder, and lateral flexion of the neck to the right occurs when you lean your head toward your right shoulder. Lateral flexion is shown in figure 2.12d and e.

Movements Around Longitudinal Axes

Internal rotation (inward or medial rotation) and external rotation (outward or lateral rotation) are joint actions occurring at the shoulder and hip joints. Starting from anatomical position, internal rotation is the joint action that occurs around the longitudinal axes through these joints and causes limb movements in the transverse plane such that the knees turn inward toward each other or the palms of the hands turn toward the body. External rotation is the joint action that occurs around the longitudinal axes through these joints and causes the opposite limb movements in transverse planes and returns the limbs to anatomical position or moves them beyond anatomical position. Figure 2.13a and b shows the internal and external rotation joint actions that occur at the hip and shoulder joints.

Pronation and supination are joint actions occurring at the radioulnar joint in the forearm. Starting from anatomical position, pronation is the joint action that occurs around the longitudinal axis of the forearm and through the radioulnar joint and causes the palm to turn toward the body. This motion is similar to internal rotation at the shoulder joint except that it occurs at the radioulnar joint. Supination is the joint action that occurs around the longitudinal axis through the radioulnar joint and causes the opposite limb movement in a trans-

verse plane and returns the forearm and hand to anatomical position or moves them beyond anatomical position. Figure 2.13d also shows the supination and pronation actions that occur at the radioulnar joint.

Horizontal abduction (horizontal extension or transverse abduction) and horizontal adduction (horizontal flexion or transverse adduction) are joint actions occurring at the hip and shoulder joints. These joint actions do not commence from anatomical position. First, hip or shoulder flexion must occur and continue until the arm or thigh is in the transverse plane. Horizontal abduction is then the movement of the arm or leg in the transverse plane around a longitudinal axis such that the arm or leg moves away from the midline of the body. The return movement is horizontal adduction. These joint actions are also shown in figure 2.13c.

Rotations of the head, neck, and trunk also occur around a longitudinal axis. Turning your trunk so that you face left is rotation to the left, and turning your trunk so that you face right is rotation to the right. These actions are shown in figure 2.13e and f.

Circumduction is a multiple-axis joint action that occurs around the transverse and AP axes. Circumduction is flexion combined with abduction and then adduction or extension and hyperextension combined with abduction and then adduction. The trajectory of a limb being circumducted forms a cone-shaped surface. If you abduct your arm at the shoulder joint and then move your arm and forearm so that your hand traces the shape of a circle, the joint action occurring at the shoulder joint is circumduction.

Just as there are diagonal planes and axes, joint actions may also occur in diagonal planes around diagonal axes. Such diagonal joint actions may be combinations of joint actions if they occur at multiple-axis joints, or they may be one of the joint actions described above if the joint or limb has been moved into a diagonal plane by the action at a more proximal joint. The joint actions and the corresponding planes and axes of motion for these actions are summarized in table 2.2.

> **Joint actions may also occur in diagonal planes around diagonal axes.**

Summary

The human body can be described as a series of linked segments that include the head, neck, trunk (thorax and abdomen), arms, forearms, hands,

Figure 2.13   Transverse plane joint actions at the hip, shoulder, shoulder and hip, radioulnar joint, neck, and trunk.

TABLE 2.2 Joint Actions and Their Corresponding Planes and Axes of Motion

Plane of motion	Axis	Joint actions
Sagittal	Transverse	Flexion
		Extension
		Hyperextension
		Plantar flexion
		Dorsiflexion
Frontal	Anteroposterior	Abduction
		Adduction
		Ulnar deviation
		Radial deviation
		Inversion
		Eversion
		Elevation
		Depression
		Upward rotation
		Downward rotation
		Lateral flexion to the left
		Lateral flexion to the right
Transverse	Longitudinal	Internal rotation
		External rotation
		Pronation
		Supination
		Horizontal abduction
		Horizontal adduction
		Rotation to the right
		Rotation to the left

thighs, legs, and feet. The bones and major superficial muscles of the body help define these segments. The shoulder, elbow, and wrist are the major joints of the upper extremity; and the hip, knee, and ankle are the major joints of the lower extremity.

Anatomical spatial and directional terms used to describe relative locations of anatomical structures are all referenced to the body when it is in anatomical position (standing erect, with arms at the sides and palms facing forward). Superior and inferior indicate above or below; anterior and posterior indicate in front of or behind; medial and lateral indicate toward the midline or toward the side; proximal and distal indicate closer to or farther from the segment's connection to the trunk; and superficial and deep indicate closer to or farther from the surface of the body (the skin).

The three principal anatomical planes are the sagittal, frontal, and transverse planes. These planes are at right angles to each other. The three principal anatomical axes that correspond to each of these planes are the transverse, AP, and longitudinal axes. The sagittal plane is an imaginary plane running from anterior to posterior and superior to inferior, cutting the body into left and right parts. Transverse axes are perpendicular to sagittal planes. The frontal plane is an imaginary plane running from superior to inferior and side to side, cutting the body into front and back parts. AP axes are perpendicular to frontal planes. The transverse plane is an imaginary plane running from anterior to posterior and side to side, cutting the body into top and bottom parts. Longitudinal axes are perpendicular to transverse planes.

Key Terms

abduction *(p. 30)*

adduction *(p. 30)*

anatomical position *(p. 24)*

anterior *(p. 24)*

anteroposterior axis *(p. 26)*

cardinal plane *(p. 25)*

circumduction *(p. 33)*

deep *(p. 24)*

depression *(p. 33)*

distal *(p. 24)*

dorsiflexion *(p. 30)*

downward rotation *(p. 33)*

elevation *(p. 33)*

eversion *(p. 30)*

extension *(p. 30)*

external rotation *(p. 33)*

flexion *(p. 30)*

frontal plane *(p. 25)*

horizontal abduction *(p. 33)*

horizontal adduction *(p. 33)*

hyperextension *(p. 30)*

inferior *(p. 24)*

internal rotation *(p. 33)*

inversion *(p. 30)*

lateral *(p. 24)*

lateral flexion *(p. 33)*

longitudinal axis *(p. 26)*

medial *(p. 24)*

plantar flexion *(p. 30)*

posterior *(p. 24)*

pronation *(p. 33)*

proximal *(p. 24)*

radial deviation *(p. 30)*

sagittal plane *(p. 25)*

superficial *(p. 24)*

superior *(p. 24)*

supination *(p. 33)*

transverse axis *(p. 26)*

transverse plane *(p. 25)*

ulnar deviation *(p. 30)*

upward rotation *(p. 33)*

REVIEW QUESTIONS

1. What muscle covers your shoulder like a shroud?

2. What is the name of the thighbone?

3. The trunk can be divided into two segments. What are the names for these two trunk segments?

4. What is the broad superficial muscle that covers much of the upper back?

5. What is the plane of motion for most of the joint actions that occur during sprint running? What is the corresponding axis of motion for these joint actions?

6. When you swing a baseball bat, what is the plane of motion for the action occurring at your leading shoulder? What is the axis of motion? What joint action occurs at the leading shoulder during the swing?

7. During the delivery phase of a baseball pitch, what joint actions occur at the shoulder and elbow joints of the throwing arm?

8. What is the plane of motion for the shoulder joint actions when you do a wide-grip pull-up? What is the axis of motion? What joint action occurs at the shoulder joint during the pulling-up phase of the pull-up?

9. How do the actions at the shoulder joint differ between a wide-grip and a narrow-grip pull-up?

10. When you spike a volleyball, what joint action occurs at the elbow joint of your hitting arm?

Fundamental Concepts and Principles of Mechanics

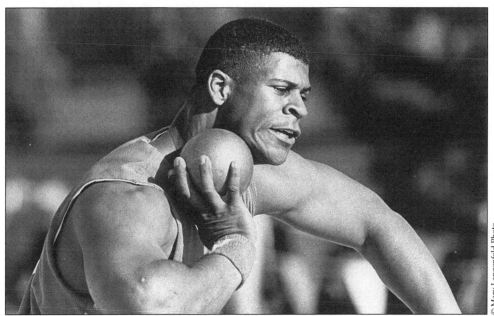

© Mary Langenfeld Photo

A shot-putter steps out of the circle after completing his put and waits for his throw to be measured. A sprinter in a 200-m dash crosses the finish line first and then looks up to the scoreboard to see what her time was. A weight lifter lifts a barbell over his head and holds it overhead while the judges determine if the lift is successful. These three examples illustrate the three basic dimensions used in mechanics: length, time, and mass. In each example, one of these dimensions was the measure of performance: length for the shot put, time for the sprint, and mass for the weight lift. Each of the skills was also influenced by other basic dimensions besides the performance measure. In the 200-m dash, the length of the race affected the finish times. In the shot put, the mass of the shot affected the length of the throw. In the weight lift, the time the lifter had to hold the barbell over his head affected how much mass he could lift. The relationships between these basic dimensions and other derived dimensions are addressed in Newton's laws of motion.

When you finish this chapter, you should be able to do the following:

▌ Define mechanics

▌ Outline the organization of mechanics

▌ Define length and the units of measurement for length

▌ Define time and the units of measurement for time

▌ Define mass and the units of measurement for mass

▌ State each of Newton's laws of motion

Before learning more about biomechanics, you should be familiar with certain basic concepts of mechanics. First, an understanding of the organization of mechanics is helpful, especially those areas of mechanics we will be studying as part of biomechanics. The basic dimensions used in mechanics to describe physical phenomena or characteristics of objects must also be understood. Finally, you should have a brief introduction to Newton's laws of motion before we go further.

The Organization of Mechanics

Our study of exercise and sport biomechanics necessitates a knowledge of mechanics. In our definition of biomechanics, we briefly defined mechanics as the analysis of forces and their effects. A more complete definition may be that mechanics is the science concerned with the effects of forces acting on objects. The objects we will be concerned with in exercise and sport biomechanics are humans and the implements they may be manipulating in sports and exercise.

Mechanics is the science concerned with the effects of forces acting on objects.

Mechanics may be divided into several branches (see figure 3.1): rigid-body mechanics, deformable-body mechanics, fluid mechanics, relativistic mechanics, and quantum mechanics. In rigid-body mechanics, objects are assumed to be perfectly rigid. This simplifies the analyses. In deformable-body mechanics, the deformation of objects is considered. These deformations complicate the analyses. Fluid mechanics is concerned with the mechanics of liquids and gases. Relativistic mechanics is concerned

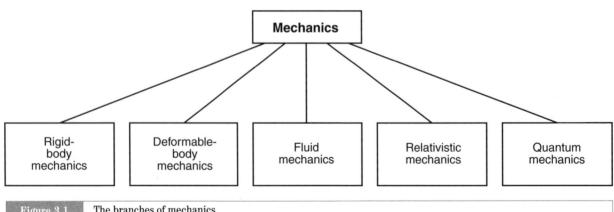

| Figure 3.1 | The branches of mechanics. |

with Einstein's theory of relativity, and quantum mechanics is concerned with quantum theory. Each branch of mechanics is best suited for describing and explaining specific features of our physical world. Rigid-body mechanics is best suited for describing and explaining the gross movements of humans and implements in sport and exercise, so the concepts of rigid-body mechanics will be important in our study of exercise and sport biomechanics. As some sport and exercise activities occur in fluid environments, we will also learn some concepts of fluid mechanics. Most of part II of this book (chapters 4 through 11) deals with concepts of rigid-body mechanics, with chapter 11 giving a brief overview of fluid mechanics.

In rigid-body mechanics, the objects being investigated are assumed to be perfectly rigid; that is, they do not deform by bending, stretching, or compressing. In describing and explaining the gross movements of the human body and any implements in sport and exercise, we will consider the segments of the human body as rigid bodies that are linked together at joints. In reality, the segments of the body do deform under the actions of forces. These deformations are small and don't appreciably affect the gross movements of the limbs or the body itself, so we can get away with considering the body as a system of linked rigid bodies. Repeated small deformations may lead to overuse injuries, however, so we will discuss the deformations of the tissues of the human body in chapter 12, which is concerned with one part of deformable-body mechanics, the mechanics of biological materials.

Rigid-body mechanics is subdivided into statics, or the mechanics of objects at rest or moving at constant velocity; and dynamics, or the mechanics of objects in accelerated motion (see figure 3.2). Dynamics is further subdivided into kinematics and kinetics. Kinematics deals with the description of motion, whereas kinetics deals with the forces that cause or tend to cause changes in motion. Our first exploration into biomechanics will be concerned with statics. The discussion of forces in the next chapter covers many of the principles of statics. Before we can proceed, though, we must first introduce some fundamental concepts of mechanics.

Basic Dimensions Used in Mechanics

Mechanics is a quantitative science, and therefore so is biomechanics. We wish to describe human movement and its causes in a quantitative manner. To do so, we must have some common units of measurement. But first, let's think about what we want to measure.

Suppose we are observing a football player returning a kickoff. What terms might we use in describing his runback? We may talk about his position on the field—where did he catch the ball and where was he tackled? How far did he run? How long did it take him to run that far? How fast was he? In what directions did he run? If you were an opposing player, you might also be concerned about how big he is and what it will feel like when you tackle him. Some of the mechanical terms we might use in our description include speed, inertia, power, momentum, force, mass, weight, distance, velocity, acceleration, and so on. (You may not be familiar with some of these terms. If you don't know what

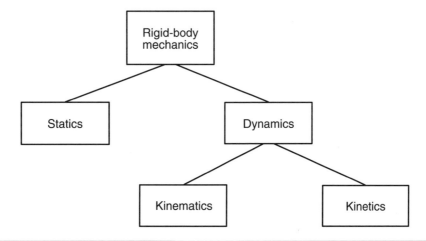

Figure 3.2   The branches of rigid-body mechanics.

they mean, don't worry about it now; they'll all be defined in this or future chapters.) Now let's sort through our descriptions and try to figure out what basic dimensions we need to describe the parameters we've listed.

Length

One basic dimension we may use is length. We need some measurement of length to describe the position of the player on the field and how far he ran. So length is used to describe the space in which movement occurs. Length is also an important dimension in many other sports. It is the most important dimension in sports such as shot putting or high jumping, where how far or how high is the actual measure of performance. In other sports, length may not be the actual measure of performance, but it may be a critical component. How far a golfer can drive the ball off the tee is one way of determining success in golf. How far a batter can hit the ball is one determinant of success in baseball. Stride length in running is one component that determines a runner's velocity.

Length is also an important dimension when we consider the anthropometry of athletes. An athlete's height may be one determinant of success in sports such as basketball or high jumping. Similarly, the length of the implements used may affect performance in sports such as golf, baseball, lacrosse, and pole vaulting. Finally, length may be a dimension of the sport specified by the rules of the activity, such as the length and width of a football field or playing court, the height of a basket, the length of a race, and so on.

Length is an important dimension, but how do we measure it? We are probably most familiar with the length measures of inches, feet, yards, or miles. But as scientists, we should be using the Systeme-Internationale d'Unités (International System of Units) or SI units. There are many systems of measurement, but the International System is by far the most widely used. This measurement system is based on the metric system. The SI unit of measurement for length is the meter, abbreviated as m, which is about 3.28 ft or 39 in. One ft is about 0.3048 m.

Time

Time is another basic dimension we may have used to describe aspects of the ballcarrier's run in our football example. Time is an important dimension of performance in almost all sports. In sports that involve races, time is the performance measure. The athlete with the shortest time from start to finish wins the race. In other sports, time is an important determinant of success. A goalie's reaction and movement times in large part determine his or her success in hockey, lacrosse, team handball, and water polo. Similar situations exist for athletes in tennis, racquetball, squash, and handball when they are attempting to return the ball. Time may also be a dimension specified by the rules of the activity, such as game time, time to begin a play in football, time-outs in basketball, time in the penalty box in hockey, time to initiate a throw or jump in track and field, and so on.

Time can be measured in seconds, minutes, hours, days, weeks, months, years, and so on. All of these units of measure are based on the second, which is the SI unit of measure for time, so we will use the second as our unit of measure for time. We will abbreviate second as s.

We now have the basic dimensions required for motion to occur: space and time—space to move in and time during which to move. With measurements of time and length, we can describe motion completely. If we consider some of the descriptors we used to describe the motion of the running back in the football example, many of them involved some measure of time and length. Speed and velocity, for instance, are derived from length and time and are expressed as some unit of length per unit of time. Acceleration is also derived from length and time measures. It is expressed as some unit of length per unit of time multiplied by itself, or squared. (If you don't know the definition of speed, velocity, or acceleration, don't worry; these terms will be clearly defined in chapter 5.)

Some of the descriptors used in our football example were not really descriptors of the motion of the running back, but rather a description of some aspects of the running back himself. These included the terms *momentum, power, force, inertia, mass* and *weight.* What are the basic dimensions used in these descriptors?

Mass and Inertia

What quality of the football running back makes it difficult for him to start running and difficult for a tackler to stop him from running? Something about the size of the football player affects how hard it is to start or stop his motion. Mechanically, we refer to the property of an object to resist changes in its motion as inertia. You may have heard this word used before or maybe even used it. Think about how you used it. You probably used inertia to indicate that something had a resistance or a reluctance to change what it was doing. This is close to the mechanical definition of inertia, but we must specify that the change being resisted is a change in motion.

How is inertia measured? Let's look at another example. Who is more difficult to start or stop moving: an Olympic shot-putter or an Olympic gymnast? Obviously, the shot-putter is more difficult to move. That athlete has more inertia. So, apparently, a larger object has more inertia than a smaller object. But what do we mean by larger? Well, maybe it's the weight of the object that determines its inertia. Not exactly; it is true that heavier objects have more inertia, but the measure of inertia is not weight. Mass is the measure of inertia, whereas weight is the measure of the force of gravity acting on an object. On the moon, the Olympic shot-putter would weigh less than he does on earth, but he would still be just as difficult to start or stop moving horizontally, so his mass is the same as on the earth.

> **Mass is the measure of inertia, whereas weight is the measure of the force of gravity acting on an object.**

How important is mass in sport and exercise? Let's think about the sports in which the athlete may have to change the motion of something. The athlete may be concerned with changing the motion of an implement such as a ball or racket; with changing the motion of another athlete, such as in football or judo; or with changing the motion of his or her own body or body parts. So the mass of the implement or athlete or body part has a great effect on the execution of the performance, because the mass of the object to be moved or stopped determines how much effort is required to get the object to move or to stop the object from moving.

The kilogram is the SI unit of measurement for mass. You may have used the kilogram as a measure for weight as well. Mechanically speaking, describing something as weighing a certain number of kilograms is incorrect. The kilogram is a unit of measure for mass only, not weight. As weight is the force of gravity acting on an object, the unit of measure for weight is the unit of measure for force. Force and its units of measurement will be further described and defined in the next chapter. The kilogram is the SI unit of measure for mass and is abbreviated as kg.

We have now defined the three basic dimensions used in mechanics as length, time, and mass. Correspondingly, the three basic SI units of measurement used in mechanics are the meter, second, and kilogram. All other mechanical quantities and all other SI units used in mechanics can be derived from these three dimensions and their units of measurement (see table 3.1). What about force and weight? Certainly force must be a basic dimension used in mechanics, and not just a derivation of length, time, and mass. Simply defined, a force is a push or a pull, but mechanically, force can be defined from the dimensions of length, time, and mass. This mechanical definition of force will be developed in the next chapter.

So only three basic dimensions are needed to define all other mechanical dimensions. This is very neat and tidy, and it makes mechanics simple. Someone must have proposed a very tight set of laws for mechanics if only length, time, and mass, or derivations of these dimensions are needed to describe and explain the motion of objects. The person responsible for this was Isaac Newton.

TABLE 3.1 Basic Mechanical Quantities and Their SI Units of Measurement				
Quantity	Symbol	SI unit	Abbreviation for SI unit	Other units
Time	t	second	s	millisecond, minute, hour, day, week, year
Length	l	meter	m	millimeter, centimeter, inch, foot, yard, kilometer, mile
Mass	m	kilogram	kg	gram, slug

Newton's Laws

Isaac Newton was an English mathematician. He was born on Christmas Day in 1642, the same year that Galileo died and three months after the death of his father (Westfall 1993, p. 7). He died on March 20, 1727. Newton was a student at Cambridge University and subsequently became a professor there. Many of his ideas about mechanics (and calculus) were conceived during a two-year retreat to his family's estate in Lincolnshire when he was in his early twenties. This retreat was prompted by the plague in England, which caused temporary closings of the university in Cambridge between 1665 and 1667. It can be said that one benefit of the plague was that it allowed Isaac Newton to establish the groundwork for a new mechanics.

More than 20 years passed before Newton shared his work with others in 1686, when his book, *Philosophiae Naturalis Principia Mathematica (Mathematical Principles of Natural Philosophy)*, or *Principia*, as it is commonly referred to, was published. *Principia* was written in Latin, the language used by scientists during that time. In *Principia*, Newton presented his three laws of motion and his law of gravitation. These laws form the basis for modern mechanics. They are explained briefly here, and a more thorough explanation of their implications for human movement is developed in chapter 6.

Newton's First Law of Motion

Corpus omne perseverare in statu suo quiescendi vel movendi uniformiter in directum, nisi quatenus a viribus impressis cogitur statum illum mutare (Newton 1686/1934, p. 644).

This is Newton's first law of motion in Latin as presented in *Principia*. It is commonly referred to as the law of inertia. Translated directly, this law states: "Every body continues in its state of rest, or of uniform motion in a straight line, unless it is compelled to change that state by forces impressed upon it" (Newton 1686/1934, p. 13). This law explains what happens to an object if no external forces act on it or if the net external force (the resultant of all the external forces acting on it) is zero. More simply stated, Newton's first law says that if no net external force acts on an object, that object will not move (it will remain in its state of rest) if it wasn't moving to begin with, or it will continue moving at constant speed in a straight line (it will remain in its state of uniform motion in a straight line) if it was moving to begin with.

> **Every body continues in its state of rest, or of uniform motion in a straight line, unless it is compelled to change that state by forces impressed upon it.**

Newton's Second Law of Motion

Mutationem motis proportionalem esse vi motrici impressae, et fieri secundum lineam rectam qua vis illa imprimitur (Newton 1686/1934, p. 644).

This is Newton's second law of motion in Latin as presented in *Principia*. It is commonly referred to as the law of acceleration. Translated directly, this law states: "The change of motion of an object is proportional to the force impressed; and is made in the direction of the straight line in which the force is impressed" (Newton 1686/1934, p. 13). This law explains what happens if a net external force acts on an object. More simply stated, Newton's second law says: if a net external force is exerted on an object, the object will accelerate in the direction of the net external force and its acceleration will be directly proportional to the net external force and inversely proportional to its mass. This can be stated in a mathematical expression as:

$$\Sigma F = ma \qquad (3.1)$$

where

ΣF = net external force
m = mass of the object
a = acceleration of the object

This is a cause-and-effect relationship; forces cause acceleration. If a net external force acts on an object, that object accelerates. If an object accelerates, a net external force must be acting on it.

> **The change of motion of an object is proportional to the force impressed and is made in the direction of the straight line in which the force is impressed.**

Later we will learn that acceleration is derived from length and time. Newton's second law gives a definition for force based on the basic dimensions of length, time, and mass. So force is a derived measure after all.

Newton's Third Law of Motion

Actioni contrariam semper et aequalem esse reactionem: sive corporum duorum actiones in se mutuo semper esse aequales et in partes contrarias dirigi (Newton 1686/1934, p. 644).

This is Newton's third law of motion in Latin as presented in *Principia*. It is commonly referred to as the law of action-reaction. Translated directly, this law states: "To every action there is always opposed an equal reaction: or the mutual actions of two bodies upon each other are always equal and directed to contrary parts" (Newton 1686/1934, p. 13). Newton used the words *action* and *reaction* to mean force, so this law explains where the external forces required to change motion arise from. More simply stated, Newton's third law says: If an object (A) exerts a force on another object (B), the other object (B) exerts the same force on the first object (A) but in the opposite direction. So forces exist in mirrored pairs. The effects of these forces are not canceled by each other because they act on different objects. Another important point is that it is the forces that are equal but opposite, not the effects of the forces. A 150-lb running back running into a 300-lb linebacker exerts the same force on the linebacker as the linebacker exerts on him. The effect of the forces on the two football players will be much different, though, due to their different masses.

> **To every action there is always opposed an equal reaction.**

Newton's Law of Universal Gravitation

Newton's law of universal gravitation was purportedly inspired by the fall of an apple on his family's farm in Lincolnshire while he was residing there during the plague years (see figure 3.3). He presented this law in two parts. First, he stated that all objects attract each other with a gravitational force that is inversely proportional to the square of the distance between the objects. Second, he stated that this force of gravity was proportional to the mass of each of the two bodies being attracted to each other. The universal law of gravitation can be represented mathematically as:

$$F = G\left(\frac{m_1 m_2}{r^2}\right) \qquad (3.2)$$

where F is the force of gravity, G is the universal constant of gravitation, m_1 and m_2 are the masses of the two objects involved, and r is the distance between the centers of mass of the two objects.

Newton's universal law of gravitation was momentous because it provided a description of the

Figure 3.3 The traditional inspiration for Newton's law of universal gravitation.

forces that act between each object in the universe and every other object in the universe. This law, when used with his laws of motion, predicted the motion of planets and stars. The gravitational forces between most of the objects in sports are very small—so small that we can ignore them. However, one object that we must be concerned with in sports is large enough that it does produce a substantial gravitational force on other objects. That object is the earth. The earth's gravitational force acting on an object is the object's weight. For an object close to the earth's surface, several of the terms in equation 3.2 are constant. These terms are G, the universal constant of gravitation; m_2, the mass of the earth; and r, the distance from the center of the earth to its surface. If we introduce a new constant,

$$g = G\left(\frac{m_2}{r_2}\right) \qquad (3.3)$$

then equation 3.2 becomes

$$F = mg \qquad (3.4)$$

or

$$W = mg \qquad (3.5)$$

where W is the force of earth's gravity acting on the object, or the weight of the object; m is the mass of the object; and g is the acceleration of the object caused by the earth's gravitational force. The acceleration due to gravity, g, has been measured experimentally and varies slightly with variations in the earth's radius, but for most purposes, we can assume that g is constant at 9.81 meters per second per second (m/s/s) downward. For quick approximations, let's round 9.81 m/s/s to 10 m/s/s and use that as our estimate of the acceleration due to gravity. It will make things easier, and our approximation won't be too far off as our estimate of g is only 2% in error. If more accuracy is required, the more precise value of 9.81 m/s/s should be used.

Summary

Exercise and sport biomechanics is concerned primarily with that branch of mechanics called rigid-body mechanics. Statics and dynamics are the subdivisions of rigid-body mechanics. Kinematics and kinetics are the further subdivisions of dynamics. In later chapters, we will discover more about each of these subdivisions of mechanics.

The fundamental dimensions used in mechanics are length, time, and mass. The SI units of measurement for these dimensions are the meter (m) for length, the second (s) for time, and the kilogram (kg) for mass. All the other dimensions we will be using in biomechanics are derived from these three fundamental units. Newton's three laws of motion and his law of universal gravitation provide us with the relationships that indicate how many other

mechanical dimensions are derived. They tell us what a force is, what the force of gravity is, how an object behaves if no net force acts on it, how an object behaves if a net force does act on it, and what the sources of external forces are. We now have a knowledge of the basic mechanical concepts and principles needed for a mechanical analysis of human movement in sport. Carrying out such an analysis may be difficult at this point, however. In the chapters that follow, we will learn techniques for applying Newton's laws to problems in human movement that will make our mechanical analysis of human movements easier. Before we can do that, though, we need to have a better understanding of forces, as forces are involved in every one of Newton's laws. The next chapter will discuss forces.

Key Terms

dynamics *(p. 41)*

force *(p. 43)*

inertia *(p. 42)*

kinematics *(p. 41)*

kinetics *(p. 41)*

mass *(p. 43)*

mechanics *(p. 40)*

rigid-body mechanics *(p. 41)*

statics *(p. 41)*

weight *(p. 43)*

REVIEW QUESTIONS

1. **Can a body be moving if no external forces are acting on it?**

2. **Can a body change its direction of motion if no external forces act on it?**

3. **Can a body be moving with constant velocity if external forces are acting on it?**

4. **If you pull on a tug-of-war rope with a force of 50 lb, how much force does the rope exert on you?**

5. **How tall are you in meters?**

6. **What is your mass in kilograms?**

II

External Biomechanics: External Forces and Their Effects on the Body and Its Movement

The movements of the body as a whole are determined by the external forces that act on the body. Part II is concerned with these forces and their effects on the body and its motion. Almost all of this part is devoted to rigid-body mechanics and its application to human movement. Part II includes eight chapters. The first four chapters, chapters 4 through 7, are concerned with linear kinematics and kinetics. Chapters 8 through 10 cover angular kinematics and kinetics. Part II concludes with a discussion of fluid mechanics in chapter 11.

Forces: Maintaining Equilibrium or Changing Motion

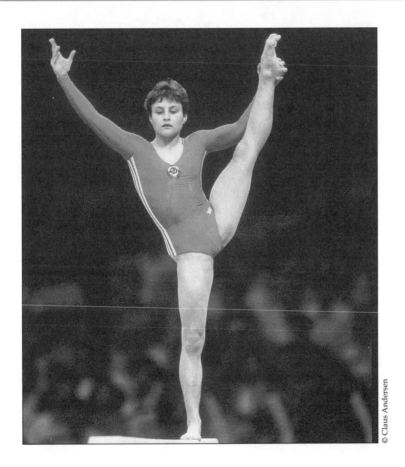

© Claus Andersen

A gymnast maintains a precarious position on one foot during a balance beam routine. A rock climber clings to the face of a cliff by his fingertips. A cyclist is motionless on her bicycle at the start of a race. A diver is supported only by his toes on the edge of the diving board before executing a back dive. What are the forces that act on each of these athletes? How do the athletes manipulate these forces to maintain balance? The information presented in this chapter will provide you with the knowledge you need to answer these questions.

OBJECTIVES

When you finish this chapter, you should be able to do the following:

▮ Define force

▮ Classify forces

▮ Define friction force

▮ Determine the resultant of two or more forces

▮ Resolve a force into component forces that act at right angles to each other

▮ Determine if an object is in static equilibrium when the forces acting on the object are known

▮ Determine an unknown force acting on an object if all the other forces acting on the object are known and the object is in static equilibrium

Forces are involved in every one of Newton's laws. Our bodies and the objects we manipulate are continually subjected to forces. Forces are important for motion, as they enable us to start moving, stop moving, and change directions. These are skills found in all sports. Forces are also important even if we aren't moving. Combinations of two or more forces enable us to maintain our balance in stationary positions. To use Newton's laws of motion analytically, we need a basic understanding of forces: how to add them to produce a resultant force, how to resolve forces into component forces, and how forces must act to maintain equilibrium.

What Are Forces?

Simply defined, a force is a push or a pull. Newton's second law provides a more precise mechanical definition: A force is something that has the ability to accelerate an object. Mechanically speaking, something accelerates when it starts, stops, speeds up, slows down, or changes direction. So a force is something that can cause an object to start, stop, speed up, slow down, or change direction.

Simply defined, a force is a push or a pull.

Our most familiar unit of measurement for force is the pound, but the SI unit of measurement for force is the newton, named in honor of Isaac Newton. The newton is abbreviated as N. One newton of force is defined as the force required to accelerate a 1-kg mass 1 m/s/s, or algebraically:

$$1.0 \text{ N} = (1.0 \text{ kg})(1.0 \text{ m/s/s}) \qquad (4.1)$$

One newton of force is equal to 0.225 lb of force, or 1 lb equals 4.448 N.

Think about how to describe a force. For instance, suppose you wished to describe the force a shot-putter exerted on a shot at the instant shown in figure 4.1. Would describing the size of the force provide enough information about it to predict its effect? What else would we want to know about the force? Some other important characteristics of a force are its point of application, its direction (line of action), and its sense (whether it pushes or pulls along this line). A force is what is known as a vector. A **vector** is a mathematical representation of anything that is defined by its size or magnitude (a number) and its direction (its orientation). To fully describe a force, you must describe its size and direction.

If we wanted to represent a force (or any other vector) graphically, an arrow makes a good representation. The length of the arrow indicates the size of the force, the shaft of the arrow indicates its line of application, the arrowhead indicates its sense or direction along that line of application, and one of the arrow's ends indicates the point of application of the force.

Force acting
on shot-putter

Force acting
on shot

Figure 4.1 The forces acting on a shot-putter and a shot at the instant before release.

Classifying Forces

Now let's consider the different types of forces and how they are classified. Newton's laws of motion make reference to external forces. This implies that there must be internal forces. Forces can thus be classified as internal or external.

Internal Forces

Internal forces are forces that act within the object or system whose motion is being investigated. Newton's third law indicates that forces come in pairs: action and reaction forces. With internal forces, the action and reaction forces act on different parts of the system (or body). Each of these forces may affect the part of the body it acts on, but the two forces do not affect the motion of the whole body because the forces act in opposition. So Newton's first and second laws of motion characterize external forces as the only forces that can change the motion of an object.

> **Internal forces are forces that act within the object or system whose motion is being investigated.**

In sport biomechanics, the objects whose motion we are curious about are the athlete's body and the implement(s) manipulated by the athlete. The human body is a system of structures—organs, bones, muscles, tendons, ligaments, cartilage, and other tissues. These structures exert forces on one another. Muscles pull on tendons, which pull on bones. At joints, bones push on cartilage, which pushes on other cartilage and bones. If pulling forces act on the ends of an internal structure, the internal pulling forces are referred to as tensile forces and the structure is under tension. If pushing forces act on the ends of an internal structure, the internal pushing forces are referred to as compressive forces and the structure is under compression. Internal forces hold things together when the structure is under tension or compression. Sometimes the tensile or compressive forces acting on a structure are greater than the internal forces the structure can withstand. When this happens, the structure fails and breaks. Structural failure occurs when muscles pull, tendons rupture, ligaments tear, and bones break.

We think of muscles as the structures that produce the forces that cause us to change our motion. Actually, as muscles can only produce internal forces, they are incapable of producing changes in the motion of the body's center of mass. It is true that muscle forces can produce motions of the body's limbs, but these motions will not produce any change in motion of the body's center of mass without external forces acting on the system. The body is only able to change its motion if it can push or pull against some external object. Imagine a defensive player in basketball jumping up to block a shot (see figure 4.2). If he has been fooled by the shooter and jumps too early, he can't stop himself in midair to wait for the shooter to shoot. He needs to touch something to have an external force act on him. So he has to get his feet back on the ground. Then he can push against the ground and create an external reaction force that causes him to jump up again. The ground provides the external force that causes the change in motion of the basketball player.

Internal forces may be important in the study of sport biomechanics if we are concerned about the nature and causes of injury, but they cannot produce any changes in the motion of the body's center of mass. External forces are solely responsible for that.

Figure 4.2 A basketball player cannot change his motion once he has jumped into the air.

External Forces

External forces are those forces that act on an object as a result of its interaction with the environment surrounding it. We can classify external forces as contact forces or noncontact forces. Most of the forces we think about are contact forces. These occur when objects are touching each other. Noncontact forces are forces that occur even if the objects are not touching each other. The gravitational attraction of the earth is a noncontact force. Other noncontact forces include magnetic forces and electrical forces.

> **External forces are those forces that act on an object as a result of its interaction with the environment surrounding it.**

In sports, the only noncontact force we will concern ourselves with is the force of gravity. In Newton's law of universal gravitation, the concept of gravitational force was developed, and the force of earth's gravity acting on an object was defined as the object's weight. Weight is a force—the force of earth's gravity. Weight is the only gravitational force we will concern ourselves with in sport biomechanics.

Contact forces are forces that occur between objects in contact with each other. The objects in contact can be solid or fluid. Air resistance and water resistance are examples of fluid contact forces, which will be further discussed in chapter 11. The most important contact forces in sports occur between solid objects, such as the athlete and some other object. For a shot-putter to put the shot, the athlete must apply a force to it, and the only way the athlete can apply a force to the shot is to touch it. To jump up in the air, you must be in contact with the ground and push down on it. The reaction force from the ground pushes up on you and accelerates you up into the air. To accelerate yourself forward as you take a running step, you must be in contact with the ground and push backward against it. The reaction force from the ground pushes forward against you and accelerates you forward.

Contact forces can be resolved into parts or components—the component of force that acts perpendicular to the surfaces of the objects in contact and the component of force that acts parallel to the surfaces in contact. We will call the first component of contact force a normal contact force, where normal refers to the fact that the line of action of this force is perpendicular to the surfaces in contact. During a running step, when the runner pushes down and backward on the ground, the normal contact force is the component of force that acts upward on the runner and downward on the ground. The second component of the contact force is called **friction**. The line of action of friction is parallel to the two surfaces in contact and opposes motion or sliding between the surfaces. So when the runner pushes down and backward on the ground during a running step, the frictional force is the component of force that acts forward on the runner and backward on the ground (see figure 4.3). The frictional force is the component of the contact force responsible for changes in the runner's horizontal motion. Frictional forces are primarily responsible for human locomotion, so an understanding of friction is important.

Friction

The frictional force described above is dry friction, which is also referred to as Coulomb friction. Another type of friction is fluid friction, which de-

Friction force

Normal contact force

| Figure 4.3 | Normal contact force and friction force acting on a runner's foot during push-off. |

velops between two layers of fluid and occurs when dry surfaces are lubricated. The behavior of fluid friction is complicated, and because it occurs less frequently in sports, we will limit our discussion to dry friction. Dry friction acts between the non-lubricated surfaces of solid objects or rigid bodies in contact and acts parallel to the contact surfaces. Friction arises as a result of interactions between the molecules of the surfaces in contact. When dry friction acts between two surfaces that are not moving relative to each other, it is referred to as static

friction or limiting friction. When dry friction acts between two surfaces that are moving relative to each other, it is referred to as dynamic friction. Other terms for dynamic friction are sliding friction and kinetic friction.

Friction and Normal Contact Force

By adding books to the pile in the self-experiment, what have we done to it to cause the static friction force to increase? We have increased the inertia of the pile by increasing its mass. This shouldn't affect the static friction force, though, as there is no apparent way an increase in mass could affect the interactions of the molecules of the contacting surfaces. It is these interactions that are responsible for friction. We also increased the weight of the pile as we added books to it. Could this affect the static friction force? Well, increasing the weight would increase the normal contact force acting between the two surfaces. This would increase the interactions of the molecules of the contacting surfaces, because they would be pushed together harder. So it is not the weight of the books that caused the increased static friction force, but the increase in the normal contact force. If we measured this normal contact force and the friction force, we would find that the friction force is proportional to the normal contact force. As one increases, the other increases proportionally. This is true for both static and dynamic friction.

SELF-EXPERIMENT

Let's do some experimentation to learn more about friction. Place a book on a flat horizontal surface such as a desk or tabletop. Now push sideways against the book and feel how much force you can exert before the book begins to move. What force resists the force that you exert on the book and prevents the book from sliding? The resisting force is static friction, which is exerted on the book by the table or desk. If the book doesn't slide, from Newton's second law, we can deduce that the static friction force acting on the book is the same size as the force you exert on the book but in the opposite direction. So the effects of these forces are canceled and the net force acting on the book is zero. Put another book on top of the original book and push again (see figure 4.4). Can you push with a greater force before the books begin to move? Add another book and push again. Can you push with an even greater force now? As you add books to the pile, the magnitude (size) of the force you exert before the books slide becomes greater and so does the static friction force.

Figure 4.4   Adding books to the stack increases static friction between the bottom book and the table.

Friction force is proportional to the normal contact force.

Friction and Surface Area

With the different orientations in the self-experiment, the surface areas in contact between the book and table vary dramatically, but friction did not change noticeably. In fact, dry friction, both static and dynamic, is not affected by the size of the surface area in contact. This statement is probably not in agreement with your previous notions about friction, but you have just demonstrated it to yourself. If that isn't enough to convince you that dry friction is unaffected by surface area, let's try to explain it.

Dry friction arises due to the interaction of the molecules at the surface areas in contact. We have seen that if we press these surfaces together with greater force, the interactions of the molecules will be greater and friction will increase. It makes sense to say that if we increase the area of the surfaces in contact, we also increase the number of molecules that can interact with each other, and thus we create more friction. But if the force pushing the surfaces together remains the same, with the greater surface area in contact, this force is spread over a greater area and the pressure between the surfaces will be less. So the individual forces pushing each of the

SELF-EXPERIMENT

Now try holding the book against a vertical surface such as a wall (see figure 4.5). Can you do this if you only push against the book with a horizontal force? How hard must you push against the book to keep it from sliding down the wall? What force opposes the weight of the book and prevents the book from falling? The force of your hand pressing against the book is acting horizontally, so it can't oppose the vertical force of gravity, which is pulling down on the book. The force acting upward on the book is friction between the book and the wall (and possibly between the book and your hand). But the force you exert against the book affects friction, because the book will slide and fall if you don't push hard enough. Again we see that friction is affected by the normal contact force, the contact force acting perpendicular to the friction and the contact surfaces.

increase in pressure decreases the magnitude of these interactions. So the net effect is zero and friction is unchanged.

Friction is not affected by the size of the surface area in contact.

Friction and Contacting Materials

One more observation about friction. When you moved the book back and forth across the table in the above experiment, was it easier to get the book started or to keep the book moving? In other words, was static friction larger or smaller than dynamic friction? It was easier to keep the book moving than to get it started moving, so static friction is larger than dynamic friction.

Let's summarize what we now know about dry friction. Friction is a contact force that acts between and parallel to the two surfaces in contact. Friction opposes relative motion (or impending relative motion) between the surfaces in contact. Friction is proportional to the normal contact force pushing the two surfaces together. This means that as the normal contact force increases, the frictional force increases as well. If the normal contact force doubles in size, the frictional force will double in size also. Friction is affected by the characteristics of the surfaces in contact. Greater friction can be developed between softer and rougher surfaces than between harder and smoother surfaces. Finally, static friction is greater than dynamic friction. Mathematically, we can express static and dynamic friction as:

$$F_s = \mu_s R \qquad (4.2)$$

$$F_d = \mu_d R \qquad (4.3)$$

| Figure 4.5 | Friction force between the book and the wall and between the book and your hand is enough to hold the book up. |

molecules together at the contact surfaces will be smaller, thus decreasing the interactions between the molecules and decreasing the friction. It looks like a tradeoff. The increase in surface area increases the number of molecular interactions, but the de-

SELF-EXPERIMENT

Let's investigate another aspect of friction. Does surface area of contact affect friction? Take a hardcover book and lay it on the table or desk (it's important that you use a hardcover book). Push the book back and forth across the table and get a feeling for how large the dynamic and static friction forces are. Try to exert only horizontal forces on the book. Now try the same thing with the book on its ends and sides (as in figure 4.6). Use a rubber band to hold the book closed, but don't let the rubber band touch the table as you're sliding the book. Are there any noticeable differences between the frictional forces you feel with the book in its different orientations? Try it with another hardcover book.

where F_s is the static friction force, μ_s is the coefficient of static friction, R is the normal contact force, F_d is the dynamic friction force, and μ_d is the coefficient of dynamic friction.

Friction in Sport and Human Movement

Friction is an important force in every sport and human movement. Locomotion requires frictional forces, so the shoes we wear are designed to provide proper frictional forces between our feet and the supporting surface. In most athletic shoes, we want large frictional forces, so the materials used for the soles have large coefficients of friction. In some activities, such as dancing or bowling, sliding is desirable, so the soles of the shoes used for these activities have smaller coefficients of friction. In snow skiing, we also want small frictional forces,

Figure 4.6   A book on its end has a smaller area of contact with the table. Does this reduced area of contact affect the frictional force between the book and the table?

SELF-EXPERIMENT

Now let's observe the difference between the frictions of a book on the table and a shoe on the table. Place the book on the table and put an athletic shoe on top of it. Push the book back and forth across the table and get a feeling for how large the dynamic and static friction forces are. Now put the shoe on the table, sole down, and place the book on top of it. Push the shoe back and forth across the table and get a feeling for how large the dynamic and static friction forces are. Which produced larger frictional forces with the table, the book or the shoe? What changed between the two conditions? In both conditions, the weight and mass of the objects being moved (the shoe and book) stayed the same. The surface area of contact changed, but we determined that friction was unaffected by that. The change that must be responsible for the changes in the observed frictional force was the difference in the type of material that was in contact with the table. Greater friction existed between the table and the softer and rougher sole of the shoe than between the table and the smoother and harder book cover.

so we wax the bottoms of our skis to decrease the coefficient of friction. In racket sports and other sports involving implements, large frictional forces are desirable so that we don't lose hold of the implement. The grips are made of material such as leather or rubber, which have large coefficients of friction. We may even alter the grips to increase their coefficients of friction by wrapping athletic tape on them, spraying them with tacky substances, or using chalk on our hands. Think about the variety of sports you have been involved in and how friction affects performance in them.

We now know about several of the various external forces that can act on us in sports activities; gravity, friction, and contact forces are the major ones. In most sport situations, more than one of these external forces will act on the athlete. Now let's go back to Newton's first two laws of motion. Newton's first law says that if no net external force acts on an object, that object will not move if it wasn't moving to begin with, or it will continue moving at constant speed in a straight line if it was moving to begin with. Newton's second law says that if a net external force acts on an object, the object will accelerate in the direction of the net external force and its acceleration will be directly proportional to the net external force and inversely proportional to the object's mass. We now know more about forces, but what is a net external force?

Addition of Forces: Force Composition

The net force acting on an object is the sum of all the external forces acting on it. This sum is not an algebraic sum; that is, we can't just add up the numbers that represent the sizes of the forces. The net force is the vector sum of all the external forces. Remember that we define a force as a push or pull, and that forces are vector quantities. This means that the full description of a force includes its magnitude (how large is it?) and its direction (which way does it act?). Visually, we can think of forces as arrows, with the length of the shaft representing the magnitude of the force, the orientation of the arrow representing its line of application, and the arrowhead indicating its direction of action along that line. When we add vectors such as forces, we can't just add up the numbers representing their sizes. We must also consider the directions of the forces. Forces are added using the process of vector addition. The result of vector addition of forces is called

the resultant force. The vector addition of all the external forces acting on an object is the net force. It is also referred to as the resultant force, because it results from the addition of all the external forces. Now we will learn how to carry out vector addition of forces.

> **The vector addition of all the external forces acting on an object is the net force.**

Colinear Forces

To begin our understanding of vector addition, let's start with a simple case that involves colinear forces. If you look closely at the word colinear, you may notice that the word *line* appears in it. Colinear forces are forces that have the same line of action. The forces may act in the same direction or in opposite directions along that line. Now here's the situation. You are on a tug-of-war team with two others. You pull on the rope with a force of 100 N and your teammates pull with forces of 200 N and 400 N. You are all pulling along the same line—the line of the rope. To find out the resultant of these three forces, we begin by graphically representing each force as an arrow, with the length of each arrow scaled to the size of the force. First, draw the 100-N force that you exerted on the rope. If you were pulling to the right, the force you exerted on the rope might be represented like this:

100 N
———>

Now draw an arrow representing the 200-N force. Put the tail of this force at the arrowhead of the 100-N force. If it is scaled correctly, this arrow should be twice as long as the arrow representing the 100-N force.

100 N   200 N
———>————>

Now draw the arrow representing the 400-N force. Put the tail of this force at the arrowhead of the 200-N force. This arrow should be four times as long as the arrow representing the 100-N force and twice as long as the arrow representing the 200-N force. Your drawing should look something like this:

100 N   200 N   400 N
———>———>————————>

An arrow drawn from the tail of the 100-N force to the tip of the 400-N force represents the resultant force, or the vector sum of the 100-N, 200-N, and 400-N forces, if we put the arrowhead on the end where it meets the tip of the 400-N force.

700 N

If we measure the length of this arrow, it turns out to be seven times as long as the 100-N force. The resultant force must be a 700-N force acting to the right. But this is what we would have found if we added the magnitudes of the three forces algebraically:

100 N + 200 N + 400 N = 700 N

Does that mean that vector addition and algebraic addition are the same? No! This is true only when the forces all act along the same line and in the same direction.

> **When forces act along the same line and in the same direction, they can be added using regular algebraic addition.**

Now let's consider the forces the opposing team exerts on the rope. That team also consists of three members. The forces they exert on the rope are to the left and are 200 N, 200 N, and 200 N, respectively. What is the resultant of these three forces?

200 N   200 N   200 N

600 N

We can determine the resultant by graphically representing the three forces as arrows and connecting the tail of the first arrow with the tip of the last arrow, as we did previously. We could also have added the magnitudes of the forces algebraically, because all three forces acted along the same line in the same direction.

200 N + 200 N + 200 N = 600 N

Now what is the resultant force acting on the rope as a result of your team pulling to the right and the opposing team pulling to the left? In this case, we have the three forces from your team pulling to the right

100 N   200 N   400 N

and the three forces from the opposing team pulling to the left

200 N   200 N   200 N

These forces are all still colinear because they act along the same line, the line of the rope in this case. If we follow the procedure we used before, we add the forces graphically by lining the vectors up tip to tail. We have done this for the three forces from your team. The tail of the 200-N force is lined up with the tip of the 100-N force, and the tail of the 400-N force is lined up with the tip of the 200-N force. We have done this for the opposing team's forces as well. Now, to add up all of these forces, we line up the tail of the 200-N force of the opposing team with the tip of the 400-N force of your team (we could have lined up the tail of your 100-N force with the tip of the 200-N force from the opposing team):

100 N   200 N   400 N

200 N   200 N   200 N

The resultant force is found by drawing an arrow from the tail of the 100-N force to the tip of the last 200-N force, with the tip of the arrow at the end meeting the tip of the 200-N force and the tail at the end meeting the tail of the 100-N force:

100 N   200 N   400 N

resultant   200 N   200 N   200 N

If we measure the length of this resultant vector, we see that it is the same length as the 100-N force. The resultant force is 100 N to the right.

We could have arrived at the same resultant if we replaced the forces exerted by your team with its 700-N resultant force to the right and the forces by the opposing team with its 600-N resultant force to the left.

700 N

resultant   600 N

Because all of the forces are acting along the same line, the resultant force could also be found through algebraic means. Now, rather than just adding up the forces algebraically, we must also consider the direction in which the forces act. Let's arbitrarily say that forces acting to the right are positive. Then forces acting to the left must be considered negative. So the resultant force acting on the rope as a result of your team pulling on it to the right and the opposing team pulling on it to the left can now be determined algebraically by adding up the positive forces from your team and the negative forces from the opposing team.

$$100 \text{ N} + 200 \text{ N} + 400 \text{ N} +$$
$$(-200 \text{ N}) + (-200 \text{ N}) + (-200 \text{ N}) = +100 \text{ N}$$

Adding a negative number is just like subtracting it, so we could also write this as:

$$100 \text{ N} + 200 \text{ N} + 400 \text{ N} -$$
$$200 \text{ N} - 200 \text{ N} - 200 \text{ N} = +100 \text{ N}$$

The positive sign associated with our answer of 100 N indicates that the resultant force acts in the positive direction. We set up our positive direction to the right, so the resultant force is a force of 100 N to the right.

If forces are colinear, they may be added using vector addition by graphically representing each force as an arrow and arranging the force arrows tip to tail. The resultant force is determined by drawing an arrow from the tail of the first force to the tip of the last force. This arrow has its tip at the tip of the last force, and it represents the resultant force. Colinear forces may also be added algebraically if the senses of the forces on the line along which they act are taken into account by assigning positive or negative signs to the magnitudes of the forces. Positive forces act in one direction along the line, and negative forces act in the opposite direction along the line.

Concurrent Forces

Now let's consider a situation in which the external forces are not colinear. Instead, the forces are concurrent. Concurrent forces do not act along the same line, but they do act through the same point. A gymnast is about to begin his routine on the high bar. He jumps up and grasps the bar, and his coach stops his swinging by exerting forces on the front and back of the gymnast's torso. The external forces acting on the gymnast are the force of gravity acting on the mass of the gymnast, a horizontal force of 20 N exerted by the coach pushing on the front

of the gymnast, a horizontal force of 30 N exerted by the coach pushing on the back of the gymnast, and an upward vertical reaction force of 550 N exerted by the bar on the gymnast's hands. The gymnast's mass is 50 kg. What is the net external force acting on the gymnast?

First, how large is the force of gravity that acts on the gymnast? If you remember, in chapter 3, we said that the force of gravity acting on an object was the object's weight. What is the gymnast's weight? Weight is defined with the equation

$$W = mg \qquad (4.4)$$

where W represented weight in newtons, m represented mass in kilograms, and g represented the acceleration due to gravity, or 9.81 m/s/s. For a good approximation, we can round 9.81 m/s/s to 10 m/s/s and make our computations easier. If we want more accuracy, we should use 9.81 m/s/s rather than 10 m/s/s. So, using the rougher approximation for g, the gymnast weighs

$$W = mg = (50 \text{ kg})(10 \text{ m/s/s}) = 500 \text{ kg m/s/s} = 500 \text{ N}$$

This weight is a downward force of 500 N. We now have all the forces that act on the gymnast. A drawing of the gymnast and all the external forces that act on him is shown in figure 4.7.

Now we can begin the process of determining the resultant of these forces. Just as we did with the colinear forces, we can represent each force graphically with an arrow, scaling the length of the arrow to represent the magnitude of the force, orienting the arrow to show its line of application, and using an arrowhead to show its sense or direction. As with the colinear forces, if we line up the forces tip to tail, we can find the resultant. Let's do that. First, draw the 20-N horizontal force acting to the right. Now draw the 550-N upward force so that its tail begins at the head of the 20-N force. Draw the 30-N horizontal force to the left so that the tail of this force begins at the head of the 550-N force. Draw the 500-N downward force of gravity so that the tail of this force begins at the head of the 30-N force. You should now have a drawing that looks something like figure 4.8.

The head of the 500-N downward force and the tail of the 20-N horizontal force do not connect. The resultant of the four forces can be represented by an arrow connecting the tail of the 20-N horizontal force (the first force in our drawing) with the head of the 500-N downward force (the last force in our drawing). Figure 4.9 shows the construction of the resultant force. This resultant force is directed from

550 N

20 N → ← 30 N

W = 500 N

Figure 4.7 Free-body diagram showing the external forces acting on a gymnast hanging from the horizontal bar.

30 N

500 N

550 N

20 N

Figure 4.8 Graphic representation of all forces acting on the gymnast.

the tail of the 20-N horizontal force to the head of the 500-N downward force. The resultant is thus directed upward and slightly to the left. The size of the resultant is indicated by the length of the arrow representing it. Using the same scale that was used to construct the other forces in figure 4.9, we can estimate that the magnitude of the resultant force is 51 N.

If we describe the direction of the resultant force as "upward and slightly to the left," we haven't provided a very precise description. Can the direction of the force be described with more precision than that? We could describe the angle that the force makes with a vertical line or a horizontal line. Measuring clockwise from a vertical line, this force is about 11° from vertical. This angular description is

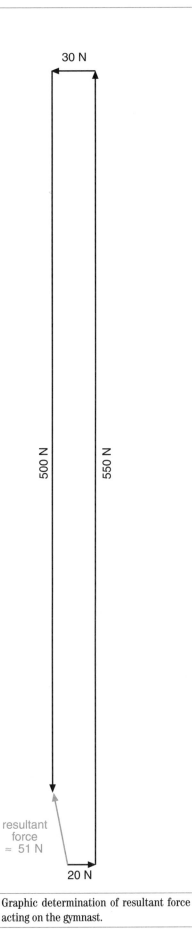

30 N

500 N

550 N

resultant
force
≈ 51 N

20 N

Figure 4.9 Graphic determination of resultant force
acting on the gymnast.

much more precise than describing the force as "upward and slightly to the left."

If vertical and horizontal forces act on a body, we can add forces graphically, as we did to determine the resultant force. Is there any way we can determine the resultant force without using graphical means? Is there a mathematical technique we can use? Let's again consider the four forces acting on the gymnast. Horizontally, there are two forces acting: a 20-N force to the right and a 30-N force to the left. Vertically, there are also two forces acting: a 500-N force downward and a 550-N force upward. Can we just add up all of these forces algebraically? If we did, we would have:

$$20 \text{ N} + 30 \text{ N} + 500 \text{ N} + 550 \text{ N} = 1100 \text{ N}$$

This is much different than what we determined graphically. Maybe we need to consider the downward forces as negative and the forces to the left as negative. Using this method, we have:

$$20 \text{ N} + (-30 \text{ N}) + (-500 \text{ N}) + 550 \text{ N} =$$
$$20 \text{ N} - 30 \text{ N} - 500 \text{ N} + 550 \text{ N} =$$
$$40 \text{ N}$$

This is much closer to the graphical result, but it still is not correct. We also don't know in which direction the resultant acts. Let's try one more method. Consider the horizontal and vertical forces separately and determine what the horizontal resultant force is and what the vertical resultant force is. Now the problem is similar to the colinear force problems we solved earlier.

Horizontally, we have a 20-N force acting to the right and a 30-N force acting to the left. Previously, we arbitrarily assigned a negative value to forces that acted to the left, so the resultant horizontal force is:

$$20 \text{ N} + (-30 \text{ N}) = 20 \text{ N} - 30 \text{ N} = -10 \text{ N}$$

The negative sign associated with this force indicates that it acts to the left. The resultant horizontal force is 10 N acting to the left.

Vertically, we have a 500-N force acting downward and a 550-N force acting upward. We arbitrarily assign a negative value to the downward force. The resultant vertical force is:

$$(-500 \text{ N}) + 550 \text{ N} = +50 \text{ N}$$

The positive sign associated with this force indicates that it acts in an upward direction. The resultant vertical force is 50 N acting upward.

Using this method, the resultant force can be expressed as a 10-N horizontal force acting to the left and a 50-N vertical force acting upward. Is this equivalent to the 51-N resultant force that acts upward and slightly to the left at 11° from vertical? How can a 51-N force be equivalent to a 50-N force and a 10-N force? Add the horizontal resultant force of 10 N and the vertical force of 50 N graphically to determine their resultant force. Draw the forces tip to tail, as shown in figure 4.10.

Now draw the resultant by connecting the tail of the 10-N horizontal force with the tip of the 50-N vertical force. How does this force compare to the resultant shown in figure 4.10? They look identical. Measure the resultant in figure 4.10, and measure the angle it makes with the vertical. The resultant force is about 51 N and makes an angle of 11° with vertical. It is identical to the resultant force shown in figure 4.10. Apparently, a 50-N force and a 10-N force can be equivalent to a 51-N force.

Take a closer look at the shape created by the three forces in figure 4.10. It's a triangle. In fact, it's a right triangle—one of the angles in the triangle is a 90° angle. The 90° angle is formed between the sides of the triangle representing the horizontal re-

sultant force and the vertical resultant force. There are special relationships among the sides of a right triangle. One of these relates the lengths of the two sides that make the right angle with the length of the side opposite the right angle. If A and B represent the two sides that make up the right angle and C represents the hypotenuse (the side opposite the right angle), then

$$A^2 + B^2 = C^2 \qquad (4.5)$$

This relationship is called the Pythagorean theorem. For our force triangle, then, we can substitute 10 N for A and 50 N for B and then solve for C, which represents the resultant force.

$$(10 \text{ N})^2 + (50 \text{ N})^2 = C^2$$
$$100 \text{ N}^2 + 2500 \text{ N}^2 = C^2$$
$$2600 \text{ N}^2 = C^2$$
$$C = 51 \text{ N}$$

This gives us an answer identical to what we got when we actually measured the graphical representation of the force.

If forces are concurrent but not colinear, we can add the forces to determine their resultant by graphically representing the forces and arranging them tip to tail. The resultant force will be represented by an arrow drawn from the tail of the first force to the tip of the last force represented. Alternatively, we can algebraically add up all the horizontal forces to determine the resultant horizontal force, then add up all the vertical forces and determine the resultant vertical force. The size of the resultant of these two forces can be determined using the Pythagorean theorem.

Resolution of Forces

What if the external forces acting on the object are not colinear and do not act in a vertical or horizontal direction? Look back at figure 4.1 and consider the forces acting on a shot during the putting action. Imagine that at the instant shown, the athlete exerts a 100-N force on the shot at an angle of 60°. The mass of the shot is 4 kg. What is the net external force acting on the shot? First, we need to determine the weight of the shot.

$$W = mg = (4 \text{ kg})(10 \text{ m/s/s}) = 40 \text{ N}$$

Now we can determine the net external force by graphically adding the 40-N weight of the shot to

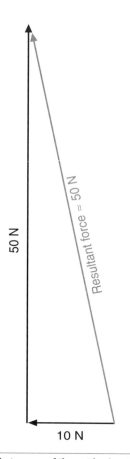

50 N

Resultant force = 50 N

10 N

Figure 4.10   Vector sum of the net horizontal force and net vertical force acting on the gymnast.

the 100-N force exerted by the athlete. Try doing this. Your graphic solution should be similar to figure 4.11. If we measure the resultant force, it appears to be about 68 N. It acts upward and to the right at an angle of about 36°.

Is there another method we could use to determine this resultant, as we did with the gymnast problem earlier? Recall that the external forces acting on the gymnast were all horizontal or vertical forces. In that problem, we could just sum the horizontal forces and the vertical forces algebraically to find the resultant horizontal and vertical forces. In the shot-putting problem, we have one vertical force, the shot's weight, but the force from the athlete is acting both horizontally and vertically. It is pushing upward and forward on the shot. As this 100-N force acts to push the shot both horizontally and vertically, perhaps it can be represented by two different forces: a horizontal force and a vertical force.

Graphical Technique

Let's start by looking at the problem graphically. We want to represent the 100-N force that acts forward and upward at 60° above horizontal as a pair of forces. The pair of forces we are trying to find are

called the horizontal and vertical components of this 100-N force. You may be familiar with the word *component*. If you have a component stereo system, the components of the system are the receiver, tape deck, compact disc player, and so on. They are the parts that make up the system. Similarly, the horizontal and vertical force components of the 100-N force are the parts that make up or have the same effect as the 100-N force. We can think of the 100-N force as the resultant of adding the horizontal and vertical components of this force. Let's draw the 100-N force as a vector, as shown in figure 4.12.

If we think about how we graphically determined the resultant of two forces, we lined up the arrows representing these forces end to end and then drew an arrow from the tail of the first force arrow to the tip of the last force arrow in the sequence. This force arrow we drew was the resultant. Now we want to work that process in reverse. We know what the resultant force is, but we want to know what horizontal and vertical forces can be added together to produce this resultant. One of the forces must end at the tip of the 100-N force and the other must start at the tail of the 100-N force, so let's arbitrarily draw a vertical line through the tip of the 100-N force and a horizontal line through the tail of the 100-N force, as shown in figure 4.13.

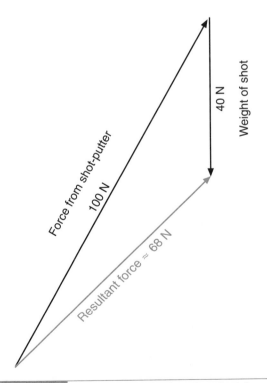

| Figure 4.11 | Graphic determination of resultant force acting on the shot. |

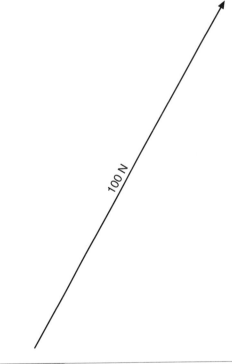

| Figure 4.12 | Vector representation of 100-N force exerted by the shot-putter on the shot. |

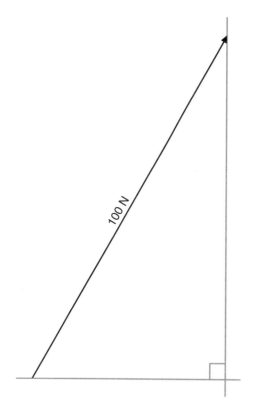

Figure 4.13 Horizontal line through tail and vertical line through tip of 100-N force vector.

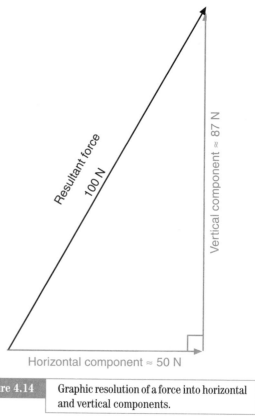

Figure 4.14 Graphic resolution of a force into horizontal and vertical components.

We know that the tail of the horizontal force component must be at the tail of the 100-N resultant and that the tip of the vertical force component must be at the tip of the 100-N resultant, so the tip of the horizontal force component must be at the tail of the vertical force component. As the horizontal and vertical lines intersect at only one point, this must be where the tip of the horizontal force component and the tail of the vertical force component meet. Let's put arrowheads on them in the proper places, as shown in figure 4.14.

If we measure the lengths of these force vectors, we find that the horizontal force component is about 50 N and the vertical force component is about 87 N. You may also notice that the shape produced by these three forces is a right triangle, so the Pythagorean theorem must apply.

$$A^2 + B^2 = C^2 \qquad (4.5)$$

For our force triangle, then, we can substitute 50 N for A, 87 N for B, and 100 N for C. Let's check to see if it works.

$$(50 \text{ N})^2 + (87 \text{ N})^2 = (100 \text{ N})^2$$
$$2500 \text{ N}^2 + 7569 \text{ N}^2 = 10{,}000 \text{ N}^2$$
$$10{,}069 \text{ N}^2 \cong 10{,}000 \text{ N}^2$$

Although 10,069 doesn't exactly equal 10,000, the difference is less than 1%.

That's pretty close, especially considering our accuracy in measuring the length of the force arrows. If the measurement accuracy were increased, the difference between the two numbers would become closer to zero.

The process of determining what two force components add together to make a resultant force is called force resolution. We resolved a force into its components. The word *resolve* sounds like *re-solve*, which is what we did. We had the resultant force and we solved the problem backward—we *re-solved* it—to determine the forces that added together to yield this resultant. But it was all done graphically. We want a nongraphical technique for doing this.

Trigonometric Technique

Let's look at the force triangle we ended up with in figure 4.14. It is a right triangle, which means that one of the angles of the triangle is 90°, or a right angle. Besides the Pythagorean theorem, there are other relationships between the sides and the angles of a right triangle. If we know the length of one side of a right triangle and the measurement of one of the angles other than the right angle, we can determine the lengths of the other sides and the measurement of the other angle using trigonometry. Trigonometry was not a prerequisite for this course, and the intent of this book is not to teach you trigonometry, but a knowledge of some of the tools of trigonometry will assist you in the study of biomechanics.

Basically, what trigonometry tells us is that a ratio exists among the lengths of the sides of right triangles that have similar angles. Look at the right triangles in figure 4.15. They are all different sizes,

but the angles are all the same and the sides all change proportionally. If you lengthened one side of any of these triangles, you would have to lengthen the other sides as well to keep the angles of the triangle unchanged. So relationships exist between the lengths of the sides of a right triangle and the angles in a right triangle.

These relationships can be expressed as ratios of one side to another for each size of angle that may exist between two sides of a right triangle. Here are the relationships that may be helpful:

$$\sin \theta = \frac{\text{opposite}}{\text{hypotenuse}} \qquad (4.6)$$

$$\cos \theta = \frac{\text{adjacent}}{\text{hypotenuse}} \qquad (4.7)$$

$$\tan \theta = \frac{\text{opposite}}{\text{adjacent}} \qquad (4.8)$$

In these equations, θ, which is pronounced as theta, represents the angle; opposite refers to the length of the side of the triangle opposite the angle theta; adjacent refers to the length of the side of the triangle adjacent to the angle theta; and hypotenuse refers to the length of the side of the triangle opposite the right angle. The term *sin* refers to the word *sine; cos* refers to the word *cosine;* and *tan* refers to the word *tangent.* For every angle between 0° and 360°, the sine, cosine, and tangent are listed in appendix A. The right triangle in figure 4.16 has these three sides labeled for you.

An easy technique for remembering these trigonometric relationships is the sentence:

Some Of His	$\mathbf{Sin}\ \theta = \dfrac{\mathbf{Opposite}}{\mathbf{Hypotenuse}}$
Children Are Having	$\mathbf{Cos}\ \theta = \dfrac{\mathbf{Adjacent}}{\mathbf{Hypotenuse}}$
Trouble Over Algebra.	$\mathbf{Tan}\ \theta = \dfrac{\mathbf{Opposite}}{\mathbf{Adjacent}}$

The first letter of each of these words matches the first letter in each of the trigonometric equations listed above.

Now let's go back to resolving the 100-N force that acts on the shot into horizontal and vertical components using the trigonometric relationships we have learned. Draw the 100-N force as an arrow acting upward and to the right 60° above horizontal, as in figure 4.17.

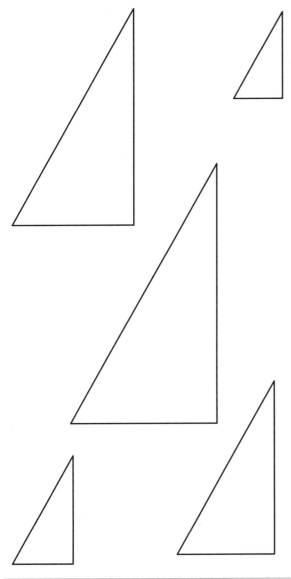

Figure 4.15 Similar right triangles. The triangles are different sizes, but the corresponding angles of each triangle are the same.

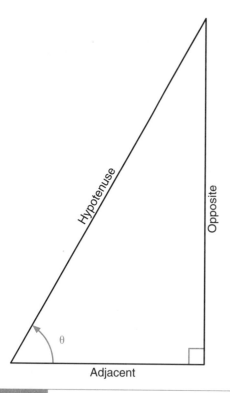

Figure 4.16 Parts of a right triangle.

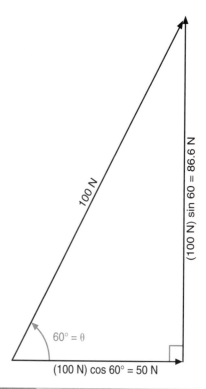

Figure 4.17 Trigonometric resolution of force into horizontal and vertical components.

Now construct a triangle by drawing a horizontal line through the tail of the force and a vertical line through the tip of the force. The 90° angle is formed by the intersection of the horizontal and vertical sides of the triangle. The 100-N force is the side opposite the 90° angle. This side is also referred to as the hypotenuse of this right triangle. The horizontal side of the triangle is the side adjacent to the 60° angle. The length of this side can be found using a trigonometric relationship:

$$\cos \theta = \frac{\text{adjacent}}{\text{hypotenuse}}$$

$$\cos 60° = \frac{\text{adjacent}}{100 \text{ N}}$$

$$(100 \text{ N}) \cos 60° = \text{adjacent} = \text{horizontal force component}$$

If we look up the cosine of 60° in appendix A, we find that it equals 0.500. Substitute this number for cos 60° in the last equation:

$$(100 \text{ N})\cos 60° = (100 \text{ N})(0.500) = \text{adjacent} = 50 \text{ N}$$

The horizontal component of the 100-N force is 50 N. Now find the vertical component of the 100-N force. The side of the triangle opposite the 60° angle represents the vertical component of the 100-N

force. The length of this side can be found by using a trigonometric relationship also:

$$\sin \theta = \frac{\text{opposite}}{\text{hypotenuse}}$$

$$\sin 60° = \frac{\text{opposite}}{100 \text{ N}}$$

$$(100 \text{ N}) \sin 60° = \text{opposite} = \text{vertical force component}$$

If we look up the sine of 60° in appendix A, we find that it equals 0.866. Substitute this number for sin 60° in the last equation:

$$(100 \text{ N}) \sin 60° = (100 \text{ N})(0.866) = \text{opposite} = 86.6 \text{ N}$$

The vertical component of the 100-N force is 86.6 N.

We can now determine the net force acting on the shot by adding up all the horizontal forces to get the resultant horizontal force and then adding up all the vertical forces to get the resultant vertical force. The only horizontal force acting on the shot is the horizontal component of the 100-N force. This resultant horizontal force is thus 50 N. Two vertical forces act on the shot: the weight of the shot acting downward (40 N) and the 86.6-N upward vertical component of the 100-N force. If we add these two forces, we get a resultant vertical force of 46.6 N acting upward. Using the Pythagorean theorem, we can then find the net force acting on the shot:

$$(50 \text{ N})^2 + (46.6 \text{ N})^2 = C^2$$
$$2500 \text{ N}^2 + 2171 \text{ N}^2 = C^2$$
$$4671 \text{ N}^2 = C^2$$
$$C = 68 \text{ N}$$

The net external force acting on the shot is 68 N. That's the same answer we got using the graphical technique.

We now have two ways to resolve a force into its components. We can do it graphically or through the use of trigonometric relationships. Resolving forces into components makes force addition easier, because we just add up the horizontal forces algebraically to determine the resultant horizontal force and add up the vertical forces algebraically to determine the resultant vertical force. In some instances, we may want to express the net force acting on an object as a pair of forces: the resultant horizontal force and the resultant vertical force. Then if we were interested in the horizontal motion of the object, we would only be interested in the horizontal component of the net force. Likewise, if we were only interested in the vertical motion of an object, we would only be interested in the vertical component of the net force.

Static Equilibrium

You should now have an understanding of what a net external force is and how to determine the net external force acting on an object if all the external forces that act on the object are known. You should now be better able to understand some aspects of Newton's first and second laws of motion. If you recall, Newton's first law of motion explained that an object remains at rest (not moving) or moving at constant (unchanging) velocity in a straight line if the net force acting on the object is zero. Newton's second law of motion explains that the acceleration of an object is directly proportional to the net force acting on it and inversely proportional to the mass of the object. If the net force acting on an object is zero, no change in velocity will occur (zero acceleration). Likewise, if an object is at rest or moving with constant velocity (zero acceleration), the net force acting on it is zero. In each of these cases, the external forces acting on the object are in equilibrium and they result in a net force of zero. If the object is at rest, the forces are in equilibrium and the object is described as being in a state of static equilibrium. The branch of mechanics dealing with the study of objects in static equilibrium is called statics.

Statics enables us to analyze two types of situations. In the first situation, we may know that an object such as an athlete is not moving, but we may want to know what external forces are acting on the athlete. An example of this could be in gymnastics. A gymnast's coach may want to know how strong a gymnast must be to hold a certain position such as an iron cross. A static analysis of the position would give the coach knowledge of the forces the athlete would have to withstand to hold that position. The coach could then test the athlete to determine if he was strong enough to react to the external forces whose magnitudes were determined by the static analysis. In the second situation, we may know what forces act on the athlete but want to know if these forces will result in static equilibrium. Let's use the gymnast example again. The coach may know how strong the gymnast is and thus how large the external forces are to which the gymnast can react. A static analysis of the position in question could then be done to determine whether or not the gymnast's strength is great enough to hold that position.

Free-Body Diagrams

Let's start with a simple situation to test our understanding. A woman in skates is standing still on the ice. The woman's mass is 50 kg. What external forces act on her? Perhaps a picture of the situation would help us visualize the problem (see figure 4.18).

Back to our question. One force that we know acts on the woman is the force of gravity, which is the woman's weight. Let's indicate that in the drawing with an arrow pointing downward through the woman's center of gravity. You probably have some notion of what center of gravity is. We'll define it in more specific terms later, in chapter 8. For now, we'll say that the center of gravity of an object is the imaginary point in space through which the force of gravity acts on that object.

Is weight the only force acting on the woman? If it is, Newton's second law (the law of acceleration; $\Sigma F = ma$) indicates that the woman should be accelerating as a result of this force. Because she isn't, another force must be acting that cancels out the effect of the weight. If you remember our discussion of forces in the last chapter, we said that when two objects touch each other, they may exert forces on each other. The woman's skates are touching the ice and exerting a force on the ice. Newton's third

Weight

Reaction
force

a   *b*

Figure 4.18   (*a*) A skater standing on the ice; (*b*) free-body diagram of the skater.

law of motion (action-reaction; for every action there is an equal but opposite reaction) leads us to the conclusion that the ice is exerting a force on the woman as well. This is the reaction force from the ice. If we drew an arrow representing this force, it would be an upward arrow whose arrowhead was just contacting the skates. But does this arrow represent a force pushing up on the skates or a force pulling up on the ice? This is confusing. As we are only concerned with the forces acting on the skater, why don't we just make a drawing of the skater alone, isolated from the rest of the environment? Any points where the skater touches something external to her are places where external forces may be acting. This is shown in figure 4.18b. In this drawing, it is clear that the reaction force from the ice is only acting on the skater and not on the ice. This type of drawing is called a **free-body diagram**. It is a mechanical representation of the skater. The free-body diagram is a valuable tool for doing mechanical analyses, so we'll be using free-body diagrams again.

The free-body diagram is a valuable tool for doing mechanical analyses.

The free-body diagram of the woman on skates now shows the two vertical forces that act on the skater: her weight and the reaction force from the ice. Do any horizontal forces act on her? What about friction? On ice, friction is very small, and because the skater is not accelerating forward or backward, the frictional force under her skates must be zero. The free-body diagram thus shows all of the external forces that act on the skater. To draw a free-body diagram of any situation, the system in question (in this case, the skater) is isolated from the environment (in this case, anything that is not the skater), and reaction forces are drawn as arrows at the points of contact of the system with the environment. The noncontact force of gravity is also drawn as an arrow acting downward from the center of gravity of the system.

Static Analysis

Now that we have shown all the external forces acting on the skater, we can finish our analysis. Newton's second law of motion led us to the conclusion that another vertical force besides weight had to act on the skater. We now use that law to determine exactly how large that other force is. Newton's second law of motion is written mathematically as: $\Sigma F = ma$. In static situations, acceleration is zero, so Newton's second law can be rewritten as: $\Sigma F = 0$. This is the equation of static equilibrium. The term ΣF represents the net external force or the vector sum of the external forces. If the only external forces acting are colinear forces, as in the situation with the ice skater, the vector sum of the external forces is found by simply adding them algebraically. But the senses of the forces must be taken into account. The sense of a force or any vector is the direction in which the arrow of the force or vector points along its line of action. For vertical forces—forces acting up or down—let's give upward forces positive signs and downward forces negative signs.

Getting back to the analysis, the upward reaction force that the ice exerts on the skater has a positive sign, but we don't know how large the force is yet. That's what we want to find out. The downward force of gravity that acts on the skater has a negative sign because it's downward. We said the skater's mass was 50 kg. What is her weight? Remember we defined weight, W, with the equation $W = mg$, where m represents mass in kilograms and g represents acceleration due to gravity, or about 10 m/s/s downward. So the skater's weight is:

$$W = mg = (50 \text{ kg})(-10 \text{ m/s/s}) = -500 \text{ kg m/s/s} = -500 \text{ N}$$

We can now write the equilibrium equation for the ice skater:

$$\Sigma F = R + (-500 \text{ N}) = R - 500 \text{ N} = 0$$

where R represents the reaction force from the ice. Solving this equation for R gives:

$$R - 500 \text{ N} = 0$$
$$R - 500 \text{ N} + 500 \text{ N} = 0 + 500 \text{ N}$$
$$R = +500 \text{ N}$$

The reaction force from the ice is +500 N, exactly equal to the weight of the skater but acting in the upward direction. This is true in all situations of static equilibrium where only two external forces act. The external forces must be equal in size but act in opposite directions.

Here's another example. An 80-kg weight lifter has lifted 100 kg over his head and is holding it still. He and the barbell are in static equilibrium. What is the reaction force from the floor that must act on the weight lifter's feet? Draw the free-body diagram first. Should the barbell be included in the free-body diagram of the weight lifter? Because the problem is concerned with the reaction force between the athlete's feet and the floor, the system can be considered to be the weight lifter alone or the weight lifter and the barbell. In both cases, the reaction force occurs at the boundary between the environment and the system.

First, let's try this analysis using a free-body diagram of the weight lifter alone (see figure 4.19).

| Figure 4.19 | Free-body diagram of a weight lifter alone. |

Three external forces act on the weight lifter: (1) the reaction force from the floor acting upward on the athlete's feet, (2) the weight of the athlete acting downward through the athlete's center of gravity, and (3) the reaction force from the barbell acting downward on the athlete's hands. With only one equilibrium equation, only one unknown force can be determined, and in this situation we have two unknown forces: the reaction force from the floor and the reaction force from the barbell. The magnitude of the athlete's weight can be determined because we know the athlete's mass. The free-body diagram of the weight lifter and its equilibrium equation must be used to determine the reaction force from the floor. So another equation must be used to determine the other unknown force, the reaction force from the barbell.

To determine the reaction force from the barbell, a free-body diagram of the barbell would be useful. The external forces acting on the barbell are the weight of the barbell and the reaction force from the hands acting upward on the barbell. The free-body diagram of the barbell is shown in figure 4.20.

The weight of the barbell is:

$$W' = mg = (100 \text{ kg})(-10 \text{ m/s/s}) = -1000 \text{ N}$$

Now use the equilibrium equation to determine the reaction force from the hands:

$$\Sigma F = R' + (-1000 \text{ N}) = R' - 1000 \text{ N} = 0$$
$$R' - 1000 \text{ N} + 1000 \text{ N} = 0 + 1000 \text{ N}$$
$$R' = +1000 \text{ N}$$

The reaction force from the hands is a 1000-N force acting upward on the barbell. From Newton's third law, the force exerted by the hands on the barbell is equal in magnitude but opposite in direction to the force exerted by the barbell on the hands. The reaction force exerted by the barbell on the hands must be a force of 1000 N acting downward. We

can now solve for the reaction force from the floor. The weight of the athlete is:

$$W = mg = (80 \text{ kg})(-10 \text{ m/s/s}) = -800 \text{ N}$$

Now use the equilibrium equation to determine the reaction force from the floor:

$$\Sigma F = R + (-1000 \text{ N}) + (-800 \text{ N}) = R - 1800 \text{ N} = 0$$
$$R - 1800 \text{ N} + 1800 \text{ N} = 0 + 1800 \text{ N}$$
$$R = +1800 \text{ N}$$

Figure 4.20   Free-body diagram of a barbell.

Figure 4.21   Free-body diagram of weight lifter with barbell.

The reaction force from the floor is an 1800-N force acting upward on the weight lifter's feet.

Now let's consider the athlete and the barbell as a system. In the free-body diagram, include the barbell. The free-body diagram is shown in figure 4.21.

What external forces are shown in the free-body diagram as acting on this system? The vertical reaction force from the floor acting upward on the athlete's feet is one. This is the force whose magnitude we wish to determine. The other forces are the noncontact forces of gravity that act vertically downward through the center of gravity of the athlete and through the center of gravity of the barbell. We have already computed these gravitational forces as the weight of the athlete, 800 N, and the weight of the barbell, 1000 N. We can now use the equilibrium equation to determine the reaction force from the floor:

$$\Sigma F = R + (-800\ N) + (-1000\ N) = R - 1800\ N = 0$$
$$R - 1800\ N + 1800\ N = 0 + 1800\ N$$
$$R = 1800\ N$$

The reaction force from the floor is a force of 1800 N acting upward on the weight lifter. This force is exactly equal to the combined weight of the athlete and the barbell. Representing the athlete and the barbell as the system in a free-body diagram was a less complicated way of solving this problem, as the reaction force between the hands and the barbell did not have to be computed.

Let's try one more problem. A 20-kg child is sitting on a swing, and the child's parent is exerting a 40-N horizontal force on the child to the right and a 10-N upward force (see figure 4.22a). The child is not moving and is in a state of static equilibrium. What force must the swing be exerting on the child? First, draw a free-body diagram of the child and show the external forces that act on the child (see figure 4.22b). Show the force of gravity, or weight, acting on the child. How large is this force?

$$W = mg = (20\ kg)(-10\ m/s/s) = -200\ N$$

Now show the reaction force from the swing and the horizontal and vertical forces exerted by the parent. Are any other forces acting? No. In this situation, both horizontal and vertical forces are acting. The equilibrium equation in the form $\Sigma F = 0$ is a little more difficult to use algebraically. We could write the equation as two equilibrium equations to represent the horizontal component of the net external forces and the vertical component of the net external forces. Instead of $\Sigma F = 0$, we have $\Sigma F_x = 0$ and $\Sigma F_y = 0$. Now ΣF_x represents the horizontal com-

Figure 4.22 (a) A parent pushes a child on a swing; (b) free-body diagram of the child on the swing.

ponent of the net external forces, and ΣF_y represents the vertical component of the net external forces. For our child in the swing problem, the equilibrium equations become:

$$\Sigma F_x = R_x + 40\ N = 0$$
$$R_x = -40\ N$$

and

$$\Sigma F_y = R_y + 10\ N + (-200\ N) = 0$$
$$R_y = +190\ N$$

where R_x represents the horizontal component of the reaction force exerted by the swing and R_y represents the vertical component. A positive sign in the horizontal direction indicates that the force is to the right, and a negative sign indicates that it is to the left. A positive sign in the vertical direction indicates that the force is upward, and a negative sign indicates that it is downward.

So the reaction force the swing exerts on the child can be represented by its component forces of 40 N to the left and 190 N upward.

Let's summarize the steps involved in determining reaction forces in a static situation. First, isolate the system in question from the environment and draw a free-body diagram. Show all the external forces that act on the system. Don't forget the gravitational forces (weights). If the external forces that act on a system are all colinear, only one equilibrium equation needs to be used; otherwise use the equilibrium equation for the horizontal direction and the equilibrium equation for the vertical direction. Resolve any forces that are not vertical or horizontal into horizontal and vertical components. Positive and negative signs associated with forces are determined by the direction of the forces. The equilibrium equations are solved using algebra. Alternatively, the reaction forces can be solved using the graphical methods used early in the chapter to demonstrate force addition and force resolution. Either method is acceptable, and both should give the same result.

Summary

Forces are pushes or pulls. They are vector quantities, so they are described by a size and a direction of action. Use arrows to represent forces graphically. Internal forces hold things together and cannot cause changes in the motion of a system. External forces may cause changes in the motion of the system. The most commonly felt external forces are gravity and contact forces. Friction and the normal reaction force are the two components of a contact force. Forces are added using vector addition. This can be accomplished using graphical techniques, or using algebra if the forces are resolved into horizontal and vertical components. If the net external force acting on an object is zero, the object is standing still or moving in a straight line at constant velocity. If the object is standing still, it is in a state of static equilibrium and the external forces acting on it are balanced and sum to a net force of zero. If the external forces acting on an object do not balance (i.e., they do not sum to zero), the object is not in equilibrium and will change its state of motion. In the next chapter, we will learn about motion and how it is described. Then we will be able to analyze objects that are not in equilibrium.

Key Terms

colinear forces *(p. 57)*

compressive force *(p. 51)*

concurrent forces *(p. 59)*

contact force *(p. 52)*

dynamic friction *(p. 53)*

external force *(p. 52)*

free-body diagram *(p. 68)*

friction *(p. 52)*

internal force *(p. 51)*

net force *(p. 57)*

resultant force *(p. 57)*

static equilibrium *(p. 67)*

static friction *(p. 53)*

tensile force *(p. 51)*

vector *(p. 50)*

REVIEW QUESTIONS

1. **A sprinter is just coming out of the starting block, and only one foot is touching the block. The sprinter pushes back (horizontally) against the block with a force of 500 N and down (vertically) against the ground with a force of 800 N. What is the resultant of these forces?**

2. The same sprinter from problem 1 is now out of the blocks and running. If the coefficient of static friction between the track shoe and the track is 0.50, and the sprinter exerts a vertical force of 2000 N downward on the track, what is the maximum horizontal force he can generate under his shoe?

3. Three athletes are tugging on a ring. Nancy pulls with a force of 150 N in the northerly direction. Evelyn pulls with a force of 100 N in the easterly direction. Sara pulls with a force of 200 N in the southwesterly direction (225° clockwise from north). What is the resultant of the three forces?

4. Dave is trying to pull Dana on a sled across a flat field. Dave pulls on the rope attached to the sled with an upward and forward force of 300 N. This force and the rope are directed at an angle of 30° above horizontal. Dana's mass is 50 kg, and the sled's mass is 8 kg. If the coefficient of static friction between the sled runners and the snow is 0.10, will Dave move the sled?

5. A 60-kg skier is in a tuck and moving straight down a 30° slope. Air resistance pushes backward on the skier with a force of 10 N (this force acts in a direction upward and parallel to the 30° slope). The coefficient of dynamic friction between the skis and the snow is 0.08. What is the resultant force that acts on the skier?

6. The coefficient of static friction between a tennis player's hand and her racket is 0.45. How hard must she squeeze the racket if she wants to exert a force of 200 N along its longitudinal axis?

Linear Kinematics: Describing Objects in Linear Motion

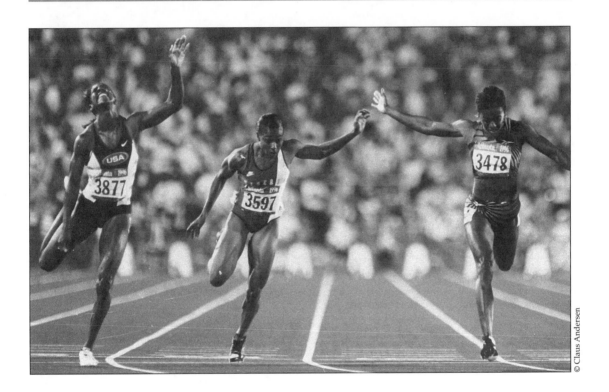

© Claus Andersen

The world's best female sprinters are lined up at the starting line in the 100-m dash finals at the Olympic Games. The winner will earn the title of "World's Fastest Woman." The starter's pistol goes off, and Mary jumps into an early lead. At 50 m she has a 3-m lead on the other runners. But during the last 40 m of the race, Jill slowly reduces that lead. At the finish, Mary finishes only 1 m ahead of Jill and wins the race. Mary wins the title of "World's Fastest Woman," but was her top speed really faster than Jill's? Whose acceleration was greater? Were both athletes accelerating during the entire race? Did either athlete decelerate? What performance parameters could be used to account for the last part of the race where Jill almost caught Mary? Can you answer these questions? They all concern the kinematic measures of performance. The information presented in this chapter will give you the means to answer these questions.

OBJECTIVES

When you finish this chapter, you should be able to do the following:

▌ Distinguish between linear, angular, and general motion

▌ Name and define distance traveled and displacement and distinguish between the two

▌ Name and define average speed and velocity and distinguish between the two

▌ Name and define instantaneous speed and velocity

▌ Name and define average acceleration

▌ Name and define instantaneous acceleration

▌ Name the units of measurement for distance traveled and displacement, speed and velocity, and acceleration

This chapter is about the subbranch of mechanics called kinematics. Dynamics is the branch of rigid-body mechanics concerned with the mechanics of moving objects. Kinematics, the topic of this chapter, is the branch of dynamics concerned with the description of motion. The outcomes of many sporting events are kinematic measures, so an understanding of these measures is important. Some of the kinematic terminology that will be introduced in this chapter may sound familiar to you (speed, velocity, acceleration, etc.). You may believe that you already know all about these terms, but we will be using them in specific ways. The precise mechanical definitions may not agree with your connotations, and there will be misunderstandings unless our definitions agree. With that in mind, let's begin.

> **Kinematics is the branch of dynamics concerned with the description of motion.**

Motion

What is motion? Can you define it? We might define motion as the action or process of a change in position. Movement is a change in position. Moving involves a change in position from one point to another. Two things are necessary for motion to occur: space and time—space to move in and time during which to move. To make the study of movement easier, we classify movements as linear, angular, or both (general).

Linear Motion

Linear motion is also referred to as translation. It occurs when all points on a body or object move the same distance, in the same direction, and at the same time. This can happen in two ways: rectilinear translation or curvilinear translation.

Rectilinear Translation

Rectilinear translation is the motion you probably would think of as linear motion. Rectilinear translation occurs when all points on a body or object move in a straight line so the direction of motion does not change, the orientation of the object does not change, and all points on the object move the same distance.

Curvilinear Translation

Curvilinear translation is very similar to rectilinear translation. Curvilinear translation occurs when all points on a body or object move so that the orientation of the object does not change and all points on the object move the same distance. The difference between rectilinear and curvilinear translation is that the paths followed by the points on an object in curvilinear translation are curved, so the direction of motion of the object is constantly changing, even though the orientation of the object does not change.

Try to think of some examples of linear motion in sports or human movement. What about a figure skater gliding across the ice in a static position? Is her motion rectilinear or curvilinear? What about a sailboarder zipping across the lake in a steady breeze? Is it possible for the sailboarder's motion to be rectilinear? What about a bicyclist coasting along a flat section of the road? (In each of these examples, it is possible for the athletes to achieve rectilinear motion.) Can you think of any examples of curvilinear motion? Can a gymnast on a trampoline experience linear motion? How? What about a diver? A ski jumper? A skateboarder rolling along a flat section of concrete? An in-line skater? (It's possible for the gymnast, diver, and ski jumper to achieve curvilinear motion. The skateboarder and in-line skater can achieve both types of linear motion.)

To determine whether a motion is linear, imagine two points on the object in question. Now imagine a straight line connecting these two points. As the object moves, does the line keep its same orientation; that is, does the line point in the same direction throughout the movement? Does the line stay the same length during the movement? If both of these conditions are true throughout the movement, the motion is linear. If both points on the imaginary line move in parallel straight lines during the motion, the motion is rectilinear. If both points on the imaginary line move in parallel lines that are not straight, the motion is curvilinear. Now try to think of more examples of linear motion in sport. Would you classify the motions you thought of as rectilinear or curvilinear?

Angular Motion

Angular motion is also referred to as rotary motion or rotation. It occurs when all points on a body or object move in circles (or parts of circles) about the same fixed central line or axis. Angular motion can occur about an axis within the body or outside of the body. A child on a swing is an example of angular motion about an axis of rotation external to the body. An ice skater in a spin is an example of angular motion about an axis of rotation within the body. To determine whether or not a motion is angular, imagine any two points on the object in question. As the object moves, are the paths that each of these points follow circular? Do these two circular paths have the same center or axis? If you imagine a line connecting the two imaginary points, does this line continually change orientation as the object moves? Does the line continually change the direction in which it points? If these conditions are true, the object is rotating.

Examples of angular motion in sports and human movement are more numerous than examples of linear motion. What about a giant swing on the horizontal bar? Are parts of this motion rotary? What about individual movements of our limbs? Almost all of our limb movements (if they are isolated) are examples of angular motion. Hold your right arm at your side. Keeping your upper arm still, flex and extend your forearm at the elbow joint. This is an example of angular motion. Your forearm rotated about a fixed axis (your elbow joint). During the flexing and extending, your wrist moved in a circular path about your elbow joint. Every point on your forearm and wrist moved in a circular path about your elbow joint. Consider each limb and the movements it can make when movement about only one joint is involved. Are these movements rotary; that is, do all the points on the limb move in circular paths about the joint?

Let's consider motion about more than one joint. Is the limb's motion still angular? Extend your knee and hip at the same time. Was the movement of your foot angular? Did it move in a circular path? Was the motion of your foot linear?

General Motion

Combining the angular motions of our limbs can produce linear motions of one or more body parts. When both the knee and hip joints extend, you can

SELF-EXPERIMENT

Hold a pencil in your hand so that it lies flat on the desk or table. Now move the pencil rectilinearly across the desk. Can you do it? You produced that motion by combining angular motions of your forearm and upper arm. The total motion of your limbs is called general or mixed motion.

produce a linear motion of your foot. Similarly, extension at the elbow and horizontal adduction at the shoulder can produce a linear motion of the hand. General motion is a combination of linear and angular motions.

General motion is the most common type of motion exhibited in sports and human movement. Running and walking are good examples of general motion. In these activities, the trunk often moves linearly as a result of the angular motions of the legs and arms. Bicycling is another example of general motion. Think of various human movements in sports and consider how you would classify them.

Classifying motion as linear, angular, or general motion makes the mechanical analysis of movements easier. If a motion can be broken down into linear components and angular components, the linear components can be analyzed using the mechanical laws that govern linear motion. Similarly, the angular components can be analyzed using the mechanical laws that govern angular motion. The linear and angular analyses can then be combined to understand the general motion of the object.

Classifying motion as linear, angular, or general motion makes the mechanical analysis of movements easier.

Linear Kinematics

Now let's examine linear motion in more detail. Linear kinematics is concerned with the description of linear motion. Questions about speed, distance, and direction are all inquiries about the linear kinematics of an object.

Position

The first kinematic characteristic we might describe about an object is its position. Our definition of motion—the action or process of change in position—refers to position. Mechanically, position is defined as location in space. Where is an object in space at the beginning of its movement or at the end of its movement or at some time during its

SELF-EXPERIMENT

How would you go about describing something that is moving? Roll a ball across the floor. Describe its movement. What words do you use? You might describe how fast or slow it is going; whether it is speeding up or slowing down; and that it is rolling and not sliding. You also might say something about where it started and where it might end up, or you might describe its direction. "It's moving diagonally across the room." Or "It's moving toward the wall or toward the door." You might say how far it traveled and how long it took to get there after it stops. The terms you used to describe the motion of the ball are all words that concern the kinematics of linear motion.

movement? This may not seem like such an important characteristic at first, but consider the importance of the position of players on the field or court in sports such as football, tennis, racquetball, squash, soccer, field hockey, ice hockey, rugby, and so on. The strategies employed often depend on where the players on each team are positioned.

Let's start with a simple example. Consider a runner competing in a 100-m dash (see figure 5.1). How would you go about describing the runner's position during the race? You might describe the runner's position relative to the starting line. "She's 40 m from the start." Or you might describe the runner's position relative to the finish line. "She's 60 m from the finish." In both cases, you have used a measure of length to describe the runner's position relative to some fixed, nonmoving reference.[*] The references were the starting line or the finish line. Some concept of direction was also implied by your description and the event itself. When you say the runner is 40 m from the start, this is usually interpreted to mean that the runner is 40 m in front of

*Are these really fixed, nonmoving references? Relative to the surface of the earth they are, but the earth itself is moving around the sun in the solar system. And the solar system is moving in the galaxy. And the galaxy is moving in the universe. So it is difficult to define a position in terms of an absolute nonmoving reference frame. For our purposes, however, we will consider anything that doesn't move relative to the earth's surface as a fixed reference.

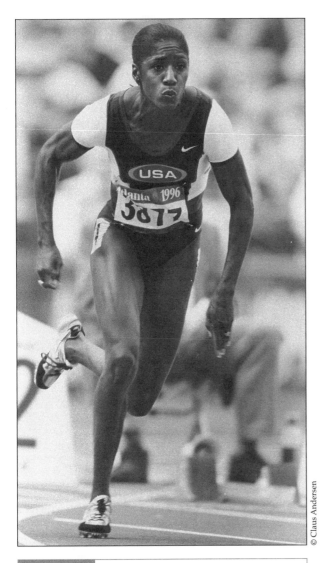

Figure 5.1 How would you describe a runner's position in a 100-m dash?

he was 20 yd from the goal line. But to fully describe his position, you would also have to give information about his location relative to the sidelines. Using the left sideline as a reference, we could then describe his position as 20 yd from the goal line and 15 yd from the left sideline. This is shown in figure 5.2.

In this situation, it might be helpful to set up a Cartesian coordinate system to help identify the location of the runner. Cartesian coordinates are named after René Descartes (1596-1650), a French philosopher and mathematician who is credited with inventing analytic geometry. You may remember this type of coordinate system from high school mathematics. First, we would need to locate a fixed reference point for our coordinate system. This fixed point is called the origin, because all our position measurements originate from it. Let's put the origin for this system at the intersection of the left sideline and the goal line. We could put the origin at any fixed point; we chose the intersection of the goal line and the sideline because it was convenient. Imagine the x-axis lying along the goal line with zero at the origin and positive numbers to the right on the playing field. Imagine the y-axis lying along the left sideline with zero at the origin and positive numbers increasing as you move toward the opposite goal. With this system, we could identify the running back's position with two numbers corresponding to his x- and y-coordinates in yards as follows: (15, 80). This situation is shown in figure 5.3. The x-coordinate of 15 indicates that he is 15 yd from the left sideline on the field, and the y-coordinate of 80 indicates that he is 80 yd from his goal line or 20 yd from scoring, because we know that the goal lines are 100 yd apart.

In three dimensions, we would need three numbers to describe the position of an object in space. For example, how would you describe the position of the ball during a game of racquetball? We might set up a three-dimensional Cartesian coordinate with one axis in the vertical direction and two axes in the horizontal plane. If we put the point of reference or origin in the lower left front corner of the court (where the front wall, left side wall, and floor intersect), the x-axis would be the line along the intersection of the front wall and floor. The y-axis would be the line along the intersection of the front wall and the left side wall, and the z-axis would be the line along the intersection of the left side wall and the floor. This is shown in figure 5.4. If the ball were 3 m to the right of the left side wall, 2 m above the floor, and 4 m away from the front wall, its x-, y-, and z-coordinates in meters would be (3, 2, 4).

the start and toward the finish line. Mechanically, if we use the starting line as our reference, we would say that the runner is at +40 m. If the runner were on the other side of the starting line, we would describe the runner's position as –40 m. We use the positive and negative signs to indicate which side of the starting line the runner is on.

This example of the 100-m dash is only one-dimensional. We were only concerned about one dimension—the dimension from the starting line to the finish line. Only one number was required to identify the position of the runner in the race. Now let's consider a two-dimensional situation. Imagine you are watching a football game. A running back has broken out of the backfield and is running toward the goal line. He is on the opposing team's 20-yd line. To describe his position, you would say

Figure 5.2 The location of a running back on a football field using the sideline and opponent's goal line as references.

> **In three dimensions, we would need three numbers to describe the position of an object in space.**

To describe the position of something in space, we need to identify a fixed reference point to serve as the origin of our coordinate system. For our purposes, any point fixed relative to the earth will do. Then we set up a Cartesian coordinate system. If we are only describing the position of objects in one dimension, only one axis is needed; for two dimensions, two axes are needed; and for three dimensions, three axes are needed. The axes of this system may point in any direction that is convenient, as long as they are at right angles to each other if we are describing the position of something in two or three dimensions. Typically, one axis will be oriented vertically (the y-axis) and the other axis (the x-axis) or axes (the x- and z-axes) will be oriented horizontally. Each of these axes will have a positive and negative direction along them. The x-coordinate of an object is the distance the object is away from the plane formed by the y- and z-axes. The y-coordinate of an object is the distance the object is away from the plane formed by the x- and z-axes, and the z-coordinate of an object is the distance the object is away from the plane formed by the x- and y-axes. Units of length are used to describe position.

Distance Traveled and Displacement

Now we have a method of describing and locating the position of an object in space. This is our first task in describing motion. If we remember how we defined motion—the action or process of change in position—our next task will be discovering a way to describe or measure change in position. How would you do this?

Distance Traveled

Let's use a football example again. Suppose a football player has received the kickoff at his 5-yd line, 15 yd from the left sideline. His position on the field (using the Cartesian coordinate system we established in the previous section) is (15, 5) when he catches the ball. He runs the ball back following the path shown in figure 5.5. He is finally tackled on his 35-yd line, 5 yd from the left sideline. His posi-

Figure 5.3 The location of a running back on the football field, as described with Cartesian coordinates.

tion on the field at the end of the play is (5, 35). If we measure the length of the path of his run with the ball, it turns out to be 48 yd. So we might describe this run as a run of 48 yd to gain 30 yd.

Another way of saying this is that the runner's displacement was +30 yd in the y-direction and –10 yd in the x-direction, or a resultant displacement of 31.6 yd toward the left sideline and goal. The distance traveled by the runner was 48 yd. We've used two different terms to describe the runner's progress: displacement and distance traveled. Distance traveled is easily defined—it's simply a measure of the length of the path followed by the object whose motion is being described, from its starting (initial) position to its ending (final) position. Distance traveled doesn't mean a whole lot in a football game, though, because the direction of travel isn't considered. Displacement does take into account the direction of travel.

Displacement

Displacement is the straight-line distance in a specific direction from initial (starting) position to final

(ending) position. The resultant displacement is the distance measured in a straight line from the initial position to the final position. Displacement is a vector quantity. If you recall from chapter 4, we said force was also a vector quantity. A vector has a size associated with it as well as a direction. It can be represented graphically as an arrow whose length represents the size of the vector and whose orientation and arrowhead represent the direction of the vector. Representation of displacement with an arrow is appropriate and communicates what displacement means as well. Figure 5.6 shows the path of the player in the kick return example. The arrow from the initial position of the player to where he was tackled represents the displacement of the back.

> **Displacement is the straight-line distance in a specific direction from starting (initial) position to ending (final) position.**

If you also recall from chapter 4, vectors can be resolved into components. In the football example, the resultant displacement of the running back does

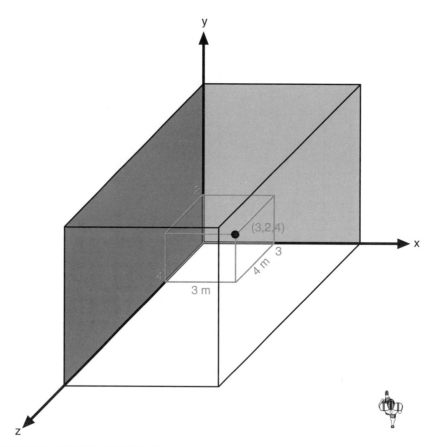

Figure 5.4   The location of a ball in a racquetball court, using Cartesian coordinates.

Figure 5.5   The locations of a running back when he receives the ball and when he's tackled, as described with Cartesian coordinates.

Figure 5.6   The running back's displacement and distance traveled are not the same.

not indicate directly how many yards the back gained. But if we resolve this resultant displacement into components in the x-direction (across the field) and y-direction (down the field toward the goal), we then have a measure of how effective the run was. In this case, the y-displacement of the running back is the measure of importance. His initial y-position was 5 yd and his final y-position was 35 yd. We can find his y-displacement by subtracting his initial position from his final position:

$$d_y = \Delta y = y_f - y_i \qquad (5.1)$$

where

d_y = displacement in the y-direction

Δ = change, so Δy = change in y-position

y_f = final y-position

y_i = initial y-position

If we put in the initial (5 yd) and final (35 yd) values for y-position, we get the runner's y-displacement:

$$d_y = \Delta y = y_f - y_i = 35 \text{ yd} - 5 \text{ yd}$$
$$d_y = +30 \text{ yd}$$

The runner's y-displacement or displacement down the field was +30 yd. The positive sign indicates that the displacement was in the positive y-direction or toward the goal (a gain in field position in this case). This measure is probably the most important measurement to the coaches, players, and

fans because it indicates the effectiveness of the kick return.

We may also be curious about the player's displacement across the field (in the x-direction). We can use the same equation to determine the x-displacement:

$$d_x = \Delta x = x_f - x_i \qquad (5.2)$$

where

d_x = displacement in the x-direction

Δx = change in x-position

x_f = final x-position

x_i = initial x-position

If we put the initial (15 yd) and final (5 yd) values in for x-position, we get the runner's x-displacement:

$$d_x = \Delta x = x_f - x_i = 5 \text{ yd} - 15 \text{ yd}$$
$$d_x = -10 \text{ yd}$$

The runner's x-displacement or displacement across the field was –10 yd. The negative sign indicates that the displacement was in the negative x-direction or toward the left sideline.

The resultant displacement of the runner could be found in a manner similar to that with which we found a resultant force. Graphically, this could be done by drawing the arrows representing the component displacements of the runner in the x- and y-directions. Look at figure 5.7. Put the tail of the

Figure 5.7   The resultant displacement of the running back, R, can be determined by vector addition of his displacement toward the goal line, Δy, and his displacement toward the sideline, Δx.

x-displacement vector at the tip of the y-displacement vector and then draw an arrow from the tip of the y-displacement vector to the tail of the x-displacement vector. This arrow represents the resultant displacement.

This resultant displacement could also be determined by starting with the y-displacement vector and drawing the x-displacement vector so that its tail ended at the tip of the y-displacement vector. The resultant would be determined by drawing an arrow from the tail of the x-displacement vector to the tip of the y-displacement vector. You should get the same resultant as determined by the previous method shown in figure 5.7.

We could also determine this resultant displacement using trigonometric relationships. The displacement vectors arranged as shown in figure 5.7 form a triangle, specifically, a right triangle with the hypotenuse represented by the resultant displacement. Recalling from chapter 4, the size of the hypotenuse can be determined as follows: if A and B represent the two sides that make up the right angle and C represents the hypotenuse, then

$$A^2 + B^2 = C^2 \qquad (5.3)$$
$$(\Delta x)^2 + (\Delta y)^2 = R^2$$

For our displacements, then, we can substitute –10 yd for Δx and +30 yd for Δy and then solve for R, which represents the resultant displacement.

$$(-10 \text{ yd})^2 + (30 \text{ yd})^2 = R^2$$
$$100 \text{ yd}^2 + 900 \text{ yd}^2 = R^2$$
$$1000 \text{ yd}^2 = R^2$$
$$R = \sqrt{1000 \text{ yd}^2} = 31.6 \text{ yd}$$

To find the direction of this resultant displacement, we can use the relationship between the two sides of the displacement triangle.

$$\tan \theta = \frac{\text{opposite}}{\text{adjacent}} \qquad (5.4)$$
$$\theta = \arctan \left(\frac{\text{opposite}}{\text{adjacent}} \right)$$
$$\theta = \arctan \left(\frac{\Delta y}{\Delta x} \right)$$

In these equations, θ, which is pronounced as theta, represents the angle between the resultant displacement vector and the x-displacement vector. To find the value of θ, substitute –10 yd for Δx and +30 yd for Δy.

$$\theta = \arctan \left(\frac{30 \text{ yd}}{-10 \text{ yd}} \right)$$
$$\theta = \arctan \ (-3)$$
$$\theta = -71.6°$$

If we look in the appendix of trigonometric functions under tangent, we find that the tangent of –71° is 2.90 and the tangent of –72° is 3.07. From this, we can determine that θ is between –71° and –72°. Using a calculator, we can get a more precise value for θ, or –71.6°.

We can now describe several qualities of movement—initial and final positions, distance traveled, and displacement. Distance traveled can be described by a single number that represents the length of the path followed by the object during its motion. Displacement, however, is a vector quantity, so it is expressed with a length measurement and a direction. The resultant displacement is the length of a straight line from the initial position to the final position in the direction of motion from the initial position to the final position. Components of the resultant displacement may also be used to describe displacement of the object in specific directions. In some situations (such as our football examples), a component displacement is more important than the resultant displacement.

Now let's see if we understand the concept of displacement. Imagine two downhill ski racers, Tamara and Cindy, competing on the same course. They start at the same starting position and finish the race at the same finish point. Tamara takes wider turns than Cindy does, so the length of the path Tamara follows is longer. Who has the greater resultant displacement from start to finish? Because they both start at the same spot and finish at the same spot, their resultant displacements are the same. Now consider a 100-m swimming race in a 50-m pool. Which measure (displacement or distance traveled) is more meaningful? In a 100-m swimming race in a 50-m pool, you have to start and finish in the same place, so your displacement is zero! Distance traveled is the more meaningful measure. What about a 400-m running race around a 400-m oval track? Or a 100-m running race on a straight section of track?

Speed and Velocity

We can now describe an object's position, and we have measures (distance traveled and displacement) for describing its change in position, but how do we describe how quickly something changes its position? When we speak of how fast or slow something moves, we are describing its speed or velocity. Both are used to describe the rate of motion. You have probably used both of these terms, perhaps interchangeably.

Speed

Are speed and velocity the same thing? Mechanically, speed and velocity are different. Speed is just rate of motion. It is described by a single number only. Velocity is rate of motion in a specific direction, so it is a vector quantity. Velocity has a magnitude (number) and a direction associated with it.

Speed is rate of motion; velocity is rate of motion in a specific direction.

Average speed of an object is distance traveled divided by the time it took to travel that distance. Mathematically, this can be expressed as

$$\bar{s} = \frac{\ell}{\Delta t} \tag{5.5}$$

where

$\bar{s}$ = average speed

ℓ = distance traveled

Δt = time taken

The units for describing speed are a unit of length divided by a unit of time. The SI unit for describing speed is meters per second. You have probably used other units of measurement for speed. If you have driven a car, you are probably more familiar with miles per hour or kilometers per hour. These are also units of measurement for speed.

Average speed is an important descriptor of performance in many sport activities. In some activities, average speed is in fact the measure of success. Consider almost any type of racing event (swimming, running, cycling, etc.). The winner is the person who completes the specified distance in the shortest time. The average speed of the winner is the distance of the race divided by the time. The winner's average speed over the race distance will always be the fastest of all the competitors if everyone raced the same distance.

This one number, average speed, doesn't tell us much about what went on during the race itself, though. It doesn't tell us how fast the racer was moving at any specific instant in the race. It doesn't tell us the maximum speed reached by the racer during the race. It doesn't indicate when the racer was slowing down or speeding up. Average speed for the whole race is just a number indicating that, on average, the competitor was moving that fast. To find out more about the speed of a competitor in a race, a coach or athlete may want to measure more than one average speed.

Let's look at a 100-m dash as an example. At the 1988 Olympics in Seoul, the men's 100-m dash was won by Ben Johnson of Canada in an apparent world record time of 9.79 s. Johnson's gold medal and world record were subsequently taken away from him after it was determined that his performance had been artificially aided by his use of illegal steroids. The second-place finisher, Carl Lewis of the United States, was then awarded the gold medal. Lewis's finishing time of 9.92 s was ultimately accepted as the world record. Comparing the average speeds of these two sprinters over the entire 100 m, we find the following:

$$\text{average speed} = \bar{s} = \frac{\ell}{\Delta t} \qquad (5.5)$$

Johnson:

$$\bar{s} = \frac{100 \text{ m}}{9.79 \text{ s}}$$
$$\bar{s} = 10.21 \text{ m / s}$$

Lewis:

$$\bar{s} = \frac{100 \text{ m}}{9.92 \text{ s}}$$
$$\bar{s} = 10.08 \text{ m / s}$$

To find out more about how the two sprinters ran this race, we might have timed them for the first 50 m of the 100 m as well. Johnson's time for the first 50 m was 5.50 s. Lewis's time for the first 50 m was 5.65 s. Their average speeds for the first 50 m of the race were:

$$\text{average speed} = \bar{s}_{0-50} = \frac{\ell}{\Delta t}$$

Johnson:

$$\bar{s}_{0-50} = \frac{50 \text{ m}}{5.50 \text{ s}}$$
$$\bar{s}_{0-50} = 9.09 \text{ m / s}$$

Lewis:

$$\bar{s}_{0-50} = \frac{50 \text{ m}}{5.65 \text{ s}}$$
$$\bar{s}_{0-50} = 8.85 \text{ m / s}$$

Their average speeds from 50 to 100 m also could be determined:

$$\text{average speed} = \bar{s}_{50-100} = \frac{\ell}{\Delta t} = \frac{100 \text{ m} - 50 \text{ m}}{\Delta t}$$

Johnson:

$$\bar{s}_{50-100} = \frac{100 \text{ m} - 50 \text{ m}}{9.79 \text{ s} - 5.50 \text{ s}}$$
$$\bar{s}_{50-100} = \frac{50 \text{ m}}{4.29 \text{ s}}$$
$$\bar{s}_{50-100} = 11.66 \text{ m / s}$$

Lewis:

$$\bar{s}_{50-100} = \frac{100 \text{ m} - 50 \text{ m}}{9.92 \text{ s} - 5.65 \text{ s}}$$
$$\bar{s}_{50-100} = \frac{50 \text{ m}}{4.27 \text{ s}}$$
$$\bar{s}_{50-100} = 11.71 \text{ m / s}$$

With two numbers to describe each runner's speed during the race, we know much more about how each runner ran the race. Ben Johnson won the race in the first 50 m. His average speed was much faster than Lewis's over this portion of the race. Both sprinters were faster over the second 50 m. Surprisingly, Lewis actually ran this part of the race slightly faster than Johnson, but not fast enough to beat him to the finish line.

If we wanted to know which athlete had the fastest top speed in the 100 m, we would have to record split times at more frequent intervals in the race. This would give us even more information about the performances of each sprinter. Sport scientists at the 1988 Olympic Games in Seoul recorded the split times at every 10 m for the finalists in the men's and women's 100-m dashes (Brüggemann & Glad 1989). They did this using high-speed video and motion picture cameras that were synchronized with each other and with the starter's pistol. The times they recorded for Carl Lewis and Ben Johnson are shown in table 5.1.

These 10-m split times can be used to determine the average speed of each sprinter during each 10-m interval. This is done by dividing the distance covered in each interval, 10 m in this case, by the time taken to run that distance, the interval time. Table 5.2 shows the values of each runner's average speed over each 10-m interval.

Now we have much more information about each sprinter's performance. From table 5.2, we can tell that Johnson was faster than Lewis over every interval up to the 50- to 60-m interval. During the 50- to 60-m interval, both Johnson and Lewis reached their maximum speed, and their average speeds during this interval were the same. After 60 m, both runners slowed down, but Johnson slowed down more, especially over the last 10 m of the race, from 90 to 100 m.

By taking more split times during the race, we can determine the runners' average speeds for more intervals and shorter intervals. This procedure also gives us a better idea of what each runner's speeds were at specific instants of time during the race. The speed of an object at a specific instant of time is its

TABLE 5.1   Elapsed and Interval Times for Each 10-m Interval for Ben Johnson and Carl Lewis in the Men's 100-m Dash Final at the 1988 Olympic Games in Seoul, South Korea

Position (m)	Ben Johnson		Carl Lewis	
	Elapsed time (s)	Interval time (s)	Elapsed time (s)	Interval time (s)
0	0		0	
10	1.83	1.83	1.89	1.89
20	2.87	1.04	2.96	1.07
30	3.80	0.93	3.90	0.94
40	4.66	0.86	4.79	0.89
50	5.50	0.84	5.65	0.86
60	6.33	0.83	6.48	0.83
70	7.17	0.84	7.33	0.85
80	8.02	0.85	8.18	0.85
90	8.89	0.87	9.04	0.86
100	9.79	0.90	9.92	0.88

Data from Brüggemann and Glad 1989.

TABLE 5.2   Interval Times and Speeds for Each 10-m Interval for Ben Johnson and Carl Lewis in the Men's 100-m Dash Final at the 1988 Olympic Games in Seoul, South Korea

Interval (m)	Ben Johnson		Carl Lewis	
	Interval time (s)	Average speed (m/s)	Interval time (s)	Average speed (m/s)
0-10	1.83	5.46	1.89	5.29
10-20	1.04	9.62	1.07	9.35
20-30	0.93	10.75	0.94	10.64
30-40	0.86	11.63	0.89	11.24
40-50	0.84	11.90	0.86	11.63
50-60	0.83	12.05	0.83	12.05
60-70	0.84	11.90	0.85	11.76
70-80	0.85	11.76	0.85	11.76
80-90	0.87	11.49	0.86	11.63
90-100	0.90	11.11	0.88	11.36

Data from Brüggemann and Glad 1989.

instantaneous speed. The speed of an object may vary with time, especially in an event such as a 100-m dash. If we are interested in the maximum or top speed a runner achieves during a race, this would be an instantaneous speed. An average speed only gives us an estimate of how fast something was moving over an interval of time—not at an instant in time. If we are told what a runner's average speed is for an interval of time, we can correctly assume that the runner's instantaneous speed was faster than the average speed during some parts of that interval and slower than the average speed during other parts of that interval.

Think about your car's speedometer. Does it measure average speed or instantaneous speed? Does it indicate how fast you were going during the past hour? During the past minute? During the past second? The speedometer on your car measures instantaneous speed. It indicates how fast you are going at the instant in time that you are looking at it. Practically speaking, we can think of instantaneous speed as distance traveled divided by the time it took to travel that distance if the time interval used in the measurement is very small. If the word *average* does not precede the word *speed,* you should assume that instantaneous speed is being referred to.

> **We can think of instantaneous speed as distance traveled divided by the time it took to travel that distance if the time interval used in the measurement is very small.**

Velocity

Now let's turn our attention to velocity. Average velocity is displacement of an object divided by the time it took for that displacement. As displacement is a vector, described by a number (magnitude) and a direction, average velocity is also a vector, described by a number (magnitude) and a direction. Mathematically, this can be expressed as:

$$\bar{v} = \frac{d}{\Delta t} \qquad (5.6)$$

where

$\bar{v}$ = average velocity

d = displacement

Δt = time taken

The units for describing velocity are a unit of length divided by a unit of time. The SI unit for describing velocity is meters per second. To measure the average velocity of an object, you need to know its displacement and the time taken for that displacement.

Sometimes we are interested in the components of velocity. So, just as we were able to resolve force and displacement vectors into components, we can also resolve velocity vectors into components. To resolve a resultant average velocity into components, we could simply determine the components of the resultant displacement. For the football player returning the kickoff in the example used earlier, the player's displacement from the instant he received the ball until he was tackled was –10 yd in the x-direction (across the field) and +30 yd in the y-direction (down the field). His resultant displacement was 31.6 yd down and across the field (or –71.6° across the field). If this kick return lasted 6 s, his resultant average velocity was

$$\bar{v} = \frac{d}{\Delta t} \qquad (5.6)$$
$$\bar{v} = \frac{31.6 \text{ yd}}{6 \text{ s}}$$
$$\bar{v} = 5.3 \text{ yd / s}$$

This resultant average velocity was in the same direction as the resultant displacement. Similarly, the running back's average velocity across the field was

$$\bar{v}_x = \frac{\Delta x}{\Delta t} \qquad (5.7)$$
$$\bar{v}_x = \frac{-10 \text{ yd}}{6 \text{ s}}$$
$$\bar{v}_x = 1.7 \text{ yd / s}$$

The running back's average velocity down the field (the most important of all these velocities) was:

$$\bar{v}_y = \frac{\Delta y}{\Delta t} \qquad (5.8)$$
$$\bar{v}_y = \frac{30 \text{ yd}}{6 \text{ s}}$$
$$\bar{v}_y = 5.0 \text{ yd / s}$$

Just as with the displacements, the resultant average velocity is larger than any of its components. And just as with the displacements, the square of the resultant average velocity should equal the sum of the squares of its components. Let's check.

$$A^2 + B^2 = C^2 \qquad (5.3)$$
$$(\bar{v}_x)^2 + (\bar{v}_y)^2 = \bar{v}^2$$
$$(-1.7 \text{ yd/s})^2 + (5.0 \text{ yd/s})^2 = \bar{v}^2$$
$$2.8 \text{ yd}^2/\text{s}^2 + 25.0 \text{ yd}^2/\text{s}^2 = \bar{v}^2$$
$$\sqrt{27.8 \text{ yd}^2/\text{s}^2} = \bar{v}$$
$$5.3 \text{ yd/s} = \bar{v}$$

This indeed matches the resultant average velocity of 5.3 yd/s we computed from the resultant displacement and elapsed time.

Average velocity and average speed would both be good descriptors to use in describing the 100-m dash, as it is in a straight line. A runner's speed and the magnitude of the velocity toward the finish line would be identical. In such a case, speed and velocity may be used interchangeably with no problem. However, if we are speaking of an activity in which the direction of motion changes, speed and the magnitude of velocity are not synonymous. Imagine a 100-m swimming race in a 50-m pool. If the first-place finisher completes the race in 50 s, the swimmer's average speed is

$$\bar{s} = \frac{\ell}{\Delta t} \qquad (5.5)$$
$$\bar{s} = \frac{100 \text{ m}}{50 \text{ s}}$$
$$\bar{s} = 2.0 \text{ m/s}$$

What is the swimmer's average velocity? If the swimmer starts and finishes in the same place, the swimmer's displacement is zero, which means the swimmer's average velocity would also have to be zero. In this case, average velocity and average speed do not mean the same thing, and the average speed measurement is a better descriptor.

What about instantaneous speed and instantaneous velocity? We haven't discussed instantaneous velocity yet. It is similar to the concept of instantaneous speed except that direction is included. If we measured average velocity over shorter and shorter intervals of time, practically speaking, we would soon have a measure of instantaneous velocity. Instantaneous velocity is the velocity of an object at an instant in time. When we speak of the magnitude of the resultant instantaneous velocity of an object, that number is the same as the instantaneous speed of the object.

A resultant instantaneous velocity can also be resolved into components in the direction of interest. For the football player running back the kickoff, we could describe his instantaneous resultant velocity, and we could also describe his instantaneous velocity in the x-direction (across the field) or the y-direction (down the field). If we were concerned about how quickly he was gaining yardage, his instantaneous velocity down the field would be important. Similarly, with the downhill ski racers, it is not their instantaneous resultant velocity that is important; it is the component of this velocity in the direction down the hill that will have the greater effect on the result of the race.

SELF-EXPERIMENT

Let's consider another example to illustrate the difference between speed and velocity. Stand in a room with four walls, facing the north wall. Let's consider north to be the direction we are interested in, so north is positive. As you begin walking forward, toward the north wall, your velocity north is positive. When you stop, your velocity north is zero. As you begin walking backward, toward the south wall, your velocity north is negative (you are moving in the negative direction). If you walk to your right or left, directly east or west, your velocity north is zero because you are not getting closer to or farther away from the north wall. Walk forward toward the north wall and begin turning right toward the east wall; your velocity north is positive and then decreases as you turn. Walk east and then turn left toward the north wall; your velocity north is zero and then increases as you turn. During all of these turns, your speed may not even be changing, but if your direction of motion changes, your velocity changes.

Importance of Speed and Velocity

Now let's make sure we realize the importance of speed and velocity in different sports activities. We've already indicated that in racing events, average speed or average velocity are direct indicators of performance. The athlete with the greatest average speed or greatest average velocity will be the winner. In what other sports is speed or velocity important? How about baseball? A good fastball pitch, which moves with a velocity of 90 to 100 mi/hr, is difficult to hit. Why? The faster the ball is pitched, the less time the batter has to react and decide whether or not to swing at the ball. For instance, in his prime, Nolan Ryan could pitch the ball at 101 mi/hr. This is equivalent to 148 ft/s or 45 m/s. The distance from the pitching rubber to home plate is 60 ft 6 in., or 60.5 ft. The ball is released about 2 ft 6 in. in front of the rubber, so the horizontal distance it must travel to reach the plate is only 58 ft (60.5 ft – 2.5 ft). How much time does a batter have to react to one of Nolan Ryan's fastballs pitched at 101 mi/hr?

$$\bar{v} = \frac{d}{\Delta t} \qquad (5.6)$$

$$148 \text{ ft/s} = \frac{58 \text{ ft}}{\Delta t}$$

$$\Delta t = \frac{58 \text{ ft/s}}{148 \text{ ft/s}}$$

$$\Delta t = 0.39 \text{ s}$$

The batter has only 0.39 s to decide whether or not to swing his bat, and if he does decide to swing it, he must do so in the time he has left. The faster the pitcher can pitch the ball, the less time the batter has to react. Speed and velocity are very important in baseball.

Are speed or velocity important in soccer, lacrosse, ice hockey, field hockey, team handball, or any other sport that involves a goalie? The speed of the ball (or puck) when it is shot toward the goal is very important to the goalie. The faster the shot, the less time the goalie has to react and block it.

Are speed or velocity important in the jumping events in track and field? Yes! Faster long jumpers jump farther. Faster pole vaulters vault higher. Speed is also related to success in the high jump and triple jump.

Can you think of any sports where speed or velocity are not important? There aren't many. Speed and velocity play an important role in almost every sport.

Acceleration

We now have a large repertoire of motion descriptors: position, distance traveled, displacement, speed, and velocity. In addition, we can use component displacements or velocities to describe an object's motion, because displacement and velocity are vector quantities. Did we use any other descriptors at the beginning of this section to describe the motion of a ball rolling across the floor? Let's try another motion of the ball. Throw the ball up in the air and let it fall back into your hand. How would you describe this motion? You might say that the ball moves upward and slows down on the way up, then it begins moving downward and speeds up on the way down. Another way to describe how the ball slows down or speeds up would be to say that it decelerates on the way up and accelerates on the way down. Acceleration is a term you are probably somewhat familiar with, but the mechanical definition of acceleration may differ from yours, so we'd better get some agreement.

Mechanically, acceleration is the rate of change in velocity. As velocity is a vector quantity, with a number and direction associated with it, acceleration is also a vector quantity, with a number and direction associated with it. An object accelerates if the magnitude or direction of its velocity changes.

When an object speeds up, slows down, starts, stops, or changes direction, it is accelerating.

Average acceleration is defined as the change in velocity divided by the time it took for that velocity change to take place. Mathematically, this is

$$\bar{a} = \frac{v_f - v_i}{\Delta t} \qquad (5.9)$$

$$\bar{a} = \frac{\Delta v}{\Delta t}$$

where

$\bar{a}$ = average acceleration

v_f = instantaneous velocity at end of an interval, or final velocity

v_i = instantaneous velocity at beginning of an interval, or initial velocity

Δv = change in velocity

Δt = time taken

From this mathematical definition of average acceleration, it is apparent that acceleration can be positive or negative. If the final velocity is less (slower) than the initial velocity, the change in velocity is a negative number and the resulting average acceleration is negative. This happens if an object slows down in the positive direction. You may have thought of this as a deceleration, but we'll just call it a negative acceleration. A negative average acceleration will also result if the initial and final velocities are both negative and if the final velocity is a larger negative number than the initial velocity. This occurs if an object is speeding up in the negative direction.

The units for describing acceleration are a unit of length divided by a unit of time divided by a unit of time. The SI units for describing acceleration are meters per second per second or meters per second squared. You may have seen some car ads that tout the acceleration capabilities of the car. An ad may say that a car can accelerate from 0 to 60 in

7 s. This would represent an average acceleration for the car of:

$$\bar{a} = \frac{v_f - v_i}{\Delta t} \qquad (5.9)$$

$$\bar{a} = \frac{60 \text{ mi/hr} - 0 \text{ mi/hr}}{7 \text{ s}}$$

$$\bar{a} = 8.6 \text{ mi/hr/s}$$

This acceleration can be interpreted as follows: In 1 s, the car's velocity increases (the car speeds up) by 8.6 mi/hr.

If we measured average acceleration over shorter and shorter time intervals, practically speaking, we would soon have a measure of instantaneous acceleration. Instantaneous acceleration is the acceleration of an object at an instant in time. Instantaneous acceleration indicates the rate of change of velocity at that instant in time.

Because acceleration is a vector (as are force, displacement, and velocity), it can also be resolved into

SELF-EXPERIMENT

Let's consider the example where you walk around the room. You are in a room with four walls, facing the north wall. Let's consider north to be the direction we are interested in, so north is positive. As you begin walking forward, toward the north wall, your velocity north is positive, and because you speed up in the northerly direction, your acceleration north is positive. When you slow down and stop, your velocity north decreases to zero and your acceleration north must be negative, because you are slowing down in the positive direction. This could also be described as an acceleration in the southerly direction. This is where you may be confused; you were moving north, but your acceleration was south! This is correct, however, because acceleration indicates your change in motion. As you begin walking backward, you speed up toward the south wall, your velocity north is negative and increasing (you are moving in the negative direction), and your acceleration is also negative (or an acceleration in the southerly direction). If you walk to your right or left, directly east or west, your velocity north is zero because you are not getting closer to or farther away from the north wall. Your acceleration is also zero because you are not speeding up or slowing down toward the north wall. If you walk forward toward the north wall and begin turning right toward the east wall, your velocity north is positive and then decreases as you turn, so your acceleration north is negative as you turn. If you are walking east and then turn left toward the north wall, your velocity north is zero and then increases as you turn, so your acceleration north is positive as you turn. During all these turns, your speed may not even be changing, but if your direction of motion changes, your velocity changes and you are accelerating. Figure 5.8 illustrates the directions of motion and acceleration for various motions along a line.

− Direction +	V (Direction of motion)	Change in motion (Speeding up +; slowing down -)	a (Direction of acceleration)
Speeding up	+	+	+
Not changing	+	0 (Constant velocity)	0
Slowing down	+	−	−
Speeding up	−	+	−
Not changing	−	0 (Constant velocity)	0
Slowing down	−	−	+

Figure 5.8   The direction of motion and direction of acceleration are the same when the object is speeding up, but opposite to each other when the object is slowing down.

SELF-EXPERIMENT

Now let's try to describe the motion of a ball thrown up into the air. Let's use the terms we have learned—displacement, velocity, and acceleration. If we set up a coordinate system with the x-axis oriented horizontally in the direction of the horizontal motion of the ball and the y-axis oriented vertically, how would you describe the vertical motion of the ball? Let's consider the positive direction along the y-axis (vertical axis) as upward. As the ball leaves your hand, it is moving in the upward direction, so its velocity is positive. Is the ball speeding up or slowing down as it goes up? The ball is slowing down in the upward direction, so its acceleration is negative or in the downward direction. When the ball reaches the peak of its flight, it isn't going up anymore and it hasn't started downward, so at this instant its vertical velocity is zero. As its velocity is changing from positive to negative (or from upward to downward), it is still accelerating downward. After the ball is past its peak, it falls downward, so its velocity is negative (downward). Because the ball speeds up in the downward direction, its acceleration is still negative. Despite the changes in the direction the ball was moving, its vertical acceleration was always downward while the ball was in the air. The direction of motion does not indicate the direction of the acceleration.

component accelerations. This is true for both average and instantaneous accelerations. The difficulty with understanding acceleration is that it is not directly observed, as are displacement and velocity.

> **The direction of motion does not indicate the direction of the acceleration.**

After completing the self-experiment on page 92, consider the motion of a ball in the horizontal direction. The ball continues to move in the direction we projected it. It does not swerve right or left. Its horizontal velocity is positive. Does the ball accelerate horizontally while it is in the air? Does it speed up or slow down horizontally? No. Does it change its direction horizontally? No. If the ball doesn't speed up or slow down or change direction, it is not accelerating in the horizontal direction.

Let's summarize some things about acceleration. If something speeds up in the positive direction, its acceleration is positive (it accelerates in the positive direction). Think of this as a double positive (+ +), which results in a positive (+). If it slows down in the positive direction, its acceleration is negative (it accelerates in the negative direction). Think of this as a negative positive (– +), which results in a negative (–). If something speeds up in the negative direction, its acceleration is negative (it accelerates in the negative direction). Think of this as a positive negative (+ –), which results in a negative (–). If something slows down in the negative direction, its acceleration is positive (it accelerates in the positive direction). Think of this as a double negative (– –), which results in a positive (+). Remember, though, the algebraic signs + and – are only symbols we use to indicate directions in the real world. Before analyzing a problem, first establish to which direction + will be assigned.

Summary

Motion may be classified as linear, angular, or a combination of both (general motion). Most examples of human movement are general motion, but separating the linear and angular components of the motion makes it easier to analyze the motion. Linear displacement is the straight-line distance from starting point to finish, whereas linear distance traveled represents the length of the path followed from start to finish. Velocity is the rate of change of displacement, whereas speed is the rate of change of distance. Acceleration is the rate of change of velocity. Displacement, velocity, and acceleration are vector quantities and are described by size and direction.

We now have the terms to describe many aspects of the linear motion of an object—distance traveled, displacement, speed, velocity, and acceleration. But what causes linear motion of objects? How do we affect our motion and the motion of things around us? We've gotten some hints in the previous chapters. In the next chapter, we will explore the causes of linear motion more thoroughly.

Key Terms

angular motion *(p. 77)*

average acceleration *(p. 90)*

average speed *(p. 85)*

average velocity *(p. 88)*

Cartesian coordinate system *(p. 79)*

curvilinear translation *(p. 76)*

displacement *(p. 81)*

distance traveled *(p. 81)*

general motion *(p. 78)*

instantaneous speed *(p. 88)*

instantaneous velocity *(p. 89)*

instantaneous acceleration *(p. 91)*

linear motion *(p. 76)*

position *(p. 78)*

rectilinear translation *(p. 76)*

resultant displacement *(p. 81)*

speed *(p. 85)*

velocity *(p. 85)*

REVIEW QUESTIONS

1. A running back in football runs 30 yd before he is tackled (i.e., his distance traveled is 30 yd). The straight-line distance between where he began the run and where he was tackled is 20 yd at an angle of 25° with the sideline. How many yards did the running back gain in this play?

2. Which shot would a hockey goalie have a better chance of blocking—a shot from 5 m away at 10 m/s or a shot from 10 m away at 40 m/s?

3. The horizontal velocity of a baseball pitcher's fastball is 40 m/s at the instant it's released from the pitcher's hand. If the horizontal distance from the pitcher's hand to home plate is 17.5 m at the instant of release, how much time does the batter have to react to the pitch and swing the bat?

4. How are the sizes of the goals in ice hockey, lacrosse, soccer, field hockey, and team handball related to the speeds of the balls (or pucks) used in these games?

5. A sailboarder is zipping northeast across the river with a velocity of 10 m/s relative to the water. The river current is moving the water north at a velocity of 3 m/s downstream. If the angle between the relative velocity of the sailboard and the river current is 30°, what is the resultant or true velocity of the sailboard?

6. Can a runner moving around a curve at constant speed be accelerating?

7. Gail Devers is running a 100-m dash. When the starter's pistol fires, she leaves the starting block and continues speeding up until she reaches her top speed of 11 m/s 6 s into the race. She holds this speed for 2 s and then gradually slows down until she crosses the finish line 11 s after she started the race.

   a. What was Gail's average acceleration during the first 6 s of the race?

   b. What was Gail's average acceleration from 6 to 8 s into the race?

   c. In which direction was Gail's acceleration during the last 3 s of the race?

8. A football is punted into the air. It has an initial vertical velocity of 15 m/s and an initial horizontal velocity of 10 m/s when it leaves the kicker's foot. If the ball experiences a constant vertical acceleration of 9.81 m/s/s downward while it is in the air, what is its vertical velocity 2 s after it leaves the kicker's foot?

9. If you run around a circle counterclockwise, in which direction (relative to the circle) is your acceleration?

Linear Kinetics: Explaining the Causes of Linear Motion

© Claus Andersen

Y ou're watching the Olympic weight lifting competition. The weights on the barbell are more than two times the lifter's weight. The athlete approaches the barbell and grips it firmly with both hands. With a sharp grunt, he snatches the bar and lifts it from the floor to overhead in one smooth motion. What forces did the athlete have to exert on the barbell to create the movement you just witnessed? An analysis using Newton's second law of motion would assist in answering this question. The contents of this chapter provide the background for completing such an analysis.

OBJECTIVES

When you finish this chapter, you should be able to do the following:

▌ Explain Newton's three laws of motion

▌ Define impulse

▌ Define momentum

▌ Explain the relationship between impulse and momentum

▌ Describe the relationship between mass and weight

▌ Identify the acceleration due to gravity

This chapter is about the subbranch of mechanics called kinetics. Dynamics is the branch of mechanics concerned with the mechanics of moving objects. Kinetics is the branch of dynamics concerned with the forces that cause motion. This chapter is specifically concerned with linear kinetics, or the causes of linear motion. In this chapter, we will learn more about Newton's laws of motion and how they can be used to analyze motion. The topics learned in this chapter will give you the basic tools to analyze and explain the techniques used in many sport skills.

Law of Inertia

In chapter 3, Newton's first law of motion was briefly introduced. Because this and Newton's other laws of motion provide the basics of mechanics, they were introduced early. Now let's study them more carefully to discover their full importance. Let's begin with Newton's first law of motion, the law of inertia.

What happens to an object if no external forces act on it or if the net external force acting on the object is zero? Newton's first law of motion provides an answer to this question. Every body continues in its state of rest, or of uniform motion in a straight line, unless it is compelled to change that state by forces impressed upon it. This law explains what happens to an object if no external forces act on it or if the net external force is zero. More simply stated, Newton's first law says: an object will not move (it will remain in its state of rest) if it wasn't moving to begin with, or it will continue moving at

constant speed in a straight line (it will remain in its state of uniform motion in a straight line) if it was moving to begin with, unless an external force (or a net external force) acts on the object.

Let's see how Newton's first law of motion applies to human movement in sport. Can you think of any situations where no external forces act on an object? Gravity is an external force that acts on all objects close to the earth. Apparently, there are no situations in sports and human movement to which Newton's first law of motion applies! Is this true? Perhaps we can find applications of Newton's first law in sports if we consider only motions of an object or body in a specific direction.

Horizontal Motion of a Projectile

Velocity is a vector quantity. The velocity of an object can thus be resolved into vector components. Perhaps Newton's first law may be applied to an object in motion if only a specific component of the object's velocity is examined. If we resolve the velocity of a projectile into horizontal and vertical components, and examine the horizontal motion of the projectile separately, we might have a situation where Newton's first law applies.

It is difficult to examine or observe the horizontal motion of a projectile separately from its vertical motion, though, because when you observe a projectile, you see both horizontal and vertical motions simultaneously as one motion. How can we view a projectile, such as the ball in the self-experiment on page 97, so that we isolate only its horizontal motion? What if we watched the projectile from above?

SELF-EXPERIMENT

Throw a ball in the air from one hand to the other. If we resolve the motion of the ball into horizontal and vertical components, we know that Newton's first law doesn't apply to the vertical motion of the ball because gravity is an external force that pulls downward on the ball and changes its motion. What about horizontally? Are there any external forces that pull or push against the ball sideways to change its horizontal motion once it leaves your hand? The only thing that could exert a force horizontally on the ball is the air through which it moves. This force will probably be very small in most cases, and its effect will be too small to notice. If we ignore air resistance, the horizontal motion of the ball does provide an example of Newton's first law. This means that the horizontal velocity of the ball will not change from the time it leaves your hand until it contacts your other hand or another object.

Imagine yourself perched on the catwalk of a gymnasium watching a basketball game. Better yet, imagine watching a football game from the Goodyear blimp. How would the motions of the football or basketball appear to you from these vantage points? If your depth perception was hindered (close one eye), could you see the vertical motion of the football during a kickoff? Could you detect the vertical motion of the basketball during a free throw? The answer is no in both cases. All you see is the horizontal motion of the balls. Does the basketball slow down, speed up, or change direction horizontally as you view it from above? How about the football? If we tried to represent the motion of the basketball as viewed from above in a single picture, it might look something like figure 6.1.

To represent the motion, we show the position of the basketball at four instants in time, each 0.10 s apart. Notice that the images line up along a straight line, so the motion of the ball is in a straight line. Also notice that the displacement of the ball over each interval of time is the same, so the velocity of the ball is constant. Newton's first law of motion does apply to the horizontal motion of a projectile. A projectile is an object that has no external forces acting on it other than gravity. Newton's first law then explains that the horizontal velocity of a projectile is constant and its horizontal motion is in a straight line.

The horizontal velocity of a projectile is constant and its horizontal motion is in a straight line.

Figure 6.1 An overhead view of a basketball free throw shows that the horizontal displacement, Δx, for each 0.10-s time interval is the same.

Examples of projectiles in sports and human movement are numerous. Can you name a few? Remember, an object is a projectile only if no force is acting on it other than gravity. If the force on an object due to air resistance is significant, then that object cannot be described as a projectile. Here are some examples of projectiles: a shot in flight during a shot put, a basketball in flight, a hammer in flight during a hammer throw, a volleyball in flight, a squash ball in flight, a lacrosse ball in flight, a football in flight, a rugby ball in flight . . . just about any ball used in sports becomes a projectile once it is thrown, released, or hit, if air resistance is negligible. So in ball sports, the horizontal path and velocity of the ball cannot be changed in flight if air resistance is negligible. The horizontal path of the ball will be in a straight line and its horizontal velocity will be constant; it won't slow down or speed up. Once a ball has left our hands and is in flight, our actions and antics cannot change its course or velocity.

Newton's first law seems pretty obvious for balls as projectiles, but what if we ourselves are the projectiles? Can the human body be a projectile? Are there situations in which the only external force acting on you is the force of gravity? Yes, of course there are! Think of some examples in sports where the human body is a projectile. How about in running? High jumping? Long jumping? Diving? Pole vaulting? Volleyball? Basketball? Soccer? Football? In each of these sports, there may be situations where the athlete is airborne and the only force acting on him or her is gravity. Does Newton's first law apply to the horizontal motion of an athlete in these situations? Yes! This means that, once an athlete's body has left the ground and become a projectile, the athlete cannot change its horizontal path or velocity. Once a volleyball player has jumped up to the left to block a shot, the direction of her body's motion cannot be changed; in other words, once she has jumped up to the left, she won't be able to change direction and block a shot to the right. And once a pole vaulter releases the pole, he cannot change his horizontal motion. Once he has let go of the pole, he no longer has control over where he falls. Once a long jumper leaves the take-off board and becomes a projectile, her actions while in the air will not affect the horizontal velocity of her body. She cannot speed up her horizontal velocity to increase the distance of the jump after leaving the ground.

Newton's first law of motion tells us that, when we throw things, we cannot change the horizontal velocity or horizontal path of objects once they leave our hands. Likewise, if we throw ourselves into the air by jumping or diving, we cannot control our horizontal velocity or direction after takeoff without any external forces acting on us horizontally.

Zero Net Force

Other than projectile motion, does Newton's first law apply to any other situations in human move-

SELF-EXPERIMENT

Hold a 10-lb dumbbell in your hand. How large a force must you exert on the dumbbell to hold it still? What external forces act on the dumbbell? Vertically, gravity exerts a force downward equal to the weight of the dumbbell, or 10 lb. Your hand exerts a reaction force upward against the dumbbell. According to Newton's first law, an object will stay at rest only if no external forces are acting on the object or if the net external force acting on the object is zero. Because the dumbbell is at rest and not moving, the net external force acting on it must be zero. Figure 6.2 shows a free-body diagram of the dumbbell.

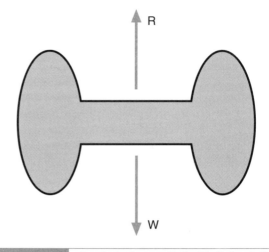

| Figure 6.2 | Free-body diagram of a dumbbell when held still in the hand. |

ment or sports? Newton's first law is applicable if *no* external forces act on an object or if the *net* external force is zero. So an object will stay at rest or moving in a straight line at constant velocity if the net external force is zero (if the external forces acting on it cancel each other out). Can you think of any examples in sport where the net external force acting on an athlete or object is zero?

The two external forces acting on the dumbbell are the force of gravity acting downward and the reaction force from your hand acting upward. The resultant of these two external forces is:

$$\Sigma F = R + (-W) = 0 \qquad (6.1)$$

$$R = W = 10 \text{ lb} \qquad (6.2)$$

where

    R = the reaction force from your hand

    W = the weight of the dumbbell = 10 lb

SELF-EXPERIMENT

Begin lifting the dumbbell, moving it upward at a constant velocity. How large a force must you exert against the dumbbell to keep it moving upward at constant velocity? What external forces act on the dumbbell? Vertically, gravity still exerts a force downward equal to the weight of the dumbbell, 10 lb, and your hand still exerts a reaction force upward against the dumbbell. According to Newton's first law, an object will move at constant velocity in a straight line only if no external forces are acting on the object or if the net external force acting on the object is zero. Because the dumbbell is moving at a constant velocity in a straight line, the net external force acting on it must be zero. Figure 6.3 shows a free-body diagram of the dumbbell. The two external forces on the dumbbell are exactly the same as when it was still: the force of gravity acting downward and the reaction force from your hand acting upward.

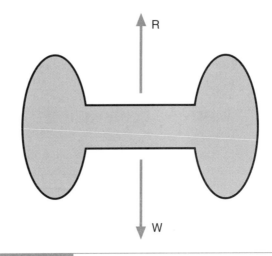

| Figure 6.3 | Free-body diagram of a dumbbell when moving upward (or downward) at a constant velocity. |

When you are holding the 10-lb dumbbell still in your hand, the force you must exert against it is 10 lb upward.

When you are moving the 10-lb dumbbell upward at a constant velocity, the force you must exert against it to keep it moving upward at a constant velocity is a 10-lb upward force.

Newton's first law of motion may be interpreted in several ways:

1. If an object is at rest and no external forces act on it, the object remains at rest.
2. If an object is in motion and no external forces act on it, the object will continue moving at constant velocity in a straight line.
3. If an object is at rest and external forces do act on it, the object remains at rest only if the resultant of the external forces is zero (if the net external force is zero).
4. If an object is in motion and external forces do act on it, the object will continue moving at constant velocity in a straight line only if the net external force is zero.

Newton's first law of motion applies to the resultant motion of an object and to the components of this resultant motion. Because forces and velocities are vectors, Newton's first law can be applied to any direction of motion. If no external forces act, or if the components of the external forces acting in the specified direction sum to zero, there is no motion of the object in that direction or the velocity in that direction is constant.

> Newton's first law of motion applies to the resultant motion of an object and to the components of this resultant motion.

Law of Acceleration

What happens to an object if the external forces acting on it do not cancel each other out; that is, if the forces do not balance and a net external force acts on the object? This is explained by Newton's second law of motion, the law of acceleration. The change of motion of an object is proportional to the force impressed and is made in the direction of the straight line in which the force is impressed. This law explains what happens if a net external force acts on an object. Stated more simply, Newton's second law says: if a net external force is exerted on an object, the object will accelerate in the direction of the net external force and its acceleration will be proportional to the net external force and inversely proportional to its mass. Mathematically, this law can be expressed as:

$$\Sigma F = ma \qquad (6.3)$$

where

ΣF = net external force

m = mass of the object

a = acceleration of the object

Newton's second law expresses a cause-and-effect relationship. Forces cause acceleration. Acceleration is the effect of forces. If a net external force acts on an object, the object accelerates. If an object accelerates, a net external force must be acting to cause the acceleration.

In chapter 5, we learned about acceleration. Newton's second law explains how accelerations occur.

> Any time an object starts, stops, speeds up, slows down, or changes direction, it is accelerating and a net external force is acting to cause this acceleration.

Vertical Motion of a Projectile

The horizontal motion of a projectile provided a good illustration of Newton's first law. Perhaps the vertical motion of a projectile provides a good illustration of Newton's second law.

SELF-EXPERIMENT

Consider the example of the ball thrown up in the air, only this time throw the ball straight up so it doesn't have any horizontal motion. Let's examine its vertical motion. What forces act on the ball? Once the ball leaves your hand, the only force acting on it is gravity (the weight of the ball) pulling down on the ball. Figure 6.4 shows a free-body diagram of the ball.

| Figure 6.4 | Free-body diagram of a ball as a projectile. The only force acting on it is gravity. |

We can now predict the motion of the ball using Newton's second law. The net force acting on the ball is the ball's weight. Mathematically, then:

$$\Sigma F_y = ma_y \qquad (6.4)$$

$$W = ma_y \qquad (6.5)$$

where

ΣF_y = net external force in the vertical direction

m = mass of the ball

a_y = vertical acceleration of the ball

W = weight of the ball

The resultant force acting on the ball is the ball's weight, which acts in the downward direction. The acceleration of the ball must therefore be downward also. A downward acceleration means that the ball speeds up in the downward direction or slows down in the upward direction. Does this explain what you observe? As the ball leaves your hand, it moves upward, but it slows down as it moves upward until it stops moving upward. The ball then begins to fall and move downward. As it moves downward, its velocity downward speeds up until it hits the ground. The acceleration of the ball throughout its flight is downward.

Does the ball's downward acceleration change during its flight? We can answer that by examining the equation:

$$W = ma_y \tag{6.5}$$

If the weight of the ball, W, doesn't change during its flight, the right-hand side of the equation doesn't change. The product of mass, m, times vertical acceleration, a_y, doesn't change. The mass of the ball, m, doesn't change during the ball's flight either. The vertical acceleration of the ball, a_y, must be constant and not change. This means that the ball slows down by the same amount over equal intervals of time while it is rising, and that the ball speeds up by the same amount while it is falling. This is true for all projectiles. Previously, we determined that the horizontal velocity of a projectile was constant. We can also say that the vertical acceleration of a projectile is constant and downward.

How large is this vertical acceleration? Is it affected by the weight of the ball? If you remember Newton's law of universal gravitation from chapter 3, the force of gravity was defined as weight:

$$F = G\left(\frac{m_1 m_2}{r^2}\right) = W \tag{6.6}$$

where

F = force of gravity

W = weight of the object

G = universal constant of gravitation

m_1 = mass of the object

m_2 = mass of the earth

r = distance between the center of mass of the earth and the object (essentially the radius of the earth for objects on the earth's surface)

Because the mass of the earth is constant, the radius of the earth is constant, and G is constant, we can represent these three constants with one constant, g:

$$g = G\left(\frac{m_2}{r^2}\right) \tag{6.7}$$

Equation 6.6 then becomes:

$$F = G\left(\frac{m_1 m_2}{r^2}\right) = m_1 G\left(\frac{m_2}{r^2}\right) = W$$
$$F = m_1 g = W \tag{6.8}$$

The force of gravity (the weight of the ball) changes proportionally to the mass of the ball, as g is a constant. If we examine equation 6.5 again:

$$W = ma_y \tag{6.5}$$

An increase in weight, W, corresponds to an increase in mass, m, and the vertical acceleration doesn't change. A heavier object experiences the same acceleration as a lighter object if the only force acting on the objects is the force of gravity. Weight does not affect the acceleration of an object due to gravity.

So the vertical acceleration of a projectile is constant, and it is also the same as the vertical acceleration of any other projectile, despite any differences in weight of the projectiles. How large is this acceleration? Look at equations 6.5 and 6.8. These equations are equivalent:

$$W = ma_y \tag{6.5}$$

$$F = m_1 g = W \tag{6.8}$$

so

$$W = ma_y = m_1 g = W$$
$$a_y = g \tag{6.9}$$

The constant, g, is equal to the vertical acceleration, a_y. This constant, g, is called the acceleration due to gravity or gravitational acceleration. Measurements of g indicate that the acceleration caused by gravity is 9.81 m/s/s downward. As suggested in chapter 3, a value of 10 m/s/s can be substituted for g for rough approximations.

The vertical acceleration of a projectile is constant.

Horizontal and Vertical Motion of a Projectile

We previously observed that Newton's first law explained the constant horizontal velocity of a projectile, and that Newton's second law explained the constant vertical acceleration of a projectile. Consider a projectile that moves up and out, so that it has horizontal and vertical motion. Does the vertical motion of the ball affect its horizontal motion or vice versa?

During flight, a projectile experiences a constant horizontal velocity and a constant vertical acceleration. This vertical acceleration is 9.81 m/s/s downward; that is, a projectile's vertical velocity is constantly reduced by 9.81 m/s for each second of its flight upward and constantly increased by 9.81 m/s for each second of its flight downward.

Other Applications of Newton's Second Law

Our examination of Newton's second law has been limited to projectiles; in those situations, the only

SELF-EXPERIMENT

Place a penny on the edge of a tabletop. Place another penny on the end of a ruler or another long, flat object. Place the ruler with the penny on it on the table next to the other penny so that the end of the ruler with the penny on it overhangs the tabletop. Strike the ruler with your hand so that it in turn strikes the penny on the table and knocks it off the table. Simultaneously, the movement of the ruler will dislodge the penny from the end of the ruler. Figure 6.5 shows the setup of the demonstration.

Which penny will strike the floor first? Try it several times to see. Both pennies hit the floor at the same time. The penny that is knocked off the table has a horizontal velocity as it begins to fall, whereas the penny that slips off the ruler does not. Both pennies fall the same vertical distance, and neither of them has a vertical velocity when it begins to fall. What force pulls the pennies toward the earth? The force of gravity pulls the pennies downward and accelerates both downward at the same rate of 9.81 m/s/s. Does that fact that one penny has a horizontal velocity affect the force of gravity acting on that penny and thus affect the vertical acceleration of that penny? No; the force of gravity acts with equal strength on the penny that was knocked off the table as it does on the penny that slid off the ruler.

Figure 6.5 The penny experiment demonstrates the independence of the horizontal and vertical components of projectile motion.

SELF-EXPERIMENT

Try to find an elevator at your school, or look for one in a tall building in your town. What happens when you ride up an elevator? How does it feel when the elevator starts up? Do you feel heavier or lighter? When the elevator is between floors, do you feel any heavier or lighter? What about when the elevator comes to a stop at an upper floor? Do you feel heavier or lighter as the elevator slows down? You probably feel heavier as the elevator starts upward and lighter as the elevator slows to a stop. Between floors, you probably feel neither heavier nor lighter. Why is this so? Did you gain and then lose weight as the elevator sped up and slowed down? Of course not, but let's try to explain what happened using Newton's second law of motion.

First, what external forces act on you as you stand in the elevator? Gravity pulls downward on you with a force equal to your weight. Do any other forces act on you? What about the reaction force under your feet? The elevator floor exerts a reaction force upward on your feet. Figure 6.6 shows a free-body diagram of the situation.

| Figure 6.6 | Free-body diagram of a person standing in an elevator. |

external force was gravity. Let's consider other applications of Newton's second law that involve contact forces as well as gravity.

The net force acting on you in the elevator self-experiment described above is

$$\Sigma F_y = R + (-W) = ma_y \qquad (6.10)$$

where

ΣF_y = net external force in the vertical direction

m = your mass

a_y = your vertical acceleration

W = your weight

R = the reaction force exerted on your feet by the elevator

If the reaction force, R, is larger than your weight, you feel heavier and the net force acts upward, resulting in an upward acceleration. This is exactly what happens when the elevator speeds up in the upward direction; it accelerates you upward and you feel heavier. If the reaction force, R, is equal to your weight, you feel neither heavier nor lighter and the net force is zero, resulting in no acceleration. If you and the elevator were already moving upward, you would continue to move upward at constant velocity. If the reaction force, R, is less than your weight, you feel lighter and the net force acts downward, resulting in a downward acceleration. This is exactly what happens when the elevator slows down; you decelerate upward (slow down in the upward direction or accelerate downward) and you feel lighter.

SELF-EXPERIMENT

You can crudely check the values of the reaction force, R, from the elevator floor that acts on your feet by taking a bathroom scale with you on your elevator ride. Stand on the scale while you are in the elevator. The scale indicates the force you exert on it, which is equal to the force it exerts on you (remember Newton's third law?). The scale reading goes up above your body weight as the elevator starts up and you accelerate upward. The scale reading then returns to your body weight as the elevator continues to move upward and you continue to move upward at constant velocity. The scale reading drops below your body weight as the elevator slows down and your acceleration is downward.

What happens when you ride the elevator down rather than up? The free-body diagram and equations used to analyze the upward elevator ride would be the same for the downward ride. When the elevator begins to descend, it speeds up downward. You feel lighter. The reaction force from the floor is less than your body weight, so the net force is downward and you accelerate downward. As the elevator continues downward, it stops speeding up downward. You feel your normal weight. The reaction force from the floor is equal to your body weight, so the net force is zero and you move at a constant velocity downward. As the elevator slows to a stop at the bottom floor, it slows down in the downward direction. You feel heavier. The reaction force from the floor is greater than your body weight, so the net force is upward and you accelerate upward (your downward movement slows down).

SELF-EXPERIMENT

Let's look at an example that has more to do with human movement in sports and consider what force you must exert against a 10-lb dumbbell to lift it. The external forces acting on the dumbbell are the pull of gravity acting downward and the reaction force from your hand acting upward. The net force is thus the difference between these two forces, just like on the elevator. When does the lift feel most difficult? When does it feel easier? To start the lift, you must accelerate the dumbbell upward, so the net force acting on the dumbbell must be upward. The force you exert on the dumbbell must be larger than 10 lb. (This is just like the elevator example.) Once you have accelerated the dumbbell upward, to continue moving it upward requires only that a net force of zero act on the dumbbell for it to move at constant velocity. The force you exert on the dumbbell must only equal 10 lb. As you complete the lift, you need to slow down the upward movement of the dumbbell, so the net force acting on the dumbbell is downward. The force you exert on the dumbbell must be less than 10 lb. When the dumbbell is held overhead, it is no longer moving, so the net force acting on the dumbbell is zero. The force you exert on the dumbbell must be equal to 10 lb.

The elevator and dumbbell examples show how a net force is needed to slow something down or speed it up. Slowing down and speeding up are examples of acceleration. Is a net force needed to change directions? Yes; because acceleration is caused by a net exertion force and a change in direction is an acceleration, a net force is needed to change direction. What happens when you run around a curve? You change your direction of motion. Because you change your direction of motion, you are accelerating. A net force must act in the horizontal direction to cause this acceleration. Where does the external horizontal force come from? Think about what happens if you try to run around a curve in an ice rink. You can't, because there isn't enough friction. Friction is the external horizontal force that causes you to change direction as you run around a curve.

Impulse and Momentum

Mathematically, Newton's second law is expressed by equation 6.3:

$$\Sigma F = ma \qquad (6.3)$$

This only tells what happens at an instant in time. The acceleration caused by the net force is an instantaneous acceleration. This is the acceleration experienced by the body or object at the instant the net force acts. This instantaneous acceleration will change if the net force changes. Except gravity, most external forces that contribute to a net external force change with time. So the acceleration of an object subjected to these forces also changes with time.

In sports and human movement, we are more concerned with the final outcome resulting from external forces acting on an athlete or object over some duration of time than with the instantaneous acceleration of the athlete or object at the time of the force application. We want to know how fast the ball was going after the pitcher exerted forces on it during the pitching actions.

Newton's second law can be used to determine this. Looking at equation 6.3 slightly differently, we can consider what average acceleration is caused by an average net force:

$$\Sigma \bar{F} = m\bar{a} \qquad (6.11)$$

where

$\Sigma \bar{F}$ = average net force

$\bar{a}$ = average acceleration

As average acceleration is the change in velocity over time:

$$\bar{a} = \frac{\Delta v}{\Delta t} \qquad (6.12)$$

Equation 6.11 becomes:

$$\Sigma \bar{F} = m\bar{a} \qquad (6.11)$$

$$\Sigma \bar{F} = \frac{m\Delta v}{\Delta t} \qquad (6.13)$$

Multiplying both sides by Δt, this becomes:

$$\Sigma \bar{F} \Delta t = m\Delta v$$
$$\Sigma \bar{F} \Delta t = m(v_f - v_i) \qquad (6.14)$$

This is the impulse-momentum relationship. **Impulse** is the product of force multiplied by the time that force acts. If the force is not constant, impulse is the average force times the duration of that average force. The impulse produced by a net force acting over some duration of time causes a change in momentum of the object upon which the net force acts.

Linear momentum is the product of an object's mass and its linear velocity. The faster an object moves, the more momentum it has. The larger a moving object's mass, the more momentum it has. To change the momentum of an object, either its mass or its velocity must change. In sports and human movement, most objects we deal with have a constant mass. A change in momentum thus implies a change in velocity.

When Newton stated his second law of motion, he really meant momentum when he said motion. The change in momentum of an object is proportional to the force impressed.

The impulse-momentum relationship described mathematically by equation 6.14 is actually just another way of interpreting Newton's second law. This interpretation may be more useful to us in studying human movement. The average net force acting over some interval of time will cause a change in momentum of an object. We can interpret change in momentum to mean change in velocity, as most objects have constant mass. If we want to change the velocity of an object, we can produce a larger change in velocity by having a larger average net force act on the object or by increasing the time during which the net force acts.

> **The average net force acting over some interval of time will cause a change in momentum of an object.**

SELF-EXPERIMENT

Take a ball in your throwing hand and see how far (or how fast) you can throw it using only wrist motions. Move only your hand, and keep your forearm, upper arm, and the rest of your body still. This is not a very effective technique, is it? Now try throwing the ball again, only this time use your elbow and wrist. Move only your hand and forearm, and keep your upper arm and the rest of your body still. This technique is better, but it still isn't very effective. Try it a third time, using your wrist, elbow, and shoulder. Move only your hand, forearm, and upper arm, and keep the rest of your body still. This technique is an improvement over the previous one, but you could do even better. Try it a fourth time, throwing as you normally would with no constraints. This throw was probably the fastest. In which throw were you able to exert a force against the ball for the longest time? For the shortest time?

Using Impulse to Increase Momentum

The task in many sport skills is to cause a large change in the velocity of something. In throwing events, the ball (or shot, discus, javelin, or Frisbee) has no velocity at the beginning of the throw, and the task is to give it a fast velocity by the end of the throw. We want to increase its momentum. Similarly, in striking events, the racket (or bat, fist, or stick) has no velocity at the beginning of the swing (or stroke or punch), and the task is to give the implement a fast velocity just before its impact. Our bodies may be the objects whose momentum we wish to increase in jumping events and other activities. In all of these activities, the techniques used may be explained in part by the impulse-momentum relationship. A large change in velocity is produced by a large average net force acting over a long time interval. As there are limits on the forces humans are capable of producing, many sport techniques involve increasing the duration of force application.

During the above self-experiment you exerted the largest impulse on the ball when you used your normal throwing technique. As a result, the ball's momentum changed very much, and the ball left your hand with the fastest velocity. The large impulse was the result of a relatively large average force being exerted on the ball for a relatively long time. You exerted the smallest impulse on the ball when you only used your wrist. The ball's momentum didn't change very much, and the ball left your hand with the slowest velocity. The small impulse was the result of a relatively small average force being exerted on the ball for a relatively short time. The normal throwing technique involved more limbs in the throwing action, and you were able to increase the time that you could exert a force on the ball (and you were probably able to exert a larger average force). The end result was a faster throw. Due to a longer period of force application, the ball had more time to speed up, and thus its velocity at release was faster.

An important thing to remember about the impulse-momentum relationship:

$$\Sigma\overline{F}\,\Delta t = m(v_f - v_i) \qquad (6.14)$$

is that the average net force, $\Sigma\overline{F}$, in the impulse term is a vector, as are the velocities, v_f and v_i, in the momentum term. An impulse will cause a change in momentum, and thus a change in velocity, in the direction of the force. If you want to change the velocity of an object in a specific direction, the force you apply, or some component of that force, must be in that specific direction.

When you tried to throw (or put) the 16-lb shot in the self-experiment on page 107, the limiting factor was not the duration of the force application but the size of the force itself. The force you exerted was definitely larger than the force exerted when throwing the Ping-Pong ball, but the amount of force you exerted was constrained by your strength. If you were stronger, the force you could exert on the shot would have been larger and it would have gone farther and faster.

SELF-EXPERIMENT

See how far (or fast) you can throw a very light object, such as a Ping-Pong ball, then a very heavy object, such as a bowling ball or 16-lb shot. What limited your throwing performance with the lighter object? Was your strength the limiting factor? Do you have to be exceptionally strong to throw a Ping-Pong ball fast? The limiting factor was your technique, not your strength. More specifically, the time during which you could exert a force on the ball was constrained. It was very short. The ball sped up so quickly that your hand had a difficult time keeping up with it while still exerting a force on it.

When you threw the Ping-Pong ball in the above self-experiment, the average net force, $\Sigma \bar{F}$, wasn't the problem, but the time of its application, Δt, was. In the shot put, the time of application was long, but the amount of force applied was limited. In both instances, maximizing both quantities, $\Sigma \bar{F}$ and Δt, will result in the fastest throw or put. But when throwing a lighter object, technique (duration of force application) is more important for success, whereas when throwing heavy objects, the force applied is more important. Compare baseball pitchers and javelin throwers to shot-putters. Shot-putters are bigger and stronger. Their training and selection have been based on their ability to produce large forces ($\Sigma \bar{F}$ in impulse). Baseball pitchers and javelin throwers are not as strong. They're successful because their techniques maximize the duration of force application (Δt in impulse).

Let's summarize what we've learned about impulse and momentum so far. The relationship is described mathematically by equation 6.14:

$$\Sigma \bar{F} \, \Delta t = m(v_f - v_i) \qquad (6.14)$$

impulse = change in momentum

where

$\Sigma \bar{F}$ = average net force acting on an object

Δt = interval of time during which this force acts

m = mass of the object being accelerated

v_f = final velocity of the object at the end of the time interval

v_i = initial velocity of the object at the beginning of the time interval

In many sport situations, the goal is to impart a fast velocity to an object. The initial velocity of the object is zero and the final velocity is fast, so we want to increase its momentum. This is accomplished by exerting a large force against the object for as long a time as possible (by exerting a large impulse). Techniques in sports activities such as throwing or jumping are largely based on increasing the time of force application to obtain a large impulse.

SELF-EXPERIMENT

Now let's try an activity in which the force element of the impulse is constrained. Fill several balloons with water so that each one is about the size of a softball. Take these balloons outside to an empty field or empty parking lot and see how far you can throw one without having it break in your hand. If you exert too large a force against the balloon, it will break. To throw the balloon far, you must maximize the duration of force application during the throw while limiting the size of the force you exert against the balloon so that it doesn't break. Don't constrain your technique to what you perceive as normal throwing styles. Remember, the best technique will be the one in which you accelerate the balloon for the longest possible time while applying the greatest (but non-balloon-breaking) force.

Techniques in sports activities such as throwing or jumping are largely based on increasing the time of force application to obtain a large impulse.

Using Impulse to Decrease Momentum

In certain other activities, an object may have a fast initial velocity, and we want to decrease this velocity to a slow or zero final velocity. We want to decrease its momentum. Can you think of any situations like this? What about landing from a jump? Catching a ball? Being struck by a punch? Does the impulse-momentum relationship apply in an analysis of these situations? Yes. Let's try another activity.

SELF-EXPERIMENT

Fill a few more water balloons and take them outside. With a friend, play a game of catch with a water balloon. See how far apart you can get from one another and still catch the balloon intact. If you can't find a friend willing to take part in this activity, play catch by yourself. Throw a balloon up in the air and try to catch it without breaking it. Try to throw it higher and higher.

What does your catching technique look like? To catch the balloon without breaking it, your arms have to "give" with the balloon. You start with your arms outstretched and then move them in the direction of the balloon's movement as you begin to catch the balloon.

This giving action increases the impact time, Δt, in the impulse-momentum equation, equation 6.14. This increase in the impact time decreases the average impact force, $\Sigma \overline{F}$, required to stop the balloon. This smaller average impact force is less likely to break the balloon.

Why does the balloon become more difficult to catch as you and your friend get farther apart or as you throw it higher and higher? As you move far-

ther apart or throw the balloon higher, the impulse you exert on the balloon must be larger to create the greater change in momentum required to stop the balloon. It becomes more difficult to create a large enough impulse to stop the balloon without allowing the impact force to exceed the point at which the balloon will break. The technique used in landing from a jump is similar to that used to catch a water balloon if you think of the jumper as the water balloon.

SELF-EXPERIMENT

Stand on a chair. Now jump off the chair and land on your feet. How did you reduce the impact force? You flexed your knees, ankles, and hips. This increased the impact time—the time it took to slow you down. This increased Δt in the impulse-momentum equation, equation 6.14, and thus decreased the average impact force, $\Sigma \overline{F}$, because the change in momentum, $m(v_f - v_i)$, would be the same whether you flexed your legs or not.

What would happen if you landed stiff-legged? **(Don't try it!)** The impact time would be much smaller and the average impact force would have to be much greater because the change in momentum would still be the same. This impact force might be great enough to injure you.

Let's examine the impulse-momentum equation to see why this is so:

$$\Sigma \overline{F} \Delta t = m(v_f - v_i) \qquad (6.14)$$

The right side of this equation is the same whether you jump from the chair and land stiff-legged or flex your legs. Your mass, m, does not change. Your final velocity, v_f, is the same for both conditions. This is your velocity at the end of the landing, which will be zero. Your initial velocity, v_i, is also the same for both conditions if you jump from the same height. This is your velocity when your feet first make contact with the ground. Your change in velocity, $(v_f - v_i)$, is the same for both conditions. So your change in momentum, $m(v_f - v_i)$, the right side of equation 6.14, is the same whether you land stiff-legged or flex your legs. But how can the average impact forces differ between conditions?

The change in momentum is the same for both landing techniques, which means that the impulse, $\Sigma\bar{F}\,\Delta t$, the left side of equation 6.14, must also be the same for both techniques. This doesn't mean that the average impact force, $\Sigma\bar{F}$, must be the same for both conditions, or that the impact time, Δt, must be the same. It only means that the product of the two, $\Sigma\bar{F}$ times Δt, must be the same for both landing techniques. If the impact time, Δt, is short, the average impact force, $\Sigma\bar{F}$, must be large. If the impact time, Δt, is long, the average impact force, $\Sigma\bar{F}$, must be small. As long as the change in momentum, $m(v_f - v_i)$, is the same for both activities, changes in impact time, Δt, are accompanied by inversely proportional changes in average impact force, $\Sigma\bar{F}$.

In some sports and activities, athletes or participants may not be able to land on their feet and flex their legs to reduce the average impact force. How do high jumpers and pole vaulters land? They land on their backs, not on the hard ground, but on soft landing pads. If they landed on their backs on the hard ground, the impact time would be short, and the large impact forces would definitely cause injury. How do landing pads prevent them from being injured? The cushioned materials the landing pads are made of slow the jumper down over a longer time. This longer impact time means that the average impact forces are lessened proportionally. Pole vault landing pads are thicker than high jump landing pads because vaulters fall from greater heights and have faster velocities at impact, so the impact time must be more greatly increased to reduce their average impact forces to safe levels.

Gymnasts don't usually land on their backs, but their landings from apparatus dismounts or floor exercise routines are worth more points if they allow less flexion at their knees, ankles, and hips. If these landings occurred on hard floors, there would be many more injuries in gymnastics. But the gymnasts don't land on hard floors; their apparatus dismounts are onto a padded surface. Likewise, their landings after floor exercise stunts are onto the elevated, padded, and sprung gymnastic floor. The gymnastic floor and the mats around gymnastic apparatus increase the time of impact during landings. This increased impact time decreases the average impact force. It should be noted that the increase in impact time is not of the same magnitude as that experienced by high jumpers and pole vaulters, even though gymnasts may be jumping, and thus falling, from similar heights. Thus, gymnasts are more likely to experience injuries due to hard landings than are high jumpers.

Now think about other equipment that is used to cushion impacts by increasing the time of impact. You may have thought of the padding and equipment worn by football and hockey players, boxing gloves, baseball gloves, the midsole material used in running shoes, wrestling mats, and so on.

The impulse-momentum relationship provides the basis for techniques used in many sports and human movement skills. In throwing events, an athlete lengthens the duration of force application to increase the change of momentum of the object being thrown. This results in the object being thrown having a faster release velocity. The same principle applies to jumping activities, but in this case, the object being "thrown" is the athlete. In catching, landing, and other impact situations, the goal may be to decrease the magnitude of the impact force. By again increasing the duration of application of the impact force (increasing the impact time), the size of the average impact force is reduced. Now let's move on to Newton's third law of motion, which provides more insight into what a force is.

Law of Action-Reaction

Where do the forces come from that cause motion? Newton's third law of motion explains this. This law states: to every action there is always an opposed equal reaction; or the mutual actions of two bodies upon each other are always equal and directed to contrary parts. Newton used the words *action* and *reaction* to mean force, so this law explains where the external forces required to change motion arise from. Stated more simply, Newton's third law says: if an object (A) exerts a force on another object (B), the other object (B) exerts the same force on the first object (A) but in the opposite direction. So forces exist in mirrored pairs acting on two different objects. The effects of these forces are not canceled by each other, because they act on different objects. Another important point is that it is the forces that are equal but opposite, not the effects of the forces.

> **Forces exist in mirrored pairs acting on two different objects.**

When you performed the self-experiment in the first column of page 110, why didn't you accelerate as a result of the force the wall exerted against you?

SELF-EXPERIMENT

What happens when you push against a wall? The wall pushes against you. The force the wall exerts against you is exactly equal to the force you exert against the wall, but it pushes in the opposite direction. Think about what you feel with your hand as you push against the wall. In what direction is the force that you feel through your hand? The force you feel is in fact the force of the wall pushing against you, not that of you pushing against the wall.

At first, you might say that the force the wall exerted against you was canceled out by the force you exerted against the wall, so the net force was zero and no acceleration occurred. Would this be a correct explanation? No. The force you exerted against the wall does not act on you, so it can't counteract the effect of the force the wall exerts on you. What other forces act on you when you push against the wall? Gravity pulls down on you with a force equal to your weight. The floor pushes up against the soles of your shoes or your feet. And . . . a frictional force from the floor also acts against your feet. This frictional force prevents you from accelerating as a result of the wall pushing against you.

Now imagine this situation. You are lined up opposite an offensive lineman from an NFL football team. He weighs twice as much as you do. Your job is to push him out of the way. When you push against him, he pushes against you. Who pushes harder? According to Newton's third law of motion, the force that he pushes against you is exactly equal to the force that you push against him. What about the effects of these forces? Newton's second law of motion tells us that the effect of a force on a body is dependent on the mass of the body and the other forces acting on the body. The larger the mass, the smaller the effect. The smaller the mass, the larger the effect. Because his mass is so large, and the friction force beneath his feet is probably large as well, the effect of your pushing force on him will be small. Because your mass is small relative to his, and the friction force beneath your feet is probably less than the friction force beneath his feet, the effect of his pushing force on you will be greater.

SELF-EXPERIMENT

Hold a ball in your hand. What happens when you push against it in a horizontal direction (like you do when you throw it)? The ball pushes back against you in the opposite direction with a force exactly equal to the force you exert against it. In this case, though, the ball accelerates as a result of the force. It still exerts a force against you while it accelerates. Again, think about what you feel with your hand as you throw the ball. The ball pushes against your hand. If you try to throw it harder, you feel a greater force against your hand. In which direction does this force act? It acts in the opposite direction of the ball's acceleration. It pushes back against your hand.

Newton's third law of motion helps explain how forces act and what they act on. It gives us the basis for drawing free-body diagrams. It does not explain what the effects of forces will be, however. It only tells us that forces come in pairs, and that each force in a pair acts on a separate object.

Summary

The basics of linear kinetics, explaining the causes of linear motion, lie in Newton's laws of motion. Newton's first law explains that objects do not move or do not change their motion unless a net external force acts on them. Newton's second law explains what happens if a net external force does act on an object. It will accelerate in the direction of the net external force, and its acceleration will be inversely related to its mass. The impulse-momentum relationship presents Newton's second law in a way that is more applicable to sports and human movement. The basis for the techniques used in many sport skills is found in the impulse-momentum relationship. Increasing the duration of force application increases the change in momentum. Newton's third law explains that forces act in pairs. For every force, there is another equal force acting in the opposite direction on another object.

Key Terms

gravitational acceleration *(p. 101)*

impulse *(p. 105)*

linear momentum *(p. 105)*

projectile *(p. 97)*

projectile motion *(p. 98)*

reaction force *(p. 98)*

vector components *(p. 96)*

REVIEW QUESTIONS

1. A high jumper is attempting to clear a crossbar set at 2.44 m (8 ft 0 in.). At the instant of takeoff (when he is no longer in contact with the ground), his vertical velocity is 5.0 m/s. What is his vertical acceleration at this instant?

2. A long jumper leaves the takeoff board with a vertical velocity of 4 m/s and a horizontal velocity of 9 m/s.

    a. What is the jumper's horizontal velocity just before she lands?

    b. If the jumper is in the air for 1.0 s, what is her horizontal displacement during her time in the air?

    c. What is her vertical velocity at the end of her 1.0-s flight?

3. You are trying to lift a 10-kg dumbbell by pulling upward on it with a 108.1-N force. What will be the dumbbell's acceleration as a result of this force?

4. What forces act on a 1.0-kg ball if its resultant acceleration is 5.0 m/s/s horizontally?

5. Tonya crashes into Nancy while they are practicing their figure skating routines. Tonya's mass is 60 kg and Nancy's is 50 kg. Just before the collision occurs, Tonya's velocity is 5 m/s and Nancy's is 6 m/s in the opposite direction. During the collision, Tonya exerts an average force of 1000 N against Nancy.

    a. How large is the average force Nancy exerts against Tonya during the collision?

    b. If this is the only horizontal force acting on each skater during the collision, what average acceleration does each skater experience during the collision?

6. You are trying to lift a stack of books off the table. The stack of books has a mass of 15 kg. What minimum upward force must you exert on the books if you want to lift them off the table?

7. A cross-country skier is gliding north across a flat section of snow at 7 m/s. The mass of the skier is 100 kg. The coefficient of dynamic friction between the skis

and the snow is 0.10. The coefficient of static friction between the skis and the snow is 0.12. The force of air resistance is 1.9 N acting backward (south) on the skier. If friction between the skis and snow and air resistance are the only horizontal forces acting on the skier, what is the skier's horizontal acceleration?

8. You are trying to stop a 5-kg bowling ball that is rolling toward you. Its horizontal velocity is 8.0 m/s when you begin to stop it. It takes you 2.0 s to stop the ball. What average force did you exert on the ball during the 2.0 s it took to stop it?

9. A 100-kg basketball player lands on his feet after completing a slam dunk and then immediately jumps up again to celebrate his basket. When his feet first touch the floor after the dunk, his velocity is 5 m/s downward, and when his feet leave the floor 0.50 s later, as he jumps back up, his velocity is 4 m/s upward. What is the average reaction force exerted upward by the floor on the player during this 0.50 s?

10. A skateboarder is rolling down a 30° slope. The mass of the skateboard and skateboarder is 75 kg. The rolling friction between the skateboard wheels and the concrete is 9 N acting backward against the skateboard. The drag force due to air resistance is 11 N acting backward against the skateboarder. What is the acceleration of the skateboarder?

Work, Power, and Energy: Explaining the Causes of Motion Without Newton

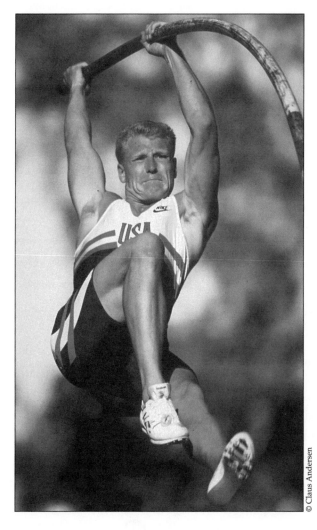

© Claus Andersen

A pole vaulter runs down the runway, slowly lowering the pole as he approaches the takeoff. He is moving as fast as a sprinter when suddenly he drops the end of the pole into the vault box and jumps off the ground. The pole bends and bends as the vaulter's forward movement slows and he swings upward. For a moment, it appears as if the pole will break, but then it begins to unbend. The unbending pole appears to fling the vaulter upward as he hangs onto it. The vaulter then swings into a handstand on the pole as the pole straightens to a vertical position. The vaulter pushes off the pole with one hand and soars up and over the crossbar to finish the vault. Wow! How is the vaulter able to convert the speed of his run-up into the height needed to get over the crossbar? The relationship between mechanical work and energy provides an answer to this question. This chapter introduces mechanical work and energy and covers material explaining their use in analysis of movement.

OBJECTIVES

When you finish this chapter, you should be able to do the following:

▌ Define mechanical work

▌ Distinguish between positive and negative work

▌ Define energy

▌ Define kinetic energy

▌ Define gravitational potential energy

▌ Define strain energy

▌ Explain the relationship between mechanical work and energy

▌ Define power

This chapter continues the study of linear kinetics begun in the previous chapter. The explanations for the causes of motion described in this chapter do not rely on Newton's laws of motion, but rather on the relationships between work, energy, and power, which were discovered and developed by several different scientists in the two centuries following Isaac Newton's achievements. Theoretically, all we need to analyze and explain linear motion are Newton's laws of motion. But some analyses and explanations are easier if based on work and energy relationships rather than Newtonian mechanics. So this chapter will provide more tools for analyzing and explaining sport skills.

Work

What is work? There are many definitions for work. *Webster's New World Dictionary* uses almost one full column (half a page) to list all the various definitions of work. Work is "the product of force and

SELF-EXPERIMENT

If you put this book on a table and push it so that it moves across the table, you have done work to the book. Try it. To quantify the amount of work you did, you would have to know what force you exerted against the book and how far you moved the book in the direction of the force (its displacement). Measuring the displacement is easy, but what about the force? When you pushed against the book, did you push with the same amount of force throughout the movement of the book, or did the force change? The force probably varied somewhat, so it wasn't constant. If the force wasn't constant, what value should we use for force, F, in equation 7.1? The force at the start of the movement? The force at the end of the movement? The peak force? How about the average force? This makes sense. The best value to describe a force that has many values during its application would be the average of these values—the average force, as in equation 7.2.

the amount of displacement in the direction of that force (it is the means by which energy is transferred from one object or system to another)" (*Webster's New World Dictionary*, p. 1538). Mathematically, this can be expressed as:

$$U = F(d) \qquad (7.1)$$

where

> U = work done on an object (The letter W would be a better abbreviation for work, but we already used it to represent weight.)
>
> F = force applied to an object
>
> d = displacement of an object along the line of action of the force

As work is the product of force and displacement, the units for work are units of force times units of length. These may be ft · lb or Nm. In the International System of Units, the **joule** (abbreviated with the letter J) is the unit of measurement for work; 1 J is equal to 1 Nm. The joule was named for James Prescott Joule (an English brewer who finally established the law of conservation of energy through practical experiments . . . more about this later).

> **Work is the product of force and displacement.**

Equation 7.1 really only describes the work done by a constant force. The work done by a force whose magnitude varies is

$$U = \bar{F}(d) \qquad (7.2)$$

where

U = Work done on an object

$\bar{F}$ = average force exerted on an object

d = displacement of an object along the line of action of the average force

To determine the amount of work done on an object, we need to know three things:

1. The average force exerted on the object
2. The direction of this force
3. The displacement of the object along the line of action of the force during the time the force acts on the object

Now let's look at an example. A discus thrower exerts an average force of 1000 N against the discus while the discus moves through a displacement of

0.6 m in the direction of this force (see figure 7.1). How much work did the discus thrower do to the discus?

$$U = \bar{F}(d)$$
$$U = (1000 \text{ N})(0.6 \text{ m}) \qquad (7.2)$$
$$U = 600 \text{ Nm} = 600 \text{ J}$$

This was easy because the average force and displacement were given. Let's try something harder. A weight lifter bench-presses a 1000-N barbell as shown in figure 7.2. He begins the lift with his arms extended and the barbell 75 cm above his chest. The lifter then lowers the barbell and stops it when it is 5 cm above his chest. He pauses there and then lifts the barbell upward away from his chest and back to the original starting position 75 cm above his chest. The average force exerted on the barbell by the lifter while lowering the weight is 1000 N upward. The average force exerted by the lifter while raising the weight is also 1000 N upward. (Could you determine this using Newton's laws?) So the average force exerted on the barbell by the lifter is 1000 N for the whole lift. How much work did the

| Figure 7.1 | A discus thrower does work on a discus by exerting an average force of 1000 N on the discus while moving it through a displacement of 0.6 m. |

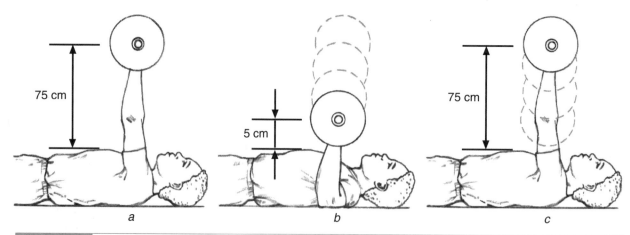

Figure 7.2   The phases of a bench press.

lifter do on the barbell from the start until the finish of the lift?

$$U = \bar{F}(d) \tag{7.2}$$
$$U = (1000 \text{ N})(d)$$

What was the displacement of the barbell? The starting and ending positions of the barbell were the same, so the displacement was zero.

$$U = (1000 \text{ N})(0) = 0$$

If the displacement was zero, the work done was also zero. Whoa! This doesn't seem correct! Certainly the lifter thinks he did work. He expended some calories in performing this lift. It is true that, physiologically, the lifter did some work, but mechanically, no work was done on the barbell because it was in the same position when the lift ended as when it started.

Was any work done during the raising of the barbell?

$$U = \bar{F}(d) \tag{7.2}$$
$$U = (1000 \text{ N})(d)$$

The displacement of the barbell during the raising of it was:

$$d = \text{final position} - \text{initial position} = y_f - y_i \tag{7.3}$$
$$d = 75 \text{ cm} - 5 \text{ cm} = 70 \text{ cm upward}$$

If we convert 70 cm into meters, we get:

$$70 \text{ cm} \times 100 \text{ cm/m} = 0.70 \text{ m} = d$$

So the work done was

$$U = (1000 \text{ N})(0.70 \text{ m}) = 700 \text{ Nm} = 700 \text{ J}$$

The lifter did do work on the barbell when he raised it. He did 700 J of work. Then how can the total work he did on the barbell for the whole lift be zero? Let's determine the work done during the lowering of the barbell.

$$U = \bar{F}(d) \tag{7.2}$$
$$U = (1000 \text{ N})(d)$$

What was the displacement of the barbell during the lowering phase? Its initial position was 75 cm above the chest and its final position after lowering was 5 cm above the chest, so:

$$d = \text{final position} - \text{initial position} = y_f - y_i \tag{7.3}$$
$$d = 5 \text{ cm} - 75 \text{ cm} = -70 \text{ cm}$$

The displacement was –0.7 m, or 0.7 m downward. In which direction was the force exerted on the barbell? The force was upward and the displacement was downward, so the work is:

$$U = \bar{F}(d) \tag{7.2}$$
$$U = (1000 \text{ N})(-0.70 \text{ m}) = -700 \text{ Nm} = -700 \text{ J}$$

The work done during the lowering of the barbell was *negative* 700 J of work. That sounds weird; how can work be negative? Mechanical work is negative if the force acting on an object is in the opposite direction of the motion (displacement) of the object.

Now it is more clear why zero work was done during the whole lift. If –700 J of work was done during the lowering of the barbell and +700 J of work was done during the raising of the barbell, the work done for the whole lift would be –700 J plus 700 J, or zero.

$$U_{\text{whole lift}} = U_{\text{lowering}} + U_{\text{raising}}$$
$$U_{\text{whole lift}} = -700 \text{ J} + 700 \text{ J}$$
$$U_{\text{whole lift}} = 0$$

Work can be positive or negative. Positive work is done by a force acting on an object if the object is displaced in the same direction as the force. A pitcher does positive work against a baseball when throwing it. A weight lifter does positive work against a weight when lifting or raising it. A gymnast does positive work when pulling up on the uneven bars. A high jumper does positive work when jumping off the ground.

Negative work is done by a force acting on an object when the object is displaced in the direction opposite the force acting on it. A first baseman does negative work against the ball when catching it. A weight lifter does negative work against a weight when lowering it. A gymnast does negative work when landing from a dismount. Friction does negative work on a skier sliding down a hill.

> **Positive work is done by a force acting on an object if the object is displaced in the same direction as the force. Negative work is done by a force acting on an object when the object is displaced in the direction opposite the force acting on it.**

Energy

What is energy? Like work, energy is a term that has many meanings. In mechanics, energy is defined as the capacity to do work. There are many forms of energy: heat, light, sound, chemical, and so on. In mechanics, we are concerned primarily with mechanical energy, which comes in two forms: kinetic energy and potential energy. Kinetic energy is energy due to motion, whereas potential energy is energy due to position.

> **Mechanical energy comes in two forms: kinetic energy, which is energy due to motion, and potential energy, which is energy due to position.**

Kinetic Energy

A moving object has the capacity to do work due to its motion. This capacity is the kinetic energy of the object. But how is kinetic energy quantified? What affects it? Try the following. Close the book and lay it flat on the desk or table. Now take another book and give it a quick push so that it slides across the desk and strikes the first book. What happened? The book you slid across the desk did work on the first book. It exerted a force on the first book, and the first book was displaced by this force. The book sliding across the desk had the capacity to do work because it had kinetic energy—energy due to its motion.

SELF-EXPERIMENT

Try this experiment again, only this time give the book a bigger push so it slides faster. Was more work done this time? Yes. Because the book was sliding faster, it had more kinetic energy and thus greater capacity to do work. Try the experiment one more time, only this time slide a heavier book (but slide it just as fast). Was more work done this time? Somehow, the larger mass of the book meant that it had more kinetic energy and thus a greater capacity to do work. (And you've also now had the satisfaction of pushing this textbook off the desk several times.)

The kinetic energy of an object is affected by the mass and velocity of the object. If we made more precise measurements, we would discover that the kinetic energy is proportional to the square of the velocity. Mathematically, we define kinetic energy as follows:

$$KE = \frac{1}{2}mv^2 \qquad (7.4)$$

where

KE = kinetic energy

m = mass

v = velocity

The units for kinetic energy are units of mass times velocity squared, or $kg(m^2/s^2)$, but this is the same as $[kg(m/s^2)]m$, which is equivalent to Nm, which is a joule. The unit of measurement for kinetic energy is the same as the unit of measurement for work. To determine the kinetic energy of an object, one must know its mass and its velocity.

How much kinetic energy does a baseball thrown at 80 mi/hr (35.8 m/s) have? A baseball's mass is

145 g (0.145 kg). To determine the ball's kinetic energy, use equation 7.4:

$$KE = \frac{1}{2}mv^2$$

$$KE = \frac{(0.145 \text{ kg})(35.8 \text{ m/s})^2}{2}$$

$$KE = 92.9 \text{ J}$$

Determining the kinetic energy of an object is easier than determining the work done by a force, because we can measure mass and velocity more easily than we can measure force.

Potential Energy

Potential energy is the energy (capacity to do work) an object has due to its position. There are two types of potential energy: gravitational potential energy, which is energy due to an object's position relative to the earth; and strain energy, which is due to the deformation of an object.

Gravitational Potential Energy

Gravitational potential energy is potential energy due to an object's position relative to the earth. The gravitational potential energy of an object is related to the object's weight and its elevation or height above the ground or some reference. Mathematically, gravitational potential energy is defined as follows:

$$PE = Wh \qquad (7.5)$$

or

$$PE = mgh \qquad (7.6)$$

where

PE = gravitational potential energy

W = weight

m = mass

g = acceleration due to gravity = 9.81 m/s/s

h = height

The units for potential energy are units of force times units of length, or Nm, which is equivalent to joules, the same unit of measure as for kinetic energy and work. To determine an object's gravitational potential energy, one must know its weight and height above the ground.

How much gravitational potential energy does a 700-N ski jumper have when taking off from the 90-m jump? We can use equation 7.5 to determine

this, but what should we use for the height, h? The 90-m ski jump is 90 m above the base of the hill, but the takeoff point is on the hill itself and is only about 3 m above the ground at the side of the hill. Potential energy is a relative term—because height is measured relative to some point that should be referred to in describing potential energy. In this case, let's use the base of the hill as our reference point. The height is then 90 m, and equation 7.5 becomes:

$$PE = Wh = (700 \text{ N})(90 \text{ m})$$

$$PE = (700 \text{ N})(90 \text{ m}) = 63,000 \text{ Nm}$$

$$PE = 63,000 \text{ J}$$

SELF-EXPERIMENT

If you have access to a hammer and nail and a block of wood, go get them. If you lift the hammer only a few inches above the nail and drop it, the hammer does not drive the nail into the wood very far. It does little work on the nail because it has little potential energy. If you lift the hammer much higher above the nail and allow it to swing down and strike the nail, it drives the nail much farther. The hammer does more work on the nail because it has greater potential energy (it is higher above the nail). Suppose a heavier hammer were used? Would it drive the nail farther if swung from the same height?

Strain Energy

Another type of potential energy is also used in sports. Strain energy is energy due to the deformation of an object. When a fiberglass vaulting pole bends, strain energy is stored in the bent pole. The greater the deformation of the object, the greater the strain energy.

The strain energy of an object is related to the object's stiffness, its material properties, and its deformation. Mathematically, the strain energy of a material with a linear stress-strain relationship is defined as:

$$SE = \frac{1}{2}k\Delta x^2 \qquad (7.7)$$

SELF-EXPERIMENT

Take a rubber band and stretch it. By stretching it, you have given the rubber band strain energy. If you stretch it farther, you increase the strain energy in the rubber band. The stiffer the object, the greater its strain energy when deformed. Stretch a wider rubber band or two rubber bands parallel to each other. The strain energy in stretching the wider rubber band or two rubber bands is greater than the strain energy in stretching the smaller, single rubber band.

where

  SE = strain energy

  k = stiffness or spring constant of material

  Δx = change in length or deformation of the object from its undeformed position

If the stiffness constant is expressed in N/m, then the strain energy is expressed in (N/m)m^2, or Nm, which are equivalent to joules—the same unit of measurement as for potential energy, kinetic energy, and work.

How much strain energy is stored in a tendon that is stretched 5 mm (0.005 m) if the stiffness of the tendon is 10,000 N/m?

$$SE = \frac{1}{2}k\Delta x^2 \qquad (7.7)$$

$$SE = \frac{(10,000 \text{ N / m})(0.005 \text{ m})^2}{2}$$

$$SE = 0.125 \text{ J}$$

In human movement and sports, energy is possessed by athletes and objects due to their motion (kinetic energy), their position above the ground (gravitational potential energy), and their deformation (strain energy). We will be concerned primarily with the first two types of energy: kinetic energy and potential energy.

The Work-Energy Relationship

The definitions of work and energy indicate that a relationship exists between them. As a reminder,

energy was defined as the capacity to do work. The definition of work included this statement: "It is the means by which energy is transferred from one object or system to another." The unit of measure for work and energy was joules—the same for each quantity. This is another indication that work and energy are related.

Demonstrating the Work-Energy Relationship

How are work and energy related? Let's look at a previous example to reveal something about the relationship. Consider the weight lifter bench-pressing the 1000-N barbell. During part of the lift, the barbell was raised 70 cm and the work done by the lifter was 700 J. How much more energy did the barbell have after it was lifted? It didn't have any more kinetic energy because it was not moving before the lift started and it was not moving at the end of the lift. But its potential energy changed because it changed heights. What was the change in potential energy?

$$\Delta PE = PE_{final} - PE_{initial}$$

$$\Delta PE = Wh_{final} - Wh_{initial}$$

$$\Delta PE = W(h_{final} - h_{initial})$$

$$\Delta PE = 1000 \text{ N } (0.70 \text{ m})$$

$$\Delta PE = 700 \text{ J}$$

The change in the potential energy of the barbell was 700 J, the same as the work done to the barbell. Perhaps work done causes a change in potential energy. Or maybe work causes a change in total mechanical energy.

In the discus-throwing example, a discus thrower exerted an average force of 1000 N against the discus while the discus moved through a displacement of 0.60 m in the direction of the force. The work done by the thrower on the discus was 600 J. Let's assume that the force exerted by the thrower was constant and the displacement was horizontal, and that the discus was not moving at the start of the throw. If work does cause a change in total mechanical energy, the work done by the thrower on the discus would cause a change in the energy of the discus. As the displacement of the discus was horizontal, there was no change in potential energy, so the work only caused a change in kinetic energy. The mass of a discus is 2 kg. Knowing the work done and that the initial velocity of the discus was zero, we can determine the velocity of the discus at release (v_f).

$$\text{Work} = \Delta E = \Delta KE + \Delta PE = \Delta KE + 0$$

$$600 \text{ J} = \Delta KE = KE_{final} - KE_{initial}$$

$$600 \text{ J} = \frac{1}{2}mv_f^2 - 0$$

$$600 \text{ J} = \frac{(2\text{kg})v_f^2}{2}$$

$$\frac{2(600 \text{ J})}{2 \text{ kg}} = v_f^2$$

$$v_f = \sqrt{\frac{2(600 \text{ J})}{2 \text{ kg}}}$$

$$v_f = 24.5 \text{ m/s}$$

Can we verify this using Newtonian mechanics? The constant 1000-N horizontal force acting on the discus is the net horizontal force acting on the discus. This force would cause the discus to accelerate at 500 m/s/s.

$$\Sigma F = ma$$

$$a = \frac{\Sigma F}{m} = \frac{1000 \text{ N}}{2 \text{ kg}} = 500 \text{ m/s/s}$$

The average velocity of the discus during the throwing action, $\overline{v}$, was the displacement of the discus during the throwing action, d, divided by the time of the throwing action, t. With a constant force acting on the discus, the velocity of the discus would increase linearly. The average velocity of the discus could also be computed by dividing the difference between the final (v_f) and initial (v_i) velocities of the discus by two:

$$\overline{v} = \frac{d}{t} = \frac{v_f - v_i}{2}$$

$$\overline{v} = \frac{0.6 \text{ m}}{t} = \frac{v_f - 0}{2}$$

We don't know the time or duration of the throwing action, but this would be the time it took to accelerate the discus from 0 m/s to its final velocity. We know the acceleration was 500 m/s/s. As average acceleration, $\overline{a}$, is change in velocity divided by time, time can be computed by dividing change in velocity by acceleration:

$$\overline{a} = \frac{v_f - v_i}{t} = \frac{v_f - 0}{t}$$

$$500 \text{ m/s/s} = \frac{v_f}{t}$$

$$t = \frac{v_f}{500 \text{ m/s/s}}$$

If we substitute this representation for time in the previous equation, we get:

$$\overline{v} = \frac{0.6 \text{ m}}{t} = \frac{0.6 \text{ m}}{\dfrac{v_f}{500 \text{ m/s/s}}} = \frac{v_f - 0}{2}$$

Solving for the final velocity gives:

$$\frac{\dfrac{0.6 \text{ m}}{v_f}}{500 \text{ m/s/s}} = \frac{v_f}{2}$$

$$v_f = 2\left(\frac{\dfrac{0.6 \text{ m}}{v_f}}{500 \text{ m/s/s}}\right)$$

$$v_f^2 = 2(0.6 \text{ m})(500 \text{ m/s/s})$$

$$v_f = \sqrt{2(0.6 \text{ m})(500 \text{ m/s/s})}$$

$$v_f = \sqrt{600 \text{ m}^2/s^2}$$

$$v_f = 24.5 \text{ m/s}$$

This is the same result we got using the work-energy principle. The computations involved in using the work-energy principle were much less complicated, though!

These examples demonstrate the work-energy principle: The work done by the external forces (other than gravity) acting on an object causes a change in energy of the object. Mathematically, this relationship is shown in equation 7.8:

$$U = \Delta E$$

$$U = \Delta KE + \Delta PE + \Delta SE \qquad (7.8)$$

$$U = (KE_f - KE_i) + (PE_f - PE_i) + (SE_f - SE_i)$$

where

U = work done on an object by forces other than gravity

ΔE = change in total mechanical energy

KE_f = final kinetic energy

KE_i = initial kinetic energy

PE_f = final gravitational potential energy

PE_i = initial gravitational potential energy

SE_f = final strain energy

SE_i = initial strain energy

> **The work done by the external forces (other than gravity) acting on an object causes a change in energy of the object.**

Doing Work to Increase Energy

Why is the relationship between work and energy so important? In sports and human movement, we are often concerned with changing the velocity of an object. Changing velocity means changing kinetic energy, and the work-energy principle shows how kinetic energy can be changed by doing work. More work is done, and thus a greater change in energy occurs, if the average force exerted is large or the displacement in line with this force is long. This sounds very similar to what we discovered using the impulse-momentum relationship in chapter 6. The impulse-momentum relationship is shown in equation 7.9:

$$\Sigma \overline{F} \, \Delta t = m(v_f - v_i) \qquad (7.9)$$

Using this relationship as the basis for technique analysis, production of a large change in velocity requires that a large force be applied over a long time. The work-energy principle indicates that production of a large change in kinetic energy (and thus a large change in velocity) requires that a large force be applied over a long distance.

Think back to the example of throwing a ball. If you only throw with your wrist, you are able to exert a force on the ball through a small displacement. The work done is small, and as a result, the ball's change in kinetic energy is small. The ball's velocity when it leaves your hand is slow.

> **A large change in kinetic energy requires that a large force be applied over a long distance.**

If the wrist and elbow are involved in the throwing motion, you are able to exert a force on the ball through a larger displacement. The work done on the ball is larger, and as a result, the ball's change in kinetic energy is larger. So the ball's velocity when it leaves your hand is faster.

When you involve your whole arm, trunk, and legs in the throwing motion, you are able to apply a force to the ball through a much larger displacement. The work done on the ball is much larger, and as a result, the ball's change in kinetic energy is much larger as well. The ball's velocity when it leaves your hand is much faster (more than 100 mi/hr for some major league pitchers).

Something similar to this actually occurred in the evolution of shot-putting technique. The rules for shot putting indicate that the put must be made from a 7-ft-diameter circle. The shot-putter must begin the shot put from a standstill, with no body part touching anything outside of this circle. The putter must complete the put without touching anything outside of the circle until the judge has ruled it a fair put. Only then is the putter allowed to step outside of the "throwing ring," but only through the rear half. The size of the ring thus limits how much work the athlete can do to the shot by constraining the distance over which the putter can exert a force on the shot.

Early in the 20th century, shot-putters began their put from the rear of the ring. The initial stance was similar to that shown in figure 7.3.

Corbis-Bettmann

Figure 7.3   Starting the shot put from the rear of the ring early in the 20th century.

The athlete's shoulders were aligned approximately 45° to the throwing direction, meaning the right-handed athlete faced slightly to the right. The putter would hop across the ring on his right leg and put the shot.

Gradually, the technique evolved, and the putter's shoulders were turned more and more toward the rear of the circle in the initial stance. The greater shoulder rotation allowed the putter to start from a position that allowed for greater displacement of the shot before release. Finally, in the 1950s, Parry O'Brien (figure 7.4) began putting the shot from an initial position in the rear of the ring facing the opposite direction of the put. This put him in a position where he could maximize the displacement of the shot in the direction of his force application. He could also involve stronger muscle groups and have a larger force act on the shot during the putting action. The work done on the shot was thus increased. This increased the change in energy (potential and kinetic), which resulted in a greater height and velocity of the shot at release. The result was a longer put.

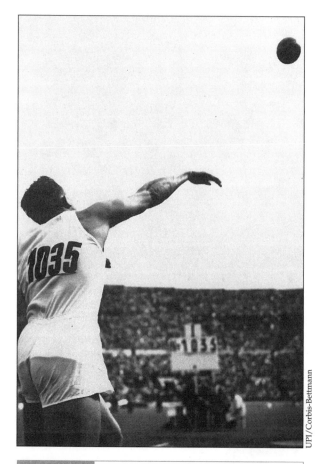

UPI/Corbis-Bettmann

Figure 7.4   Parry O'Brien putting the shot.

Doing Work to Decrease (or Absorb) Energy

The work-energy principle can also be used to explain the techniques used in transferring (or absorbing) energy from an object. When you catch a ball, its kinetic energy is reduced (or absorbed) by the negative work you do on it. Similarly, your muscles do negative work on your limbs and absorb their energy when you land from a jump or fall. The average force you must exert to absorb energy in catching a ball or landing from a jump or fall depends on how much energy must be absorbed and over how long a distance you can apply the force. If this force is too great, it may injure you. You attempt to decrease it by "giving" with a ball when you catch it or by flexing at your knees, ankles, and hips when you land from a jump or fall. These actions increase the distance over which the force acts, thus decreasing the average value of the force.

The safety and protective equipment used in many sports utilizes the work-energy principle to reduce potentially damaging impact forces. The landing pads used in gymnastics, high jumping, and pole vaulting all increase the displacement of the athlete during the impact period, as the kinetic energy of the athlete is decreased (absorbed). The impact force is thus decreased because the displacement during the impact was increased. The sand in a long jumping pit does the same thing when you jump into it, as does the water in a pool when you dive into it, the midsole material in your running shoes when you run on them, the padding in a boxing glove when you punch with it, the air bag in a car when you crash into it, and so on. All of these materials may be referred to as "shock absorbing," but they are actually energy-absorbing materials.

Conservation of Mechanical Energy

The work-energy relationship is also useful when examining situations in which no external forces act other than gravity. In these situations, no work can be done because no external forces act. If no work can be done, the total mechanical energy of the object in question is conserved; it cannot change, and equation 7.8 becomes:

$$U = \Delta E \qquad (7.8)$$
$$U = 0 = \Delta KE + \Delta PE + \Delta SE$$
$$0 = (KE_f - KE_i) + (PE_f - PE_i) + (SE_f - SE_i)$$
$$(KE_i + PE_i + SE_i) = (KE_f + PE_f + SE_f)$$

$$E_i = E_f \qquad (7.10)$$

The total mechanical energy of the object is constant if no external forces other than gravity act on the object. This principle may be useful in examining projectile motion. Gravity is the only external force that acts on a projectile. If this is true, the total mechanical energy of a projectile does not change during its flight. Let's consider dropping a ball as an example. Just before you let go of it, the ball has potential energy but no kinetic energy. During the ball's fall, its potential energy decreases because its height decreases. At the same time, though, its kinetic energy increases as it is accelerated downward by gravity. This increase in kinetic energy is exactly matched by the decrease in potential energy, so the total mechanical energy of the ball remains the same.

> **The total mechanical energy of an object is constant if no external forces other than gravity act on the object.**

Let's try some numbers in this example. Suppose you drop a 1-kg ball from a height of 4.9 m (that's about 16 ft). When you first let go of the ball, its potential energy (measured relative to the ground) would be:

$$PE_i = Wh = mgh$$
$$PE_i = (1 \text{ kg})(9.81 \text{ m/s/s})(4.9 \text{ m})$$

When you first let go of the ball, its velocity is zero, so its kinetic energy would be zero. The instant before the ball strikes the ground, it has no height above the ground, so its potential energy would be zero. The kinetic energy of the ball at this time would be:

$$KE_f = \frac{1}{2}mv_f^2$$
$$KE_f = \frac{1}{2}(1 \text{ kg})v_f^2$$

If we use equation 7.10, we can determine how fast the projectile is going just before it hits the ground:

$$E_i = E_f \qquad (7.10)$$
$$(KE_i + PE_i + SE_i) = (KE_f + PE_f + SE_f)$$
$$(0 + PE_i + 0) = (KE_f + 0 + 0)$$
$$PE_i = KE_f$$
$$(1 \text{ kg})(9.81 \text{ m/s/s})(4.9 \text{ m}) = \frac{1}{2}(1 \text{ kg})v_f^2$$

$$\frac{2(1 \text{ kg})(9.81 \text{ m/s/s})(4.9 \text{ m})}{1 \text{ kg}} = v_f^2$$
$$v_f = \sqrt{\frac{2(1 \text{ kg})(9.81 \text{ m/s/s})(4.9 \text{ m})}{1 \text{ kg}}}$$
$$v_f = \sqrt{2(9.81 \text{ m/s/s})(4.9 \text{ m})}$$
$$v_f = 9.81 \text{ m/s}$$

The conservation of mechanical energy principle gives us another tool for analyzing projectile motion. It can also allow us to analyze other situations where no work is done. For instance, in the pole vault, if the vaulter does no work during the vault itself, his total mechanical energy at the instant of takeoff should equal the total mechanical energy at bar clearance. In this case, the vaulter's kinetic energy at takeoff is transformed into strain energy as the pole bends, and this strain energy is in turn transformed into potential energy as the vaulter is lifted by the unbending pole. How high a pole vaulter can vault thus has a lot to do with how fast the vaulter can run.

Power

The ability of an athlete to increase the displacement of an object (or body part) while exerting a force will affect performance in many skills. Success in these skills thus requires the athlete to exert a large amount of work on an object (or body part). In some sports, excelling requires not just the ability to do a large amount of work, but also the ability to do that work in a short time. Power is the mechanical term that describes this ability. Like work and energy, power is another word that you have some familiarity with and that has numerous meanings. In mechanics, power is the rate of doing work, or how much work is done in a specific amount of time. Mathematically, power is defined as:

$$P = \frac{U}{t} \qquad (7.11)$$

where

P = power

U = work done

t = time taken to do the work

Power can be thought of as how quickly or slowly work is done. The SI units for power are watts (abbreviated with the letter W), named after the Scottish inventor James Watt; 1 W equals 1 J/s. You may be familiar with watts because light bulbs, amplifiers, and other electrical devices are rated in watts. Another unit of measurement for power is horsepower, but the watt is the unit of measure for power in the International System of Units.

> **Power can be thought of as how quickly or slowly work is done.**

If we examine equation 7.11 more closely, another way of defining power can be derived:

$$P = \frac{U}{t} \qquad (7.11)$$

$$P = \frac{\overline{F}(d)}{t}$$

$$P = \overline{F}\left(\frac{d}{t}\right)$$

$$P = \overline{F}\,\overline{v} \qquad (7.12)$$

Power can be defined as average force times average velocity along the line of action of that force.

The concept of power is useful in biomechanics for several reasons. The best way to explain one use of power is by example. Suppose you have to move a stack of books from one table to another, and you want to finish this task as quickly as possible. This means that you want to maximize your power output. Numerous strategies are available to you, from moving the books one at a time to moving them all at once. The amount of work done to the books will be the same, but the time it takes (and thus the power output) may differ. Carrying the entire stack at once would require a large force, and the movement would be slow. Carrying a few books at a time in several trips wouldn't require as much force, and each trip would be quicker. In the first case, you are exerting large forces but moving with a slow velocity. In the second case, you are exerting smaller forces but moving with a faster velocity. The combination of force and velocity determines the power output. Does the larger force in the first case make up for its decreased velocity, or does the larger velocity in the second case make up for its decreased force? What is the tradeoff between force and velocity?

You are faced with similar questions in certain sports and activities. How do you choose which gear to use while pedaling your bicycle? Do you use a high gear, which requires larger pedal forces and a slower pedaling rate, or do you use a lower gear, which requires smaller pedal forces but a faster pedaling rate? When you're running, how do you choose your stride length and stride rate? Do you use a long stride, which requires larger forces and a slower stride rate, or do you use a short stride, which requires smaller forces and a faster stride rate? These questions are difficult to answer because of the number of variables involved, so you have to experiment. One clue to answering these questions may come from studying muscles.

Because the power we produce in our movements ultimately originates in our muscles, the power-production characteristics of muscles may provide some insight into the questions raised above. As a muscle's velocity of contraction increases, its maximum force of contraction decreases. So a muscle contracting slowly can produce greater force than the same muscle contracting at a faster rate. If the muscle's velocity of contraction is multiplied by its maximum force of contraction for that velocity, the muscle's power output for each velocity can be determined. The maximum power output occurs at a velocity approximately one-half the muscle's maximum contraction velocity. This would mean that the best gear to use in cycling may not be the highest or the lowest, but one in between. The best stride length may not be the longest or the shortest, but one in between. The best way to move that stack of books may not be to carry them all at once or one at a time, but several at once. The best choice of bicycling gear, stride length, and so on, may be the one in which your muscles contract at a velocity corresponding to their velocity of maximal power output. The mechanics of muscular contraction are discussed further in chapter 14.

Another reason that power is an important topic in the study of human movement is because it is actually a constraint on human movement. What does that mean? Consider an Olympic weight lifter performing a clean and jerk. The forces he exerts on the barbell and the fast movement of the barbell indicate that the power output of the lifter is quite large—but only for a brief interval of time. If the time interval were longer, would the lifter be able to produce as much power? The duration of the activity influences the power output that an individual can sustain. A sprinter can maintain a high power output for only a short time (0-60 s). A middle-distance runner's power output is smaller but sustained for a longer time (1-7 min). A marathon runner's power output is smaller still, but it is sustained for a much longer time (2-4 hr). Figure 7.5 shows the theoretical relationship between maximum power output and the duration of that power output for humans. This relationship shows the mechanical constraint placed on humans by their power-generating system (their metabolic system).

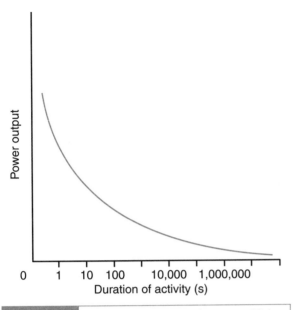

Figure 7.5	Maximal power output decreases with increased duration of activity.

motion. Potential energy could be due to the position of the object in a gravitational field, called gravitational potential energy, or to its bending, stretch, or deformation, called strain energy. The work done by a force (other than gravity) on an object causes a change in energy of the object. If a force does work, a change in energy results. Likewise, if a change in energy is observed, a force did work to cause this change in energy.

Power is defined as the rate of doing work. An alternative definition of power is average force times average velocity. In human movement, the maximum power output a human is capable of producing is related to the duration of the activity involved. This has to do with the metabolic capabilities of the human body.

Summary

In this chapter, we learned the mechanical definitions of work, energy, and power. Work done by a force is the force times the displacement of the object along the line of action of the force acting on it. Energy was defined as the capacity to do work. Mechanically, energy takes two forms: potential energy, which is energy due to position or deformation, and kinetic energy, which is energy due to

Key Terms

energy *(p. 117)*

gravitational potential energy *(p. 118)*

joule *(p. 115)*

kinetic energy *(p. 117)*

potential energy *(p. 117)*

power *(p. 123)*

strain energy *(p. 118)*

watt *(p. 123)*

work *(p. 114)*

REVIEW QUESTIONS

1. How much kinetic energy does a 2-kg discus have if its velocity is 20 m/s?

2. At the top of a giant swing on the high bar, a 50-kg gymnast's velocity is 1 m/s and he is 3.5 m high. What is the gymnast's total mechanical energy at this instant?

3. You are slowly pulling your opponent's wrist to the table in an arm wrestling match.
   a. Are you doing any mechanical work against your opponent? If so, is the work positive or negative?
   b. Is your opponent doing any work against you? If so, is the work positive or negative?

4. During the pitching motion, a baseball pitcher exerts an average horizontal force of 60 N against the 0.15-kg baseball while moving it through a horizontal displacement of 2.0 m before releasing it.

   a. How much work did the pitcher do to the baseball as a result of this force?

   b. Is the work done positive or negative?

   c. If the baseball's velocity at the start of the pitching action was zero, how fast was the ball moving at the instant of release?

5. A baseball strikes the catcher's glove with a horizontal velocity of 40 m/s. The mass of the baseball is 0.15 kg. The displacement of the baseball due to the deformation of the catcher's glove and the movement of the catcher's hand is 8 cm from the instant it first makes contact with the glove until it stops.

   a. How much kinetic energy does the baseball possess just before it strikes the glove?

   b. How much work does the catcher do to the baseball during the catch?

   c. Is the work done positive or negative?

   d. What is the average impact force exerted by the glove on the baseball?

6. Which bowling ball has more energy, a 5-kg ball rolling at 4 m/s or a 6-kg ball rolling at 3 m/s?

7. An Olympic weight lifter snatches 100 kg. In a snatch, the barbell is moved from a stationary position on the floor to a stationary position over the athlete's head. Only 0.5 s elapsed from the first movement of the barbell until it was overhead, and the barbell moved through a vertical displacement of 2.0 m. What was the weight lifter's average power output during the lift?

8. In a vertical jump-and-reach test, a 60-kg student jumps 60 cm, whereas a 90-kg student jumps 45 cm. Assuming both jumps took the same time, which jumper was more powerful?

9. The sand in a long jump pit serves two purposes. First, it enables the distance of the jump to be measured by marking the jumper's landing. Second, it cushions the landing of the jumper. Give a mechanical explanation for how the sand "cushions" the landing of the jumper.

10. Why is a follow-through important in throwing and striking activities?

11. What advantage does a longer-limbed individual have in throwing and striking activities?

Torques and Moments of Force: Maintaining Equilibrium or Changing Angular Motion

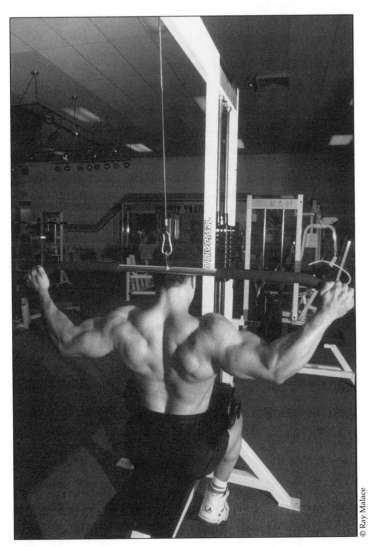

© Ray Malace

You're sitting in the gym contemplating your next lat pull on the weight-lifting machine when you notice that the stack of weight you're about to lift doesn't look heavy enough to be the 90 lb indicated by the label on the weights. But when you sit down on the bench and pull down on the bar, it feels like 90 lb. How do the pulleys, levers, and cables of the machine make a relatively small stack of metal plates feel like 90 lb? The answer has to do with the torques created by the weight stack and those pulleys, levers, and cables. This chapter introduces the concepts of torque and center of gravity and adds to the concept of static equilibrium.

OBJECTIVES

When you finish this chapter, you should be able to do the following:

▌ Define torque (moment of force)

▌ Define static equilibrium

▌ List the equations of static equilibrium

▌ Determine the resultant of two or more torques

▌ Determine if an object is in static equilibrium when the forces and torques acting on the object are known

▌ Determine an unknown force or torque acting on an object if all the other forces and torques acting on the object are known and the object is in static equilibrium

▌ Define center of gravity

▌ Estimate the location of the center of gravity of an object or body

This chapter discusses torques. Torques cause changes in angular motion. The movements of our limbs at joints are controlled by the torques produced by muscles. Muscles create torques about joints, and these torques control and cause the movements of the limbs and the entire body. Torques are important even if you aren't moving. Equilibrium and balance are affected not only by forces but by torques as well. In this chapter, you will learn about torques, equilibrium of forces and torques, center of gravity, and stability. All of these concepts are related to torque in some way.

What Are Torques?

The turning effect produced by a force is called a torque. The torque produced by a force may also be called a moment of force. Occasionally, this term is simplified further and shortened to moment. One way to think of torque is to think of it as an angular or rotary force. To get a greater understanding of how a force can produce a torque, let's try the self-experiments on pages 129 and 130.

> **The turning effect produced by a force is called a torque.**

An examination of the results of these self-experiments leads to some generalizations. In the first case, the resultant force acting on the book was directed through the center of gravity of the book. An external force directed through the center of gravity of an object is called a centric force. The effect of a centric force is to cause a change in the linear motion of the object, as predicted by Newton's second law and as shown in this first experiment.

In the second experiment, the resultant force acting on the book was not directed through the center of gravity of the book. An external force not directed through the center of gravity of an object is called an eccentric force. (Eccentric in this case refers to a type of force, not a type of muscular contraction.) The effect of an eccentric force is to cause a change in the linear and angular motions of an object. Both motions were observed in this second experiment. The change in linear motion is explained by Newton's second law. The torque produced by the eccentric force caused the book to rotate.

In the third experiment, a pair of forces acted on the book. These forces were equal in size but opposite in direction and noncolinear. A pair of such forces is called a force couple. The effect of a force couple is to cause a change only in the angular motion of an object. The resultant of the two forces

SELF-EXPERIMENT

Place a book flat on a table or desk. Using two fingers (or a pencil), strike the book on its side to create a force directed through the center of gravity of the book (see figure 8.1a). (If you could balance the book on a pencil point, the book's center of gravity would lie vertically above this point of balance. A more complete discussion of center of gravity occurs later in this chapter. For now, consider the center of gravity of the book as the center of the book.) What happens as a result of the force you exerted on the book?

The book moved linearly. The net force acting on the book (your pushing force minus friction) was directed through the book's center of gravity, and this net force caused a linear acceleration of the book. Little if any turning or rotation of the book occurred. Evidently, this type of force does not produce a turning effect or torque on the book.

Now try the experiment once again, only this time hit the book with your fingers or a pencil so that the force is not directed through the book's center of gravity (see figure 8.1b). What happens?

The linear motion of the book changed (the book's center of gravity translated), but the book rotated also. In this case, the net force acting on the book was not directed through the book's center of gravity, and this net force caused a linear acceleration of the book as well as a rotation of the book. This type of force did create a torque on the book.

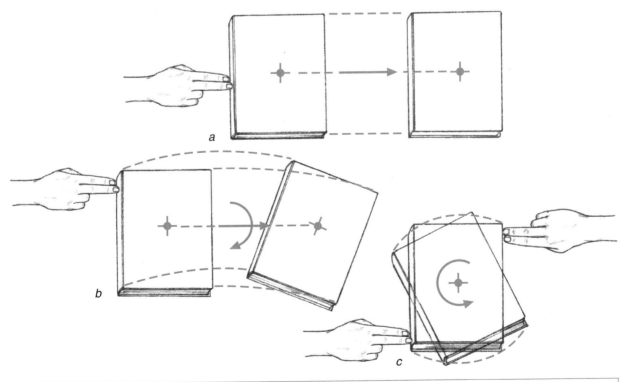

Figure 8.1   Sliding a book on the table so that it (*a*) translates; (*b*) translates and rotates; and (*c*) rotates.

in a force couple is a force of zero, so according to Newton's first and second laws, no change in linear motion occurs, and none was observed in this third experiment. (Any translation of the book that occurred was small and due to the fact that the forces you exerted on the book were not exactly equal in size and opposite in direction.) The torque produced by the force couple caused the book to rotate.

In general, then, a centric force will cause or tend to cause a change in the linear motion of an object; an eccentric force will cause or tend to cause changes in the linear and angular motions of an object; and a force couple will only cause or tend to cause a change in the angular motion of an object. The turning effect or torque produced by the eccentric force in the second case and by the force couple in the third case will be examined further.

Mathematical Definition of Torque

What affects the size of a torque and how is it quantified? Intuitively, you would think that the size of the force that produces the turning effect would influence the size of the torque. To verify this, try the second experiment again several times and increase the size of the eccentric force you exert on the book each time. As you increased the size of the force, the turning effect increased, as evidenced by the increase in the amount of rotation caused. Do the same with the third experiment. Try the experiment several times and increase the size of the forces in the force couple each time. Again, as the size of the forces in the force couple increased, the amount of rotation increased, so the turning effect increased. Intuition was correct in this case; torque is directly related to the size of the force that creates it.

The torque produced by a force does not depend on the force alone, however. If this were the case, a turning effect would have been produced by the centric force you exerted on the book in the first experiment. The only difference between the first and second experiments was the line of action of the force you applied to the book.

As you strike the book farther and farther from its center of gravity, the torque created by the force becomes larger and larger, causing the book to rotate more and more. Torque is influenced by the position and orientation of the line of action of the force as well as by its size.

As the distance between the lines of action of each of the forces gets shorter and shorter, the torque produced by the force couple gets smaller and

SELF-EXPERIMENT

Now repeat the first experiment. Strike the book so that the force your fingers create is directed through the center of gravity of the book. Do it again, only this time direct the force so its line of action is just off center, causing the force to be an eccentric force. Repeat the experiment several times, each time striking the book so the line of action of the force is farther from the book's center of gravity. Try to keep the size of the force the same in each trial. What happens?

SELF-EXPERIMENT

To reinforce this concept, try the third experiment again. Apply a pair of forces (a force couple) to the book at either end of the book and in opposite directions. Do it again, only this time move your hands so the distance between them and the forces they apply to the book are not as great. Repeat this a number of times, so that each time the lines of application of the forces move closer and closer together until finally the forces are colinear. What happens?

smaller and disappears completely when the forces become colinear.

We can now conclude that the torque produced by a force is directly proportional to the size of the force as well as the distance between the line of action of the force and the point about which the object tends to rotate (the axis of rotation). In the case of the force couple, the turning effect is again directly proportional to the size of the forces and the distance between the lines of action of these forces. Now we can expand our definition of torque. A torque is the turning effect produced by a force and is equal to the product of the magnitude of the force and the distance between the line of action of the force and the axis of rotation of the object (or the axis about which the object will tend to rotate). This distance between the line of action of the force and the axis of rotation is the perpendicular distance between the line of action of the force and a line parallel to it that passes through the axis of rotation. This distance is sometimes referred to as the perpendicular distance and is more often termed the moment arm of the force. Figure 8.2 shows the force and the moment arm of the force for the second experiment.

Mathematically, then, torque is defined as follows:

$$T = F \times r \qquad (8.1)$$

where

T = torque (or moment of force)

F = force

r = moment arm (or perpendicular distance)

The units of measurement for torque are thus units of force (newtons in the SI system) times units of length (meters in the SI system). So torque is measured in newton meters, which are abbreviated Nm.

To fully describe a torque, you must describe the size of the torque, the axis about which the turning effect is created, and the sense of the turning effect about that axis (clockwise or counterclockwise). So torque is a vector quantity, because the turning effect is around a specific axis that is directed in a specific direction. As with forces, once a direction is specified (or axis, in the case of a torque), a posi-

tive (+) or negative (−) sign is used to indicate the sense of the force or torque along (or around) that line (or axis). The conventional approach is to indicate counterclockwise torques as positive (+) and clockwise torques as negative (−). As with forces, torques that act about the same axis may be added or subtracted algebraically. Before getting further into the mathematics, let's look at examples of how we use torques.

Examples of How Torques Are Used

First, let's consider a door. How do you open or close a door? You probably grab its handle (or doorknob) and push or pull. This force you exert on the door creates a torque that acts about the axis of rotation of the door, an axis passing vertically through the door's hinges. This torque causes the door to start swinging open or closed, as shown in figure 8.3.

The same size torque can be created with a large force and a small moment arm or with a small force and a large moment arm. Because the amount of force humans can exert is generally limited, we use large moment arms when we want to create large torques. How do the tools shown in figure 8.4 increase the torque we are capable of producing?

All of these tools have handles that increase the length of the moment arm of the force, thus increasing the torque applied to the screw or nut, and so on. In the case of the wrench or pliers, holding these tools farther out on the handle increases the torque in the same way, by increasing the moment arm of the force. Other everyday things we turn and thus apply torques to include steering wheels (why do

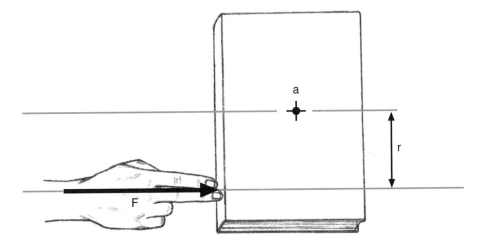

| Figure 8.2 | The moment arm (r) of a force (F) is the perpendicular distance between the line of action of the force and a parallel line passing through the axis of rotation (a). |

Axis

F

r

Figure 8.3   The torque created by the pulling force on the doorknob causes the door to swing open.

heavy trucks have larger-diameter steering wheels than cars?), bicycle handlebars, jar tops, knobs and dials on appliances, bicycle cranks, light switches, clothespins, staplers, and so on.

How is torque used in sport? In rowing, canoeing, and kayaking, torque is applied by the athlete to the oar or paddle to cause it to rotate. In golf, baseball, and tennis, torque is applied to the club, bat, or racket to swing these implements. In any sport where we turn, spin, or swing something (including our bodies), torque must be created to initiate these turns, spins, and swings. The holds used in wrestling provide great examples of torques that are used to turn an opponent. Consider the half nelson, as shown in figure 8.5.

In this hold, your opponent is prone, and you try to turn him over onto his back by putting your hand under his shoulder and onto the back of his head. If you then push down on his head with your hand and use your arm to lift up under his shoulder, you produce a force couple that creates a torque about a longitudinal axis through your opponent.

SELF-EXPERIMENT

Why do you suppose doorknobs or door handles are located on the opposite side of the door from the hinges? This location makes the moment arm (r in the torque equation) large so that the force required to create a large enough torque to swing open the door is small. Try opening or closing the door by pushing on it with a force directed closer to the hinges. You have to push with a greater force to create the same torque. As you move your hand closer and closer to the hinges, the force required to move the door gets greater and greater because the moment arm of the force is getting smaller and smaller.

This torque creates a turning effect that turns your opponent over. To counter this, your opponent could create a torque with his arm by abducting it so that it is perpendicular to his body. Pushing down onto the mat with this arm and hand creates a torque about the same axis but in the opposite direction.

Muscular Torque

The examples given so far are examples of external torques that act on the body or other objects. What about within the body? What creates the turning effects that cause our limbs to rotate about our joints? Muscles create the torques that turn our limbs. A muscle creates a force that pulls on its points of attachment to the skeletal system when it contracts. The line of action (or line of pull) of a muscle force is along a line joining its attachments and is usually indicated by the direction of its tendons. The bones that a muscle attaches to are within the limbs on either side of a joint, or two or more joints in some cases. The force of contraction of a muscle pulls on these limbs. Because the line of action of the muscle force is some distance from the joint axis, a moment arm exists, and torques about the joint axis are produced by the muscle force on both limbs to which it attaches. The torque produced by the muscle on the distal limb will tend to rotate that limb in one direction about an axis through the joint, and the torque

Figure 8.4 Common tools that you exert forces (F) on, thus creating torques around axes (a).

The forces exerted by a wrestler using a half nelson on his opponent create a torque that tends to turn the opponent over. The opponent can counter this torque by pushing down on the mat with his outstretched arm.

produced by the muscle on the proximal limb will tend to rotate that limb in the opposite direction about the same axis. Figure 8.6 shows how the force produced by the biceps brachii muscle creates a torque that tends to rotate both the forearm and the upper arm about the elbow joint.

What happens to the torque on the forearm produced by the biceps brachii muscle as the forearm is moved from full extension to 90° of flexion at the elbow joint? Can the muscle create the same torque throughout this range of motion? If the muscle produces the same force and the moment arm of the muscle stays the same throughout the range of motion, the torque produced will not change. Let's try the self-experiment on page 135 to see if the moment arm of the muscle stays the same length.

Strength-Training Devices and Torque

Strength-training exercises with free weights or weight-lifting machines provide many examples of torques that are exerted on our limbs by non-muscular forces. Consider an arm curl exercise. In this exercise, you hold a dumbbell in your hand and lift it by flexing at the elbow joint. We already saw how contraction of the biceps brachii muscle produces a torque around the elbow joint. As you lift the dumbbell, its weight produces a torque around the elbow joint as well, but this torque tends to rotate the forearm in the opposite direction. It creates a

The biceps brachii exerts a torque around the axis of the elbow joint by producing a force (F_m) with a moment arm (r) around the joint.

SELF-EXPERIMENT

Hold a heavy book in your hand and extend your right arm. Now use the thumb and index finger of your left hand to pinch your right elbow. Your index finger should be able to feel the tendon of your right biceps brachii, and your thumb should feel the back of your elbow, the olecranon process of your ulna. Now flex your forearm at the elbow. Does the distance between your thumb and forefinger change? It gets larger as the degree of elbow flexion approaches 90°. This distance is a crude measurement of the moment arm of the biceps brachii muscle about the elbow joint. The moment arm of the biceps brachii is largest when the elbow is at 90° and gets smaller as it is flexed or extended away from this position. Figure 8.7 shows how the moment arm of the biceps brachii muscle changes with elbow joint position. The biceps brachii's ability to create a torque about the elbow joint is dependent on the position of the elbow joint because the moment arm of the muscle changes as the elbow is flexed and extended. This is true for many muscle and joint combinations. It partially explains why our muscles are apparently stronger in some joint positions than in others.

Figure 8.7 The moment arm of the biceps brachii muscle decreases from r to r' as the elbow extends from 90°.

torque that would tend to extend your forearm at the elbow joint.

We already saw how the biceps brachii torque could change as the arm moved through its range of motion at the elbow joint, because the moment arm of the biceps brachii changed with changing joint position. What happens to the torque produced around the elbow joint by the barbell when an arm curl exercise is performed? The barbell doesn't get heavier, but the torque gets larger as the elbow is flexed. This is because the line of action of the barbell's weight moves farther from the elbow joint as the exercise is performed, thus increasing the moment arm. When the forearm is horizontal and the elbow is at 90°, the moment arm of the dumbbell is at maximum and the torque is greatest. Flexion beyond this position results in a smaller moment arm and thus a smaller torque, as does extension beyond this position. For most exercises involving free weights, the torques produced by the weights vary as the moment arms of these weights change during the movement.

Weight-lifting machines may or may not have this characteristic. To analyze the torques a weight-lifting machine exerts on you, you must first identify the resistance force. Often cables or chains are used to redirect the line of action of the force of gravity acting on the weight stack. The direction of this cable (where it attaches to the arm or cam of the part of the machine you actually move) indicates the line of action of the resistance force. The moment arm is then determined by measuring the perpendicular distance between this line of action and the axis of rotation of the arm or cam of the device.

Let's examine a leg extension machine, which is common in many weight-training facilities (see figure 8.8). The resistance force is provided by a stack of weights. This force is then transmitted to the arm of the machine via a cable attached to the middle of the arm. As the exercise is performed, the cable pulls on the machine arm. A torque is created about the axis of the machine arm because the cable pulls some distance away from this axis. As the exercise is performed, the line of action of the cable force changes, and the moment arm gets smaller as the leg reaches full extension at the knee. The resistance torque produced by this machine is thus largest at the starting position with the knee at 90° and gets smaller as the leg extends. Could you think of a way to redesign this machine so that the torque remains constant throughout the exercise? The design goal of some exercise machines on the market is to provide a constant resistance torque throughout the range of motion of the exercise. Other machines are designed to provide a resistance torque that varies in proportion to the changes in the moment arm of the muscle being exercised as the exercise is performed. This is the premise behind the design of the Nautilus-type weight machine.

By now you should have an understanding of what a torque is and how it is quantified. The fact that torques can be added (and subtracted) was briefly noted, but we have not discussed the effects that multiple torques have when acting on the same object. Can torques be added together to form an equivalent net torque? What if the torques act in opposite directions? Must there be equilibrium of torques (as is the case for forces) for an object to be in equilibrium?

Forces and Torques in Equilibrium

In chapter 4, the concept of static equilibrium was introduced during the discussion of forces. If an object is at rest, it is described as being in a state of static equilibrium. For an object to be in static equilibrium, the external forces acting on it must sum to zero (i.e., the net external force must equal zero). Our experiment with a force couple acting on a book earlier in this chapter indicated that a net force of zero is not the only condition of static equilibrium. A net force of zero ensures that no change will occur in the linear motion of an object, but it does not put restrictions on the object's angular motion. The net torque acting on an object must be zero to ensure that no changes occur in the angular motion of an object. For an object to be in static equilibrium, the external forces must sum to zero and the external torques (about any axis) must sum to zero as well. Mathematically, these conditions are expressed as:

$$\Sigma F = 0$$
$$\Sigma T = 0 \qquad (8.2)$$

where

ΣF = net external force
ΣT = net torque

Figure 8.8　A leg extension machine. The torque varies with position due to the change in the size of its moment arm (r).

For an object to be in static equilibrium, the external forces must sum to zero and the external torques (about any axis) must sum to zero as well.

Net Torque

Earlier in this chapter, when we came up with a mathematical definition of torque, we also said that torques which acted around the same axis could be added or subtracted algebraically, just like forces that act in the same direction. In a planar situation, then, a net torque is computed by summing the torques that act on an object.

SELF-EXPERIMENT

To investigate net torques, let's use a simple system. Find a ruler, a rubber eraser, and 10 pennies. Balance the ruler on the edge of the eraser. If you don't have an eraser, find something with a flat surface about 1/4 in. wide on which you can balance the ruler. If the ruler is 12 in. long, it probably balances with the eraser at 6 in. Now place a penny on the ruler 5 in. to the left of the eraser. Does the ruler stay balanced? Is it in a state of static equilibrium? No. Why not?

The penny in the preceding self-experiment created a counterclockwise torque about the eraser that caused the ruler to rotate counterclockwise. The moment arm of the penny was 5 in. The force created by the penny was the weight of the penny. The torque created about the eraser was one penny-weight ("p") times 5 in., or 5 penny inches of torque. In this case, the net torque acting on the ruler about an axis through the eraser was caused only by the weight of the penny:

$$\Sigma T = \Sigma(F \times r)$$
$$\Sigma T = (-1 \text{ penny})(-5 \text{ in.})$$
$$\Sigma T = +5 \text{ penny inches (p·in)}$$

This torque is positive because it tended to cause a rotation of the ruler in the counterclockwise direction.

SELF-EXPERIMENT

Now place a penny on the right side of the ruler 3 in. from the eraser. What is the net torque acting on the ruler about an axis through the eraser in this case?

The net torque is

$$\Sigma T = \Sigma(F \times r)$$
$$\Sigma T = (-1 \text{ p})(-5 \text{ in.}) + (-1 \text{ p})(+3 \text{ in.})$$
$$\Sigma T = +5 \text{ p} \cdot \text{in} + (-3 \text{ p} \cdot \text{in})$$
$$\Sigma T = +5 \text{ p} \cdot \text{in} - 3 \text{ p} \cdot \text{in}$$
$$\Sigma T = +2 \text{ p} \cdot \text{in}$$

The net torque is still counterclockwise (+), but it has been reduced to 2 p · in by the clockwise torque created by the penny on the right side of the ruler.

SELF-EXPERIMENT

Now place another penny on the right side of the ruler 2 in. to the right of the eraser. What happens? The ruler balances on the eraser. It is in static equilibrium.

This seems odd, though, as there are twice as many pennies on the right side of the ruler. Maybe an analysis of the free-body diagram shown in figure 8.9 will help explain the situation.

What is the net torque acting on the ruler about an axis through the eraser? The net torque is:

$$\Sigma T = \Sigma(F \times r)$$
$$\Sigma T = (-1 \text{ p})(-5 \text{ in.}) + (-1 \text{ p})(+3 \text{ in.}) + (-1 \text{ p})(+2 \text{ in.})$$
$$\Sigma T = +5 \text{ p} \cdot \text{in} + (-3 \text{ p} \cdot \text{in}) + (-2 \text{ p} \cdot \text{in})$$
$$\Sigma T = +5 \text{ p} \cdot \text{in} - 3 \text{ p} \cdot \text{in} - 2 \text{ p} \cdot \text{in}$$
$$\Sigma T = 0$$

The ruler balances (it is in static equilibrium) because the net torque acting on it is zero.

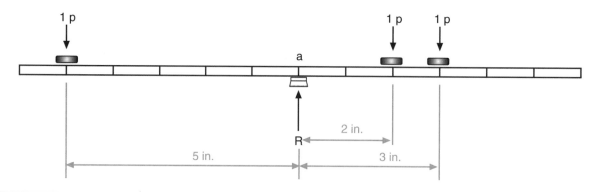

Figure 8.9   Free-body diagram of a ruler balanced on an eraser with pennies on the ruler. R is the reaction force exerted by the support of the ruler and a is the axis. (The weight of the ruler is omitted in this diagram.)

$\mathbb{S}$ELF-EXPERIMENT

Let's try another example. Clear the pennies off the ruler. Now stack four pennies on the ruler 3 in. to the right of the eraser. What net torque do these pennies create about the eraser? Four pennies times 3 in. is 12 p · in of torque in the clockwise direction about an axis through the eraser. If you only had two pennies left to use to balance the ruler, where would you stack them?

As we now know that the net torque has to be zero if the ruler is balanced (and in static equilibrium), we can solve this problem mathematically as follows:

$$\Sigma T = \Sigma(F \times r) = 0$$
$$0 = (-2\ p)(r) + (-4\ p)(+3\ \text{in.})$$
$$(-2\ p)(r) = (4\ p)(3\ \text{in.})$$
$$r = \frac{(4\ p)(3\ \text{in.})}{-2\ p} = -6\ \text{in.}$$

Two pennies will create a counterclockwise torque of 12 p · in (to counter the 12 p · in of clockwise torque created by the four pennies) if you stack them 6 in. to the left of the eraser (the negative sign before the 6 in. indicates that the moment arm is to the left of the axis). Two pennies times 6 in. equals 12 p · in of torque in the counterclockwise direction.

Muscle Force Estimates Using Equilibrium Equations

Now let's examine how the conditions for static equilibrium allow us to estimate the forces our muscles produce to lift or hold up objects. Suppose you are holding a 20-lb dumbbell in your hand and your elbow is flexed 90° so that your forearm is parallel to the floor. If the moment arm of this dumbbell is 12 in. about the elbow joint axis, what torque is created by this dumbbell about your elbow joint axis? The torque is:

$$T = F \times r \qquad\qquad (8.1)$$
$$T = (-20\ \text{lb})(-12\ \text{in.})$$
$$T = +240\ \text{lb} \cdot \text{in.}$$

The dumbbell creates a torque of 240 lb · in. about the elbow joint axis in the counterclockwise direction.

To hold up this dumbbell, your elbow flexor muscles must create a clockwise torque equal to the counterclockwise torque created by the dumbbell (if the weight of the forearm and hand are ignored). If the moment arm of these muscles is 1 in., what force must they pull with to hold the dumbbell in the position described? Because the forearm must be in static equilibrium, the problem is solved as follows (let F_m represent the muscle force):

$$\Sigma T = \Sigma(F \times r) = 0$$
$$0 = (-20\ \text{lb})(-12\ \text{in.}) + F_m(-1\ \text{in.})$$
$$-[(-20\ \text{lb})(-12\ \text{in.})] = F_m(-1\ \text{in.})$$
$$\frac{-[(-20\ \text{lb})(-12\ \text{in.})]}{-1\ \text{in.}} = F_m = +240\ \text{lb}$$

Wow! Your elbow flexor muscles must create a force of 240 lb just to hold up a 20-lb dumbbell! This seems to be too large a force. But our muscles are arranged so that their moment arms about the joints are short. This means that they must create relatively large forces to produce practically effective torques about the joints.

More Examples of Net Torque

Now let's look at some examples of net torques in sport. Consider a pole vaulter. What external forces act on a pole vaulter? Gravity pulls downward on the vaulter with a force equal to his weight, and the pole exerts forces on each of the vaulter's hands where he grips it. Figure 8.10 shows a free-body diagram of the pole vaulter with values of the external forces estimated as well.

What net torque acts on this vaulter around an axis through his center of gravity? Is the vaulter in equilibrium? The 500-N force acting at the vaulter's left hand has a moment arm of 0.50 m about his center of gravity. This force creates a clockwise torque about the vaulter's center of gravity. The 1500-N force acting at the vaulter's right hand has a moment arm of 1.00 m about his center of gravity. The torque created by this force about the vaulter's center of gravity is also clockwise. The vaulter's weight of 700 N acts through his center of gravity, and thus

the moment arm of this weight is zero, so it does not create any torque about the vaulter's center of gravity. Mathematically, the net torque about a transverse axis through the vaulter's center of gravity is:

$$\Sigma T = \Sigma(F \times r) = (-500\text{ N})(0.50\text{ m}) - (1500\text{ N})(1.00\text{ m})$$

$$\Sigma T = -250\text{ Nm} - 1500\text{ Nm}$$

$$\Sigma T = -1750\text{ Nm}$$

The negative sign on this net torque indicates that it acts in a clockwise direction. The 1750-Nm net torque produces a turning effect that tends to rotate the vaulter clockwise onto his back, as if he were doing a backward somersault.

Now let's consider what happens to the vaulter later in the vault, when he is in the position shown in the free-body diagram in figure 8.11.

What is the net torque acting on the vaulter in this situation? The 300-N force acting on the vaulter's left hand has a moment arm of 0.50 m and still creates a clockwise (–) torque about the vaulter's center of gravity. The 500-N force acting on the vaulter's right hand also has a moment arm of 0.50 m, but it now creates a counterclockwise (+) torque about the vaulter's center of gravity. The net torque is thus:

| Figure 8.10 | Free-body diagram of a pole vaulter just after takeoff. The reaction forces from the pole create clockwise torques on the vaulter. |

| Figure 8.11 | Free-body diagram of a pole vaulter midway through the vault. The reaction forces from the pole create torques in opposite directions on the vaulter. |

$\Sigma T = \Sigma(F \times r) = -(300 \text{ N})(0.50 \text{ m}) + (500 \text{ N})(0.50 \text{ m})$

$\Sigma T = -150 \text{ Nm} + 250 \text{ Nm}$

$\Sigma T = +100 \text{ Nm}$

The positive sign on this net torque indicates that it acts in the counterclockwise direction. The turning effect produced by the forces acting on the vaulter thus tends to rotate the vaulter counterclockwise, as if he were doing a forward somersault.

What Is Center of Gravity?

By now you should understand what a torque is and how it is created by external forces as well as muscular forces; how to determine a net torque; and what the conditions of static equilibrium are. The expression "center of gravity" has been used several times in this chapter. You probably have heard this expression before and already have some ideas about what it means. In this section, the concept of center of gravity will be defined and explained.

Webster's New World Dictionary defines center of gravity as "that point in a body or system around which its mass or weight is evenly distributed or balanced and through which the force of gravity acts" (p. 227). Center of mass is defined as "the point in a body or system of bodies at which, for certain purposes, the entire mass may be assumed to be concentrated: for bodies near the earth, this coincides with the center of gravity" (p. 227). As all of the human movement activities we are concerned with occur on or near the earth, center of gravity and center of mass are equivalent terms and may be used interchangeably. The center of gravity is an imaginary point in space. It is not a physical entity; its location is not marked on an object. The center of gravity is a useful concept for analysis of human movement because it is the point at which the entire mass or weight of the body may be considered to be concentrated. So the force of gravity acts downward through this point. If a net external force acts on a body, the acceleration caused by this net force is the acceleration of the center of gravity. If no external forces act on an object, the center of gravity does not accelerate. When interpreting and applying Newton's laws of motion, it is the center of gravity of a body whose motions are ruled by these laws. Is it thus important to know how to locate or estimate the location of the center of gravity of a body or object.

> The center of gravity is the point at which the entire mass or weight of the body may be considered to be concentrated.

Locating the Center of Gravity of an Object

The definition of center of gravity gives an indication of how to find its location: the center of gravity is the point in the body around which its weight is balanced. The center of gravity is the point of balance. What does this mean? Every object can be considered to be composed of many smaller elements. These smaller elements may represent the parts that make up the object. In the human body, these elements might be represented by the limbs, trunk, and head (i.e., two hands, two forearms, two upper arms, two feet, two shanks, two thighs, one torso, one neck, and one head make up the human body). At the most basic level, these elemental parts could represent molecules or atoms. The force of gravity pulls downward on each of these smaller elements. The sum or resultant of these forces is the total weight of the object. This weight acts through a point about which the moments of the weights of each of the elemental parts (the torques created by their weights) sum to zero, no matter what position the object is in.

Mathematical Determination of the Center-of-Gravity Location

If the weights and locations of the elemental parts that make up an object are known, the center-of-gravity location can be computed mathematically. The definition of center of mass indicated that it is the point at which the entire mass (or weight) may be considered to be concentrated. By this definition, a ruler with six pennies distributed on it at 2-in. intervals is equivalent to a ruler with six pennies stacked on it at one location if that location is the center of gravity of the first ruler. Let's look more closely at this. Suppose the first ruler had pennies placed on it at the 1-, 3-, 5-, 7-, 9-, and 11-in. marks. This ruler and its equivalent are shown in figure 8.12.

If you closed your eyes and picked up either of these rulers by the end, they would feel identical. You wouldn't be able to tell which ruler had the pennies distributed on it at 2-in. intervals and which ruler had the pennies all stacked in one place. The

SELF-EXPERIMENT

Take out the ruler, eraser, and pennies again. First, balance the ruler on the edge of the eraser. By our definition of center of gravity as "the point of balance," the center of gravity of the ruler must lie above the points of support provided by the edge of the eraser. The counterclockwise torque created by the weight of the ruler to the left of the eraser balances the clockwise torque created by the weight of the ruler to the right of the eraser. Now make two stacks of pennies with four pennies in each stack. Place one stack 1 in. to the left of the eraser and the other stack 1 in. to the right of the eraser so that the ruler remains balanced. The center of gravity of the pennies and the ruler is still above the eraser.

Now slide the right stack of pennies 1 in. to the right and move the eraser so that the ruler remains in equilibrium. Which way did you have to move the eraser? Which way did the center of gravity move? To keep the ruler in balance, you had to move the eraser to the right a little. The centers of gravity of the ruler and pennies also moved to the right.

Now move the right stack of pennies all the way to the right end of the ruler and move the eraser so that the ruler remains balanced. Again you had to move the eraser to the right, so the center of gravity of the system (ruler and pennies) moved to the right. You didn't have to move the eraser as far as you had to move the pennies, though. If some of the elemental parts of an object move or change position, the center of gravity of the object moves as well, although not as far.

Now take three pennies off the left stack and move the eraser so that the system remains in balance. You had to move the eraser to the right again, so the center of gravity moved to the right as well. If an elemental part of an object is removed from the object, the center of gravity of the object moves away from the point of removal.

Now add three pennies to the right stack and move the eraser so that the ruler remains balanced. The eraser is now farther to the right, and the center of gravity of the system has also moved farther to the right. If mass is added to an object, the center of gravity of the object moves toward the location of the added mass. Observe how many pennies are on either side of the center of gravity of the system now. The weights on either side of the center of gravity do not have to be equal; one penny does not equal seven pennies, but the torques created by these weights about the center of gravity must equal each other.

rulers would weigh the same, and you would have to create the same torque on the end of each ruler to hold it up with one hand. This is the key to determining the center-of-gravity location mathematically—both rulers create the same torque about the end of the ruler. The sum of the moments of force created by each of the elemental weights, the pennies in this case, equals the moment of force created by the total weight stacked at the center-of-gravity location. Mathematically, this can be expressed as:

$$\Sigma T = \Sigma (W \times r) = (\Sigma W) \times r_{cg} \qquad (8.3)$$

where

W = the weight of one element

r = the moment arm of an individual element

ΣW = the total weight of the object

r_{cg} = the moment arm of the entire weight of the object (the location of the center of gravity of the object relative to the axis about which the moments of force are being measured)

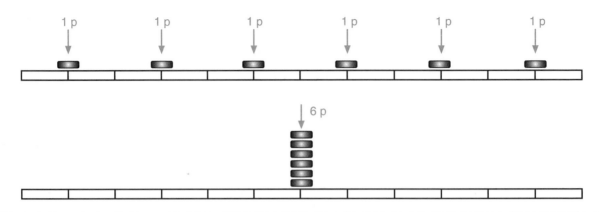

Figure 8.12   A ruler with six pennies on it, one placed every two inches, feels the same as and is equivalent to a ruler with six pennies stacked at the center of the ruler.

Equation 8.3 can be used to solve for the location of the center of gravity, r_{cg}. Let's use the end of the first ruler shown in figure 8.12 as the axis about which the moments of force will be measured (any axis may be used in this calculation, so long as the same one is used on either side of the equation).

$$\Sigma(W \times r) = (\Sigma W) \times r_{cg} \qquad (8.3)$$

$$\Sigma(W \times r) = (1\ p)(1\ in.) + (1\ p)(3\ in.) + (1\ p)(5\ in.)$$
$$+ (1\ p)(7\ in.) + (1\ p)(9\ in.) + (1\ p)(11\ in.)$$

$$\Sigma(W \times r) = 36\ p \cdot in.$$

$$(\Sigma W) \times r_{cg} = (1\ p + 1\ p + 1\ p + 1\ p + 1\ p + 1\ p)$$
$$\times r_{cg} = (6\ p) \times r_{cg}$$

$$\Sigma(W \times r) = 36\ p \cdot in. = (6\ p) \times r_{cg} = (\Sigma W) \times r_{cg}$$

$$\frac{36\ p \cdot in.}{6\ p} = r_{cg}$$

$$r_{cg} = 6\ in.$$

So the center of gravity of the ruler is 6 in. from the end of the ruler. If all six pennies were stacked at this point rather than distributed across the ruler

at 2-in. intervals, this ruler would feel the same as the first ruler.

To determine the location of the center of gravity mathematically, the relationship used is that the sum of the moments of force created by the elemental weights equals the moment of force created by the sum of the elemental weights. Stated more simply, the sum of the moments equals the moment of the sum.

Mathematically, we can solve for the location of the center of gravity as we did in the previous example. Let's measure the moments about the end of the ruler for convenience (remember, we can measure these moments about any axis, so we choose an axis that is convenient).

$$\Sigma(W \times r) = (\Sigma W) \times r_{cg} \qquad (8.3)$$

$$\Sigma(W \times r) = (3\ p)(1\ in.) + (7\ p)(8\ in.)$$
$$= (3\ p + 7\ p) \times r_{cg} = (\Sigma W) \times r_{cg}$$

$$3\ p \cdot in. + 56\ p \cdot in. = (10\ p) \times r_{cg}$$

$$59\ p \cdot in. = (10\ p) \times r_{cg}$$

$$\frac{59\ p \cdot in.}{10\ p} = r_{cg}$$

$$r_{cg} = 5.9\ in$$

The center of gravity is 5.9 in. from the end of the ruler. Pick up the ruler 5.9 in. from its end and see if it will balance on your finger at that location. If it does, you've verified that 5.9 in. is the center-of-gravity location.

With the pennies and ruler, we found the center of gravity along one dimension only. For more complex objects, the center-of-gravity location is defined by three dimensions, because most objects occupy space in three dimensions. To determine the center-of-gravity location for an object in three dimensions, the procedure we used in the previous two ex-

SELF-EXPERIMENT

Let's try another example. Place three pennies on the ruler at the 1-in. mark and seven pennies on the ruler at the 8-in. mark. Without picking up the ruler, can you determine where its center of gravity is located?

amples is repeated for each dimension, with gravity assumed to be acting in a direction perpendicular to that dimension.

Center of Gravity of the Human Body

Now let's consider the human body and the location of its center of gravity. The human body is not a rigid object, so the location of its center of gravity depends on the position of its limbs, just as the location of the center of gravity of the ruler and pennies depended on the position of the pennies on the ruler. Let's assume you are standing up with your arms at your sides, as shown in figure 8.13.

Because your left and right sides are symmetrical, your center of gravity lies within the plane that divides your body into left and right halves (the midsagittal plane). If you lift your left arm away from your side, your center of gravity shifts to the left. Although your body is not symmetrical from front to back, the center of gravity lies within a plane that divides your body into front and back halves (the frontal plane). This plane passes approximately through your shoulder and hip joints and slightly in front of your ankle joints. If you raise your arms

in front of you, your center of gravity will be moved forward slightly.

The center-of-gravity location in these two dimensions, front to back and side to side, is easy to approximate if you are in anatomical position. The location of the center of gravity in the vertical dimension is more difficult to estimate. The center of gravity from top to bottom lies within a plane that passes horizontally through your body 1 to 2 in. below your navel, or about 6 in. above your crotch. This plane is slightly higher than half of your standing height, about 55% to 57% of your height. If you reach overhead with both arms, your center of gravity will move upward slightly (about 2 or 3 in.). Someone with long legs and muscular arms and chest will have a higher center of gravity than someone with shorter, stockier legs. A woman's center of gravity is slightly lower than a man's because women have larger pelvic girdles and narrower shoulders relative to men. A woman's center of gravity is approximately 55% of her height from the ground, whereas a man's center of gravity is approximately 57% of his height from the ground. Infants and young children have higher centers of gravity relative to their height because of their relatively larger heads and shorter legs.

The center of gravity of the human body will move from the position described above if the body parts change their positions. If the limbs, trunk, head, and neck are considered the elemental units of the body, then their positions determine the center-of-gravity location of the body. The center-of-gravity location of the body can be estimated for any body position using the following procedure. First, imagine the body in the standing position shown in figure 8.13 and locate the center of gravity for this position using the information described in the previous paragraphs. Then consider the movement each limb had to make to get into the position being described. For each limb movement, the center of gravity of the entire body shifts slightly in the direction of that movement. How much it moves depends on the weight of the limb that moved and the distance it moved. Raising one leg out in front of you will move your center of gravity forward and upward farther than will raising one arm out in front of you. Practice estimating the location of the body's center of gravity for each of the positions shown in figure 8.14.

For the high jumper in the arched position over the crossbar shown in figure 8.14a, her center of gravity is actually outside of her body, as it may be for the diver in the piked position shown in figure 8.14b and the pole vaulter shown in figure 8.14c.

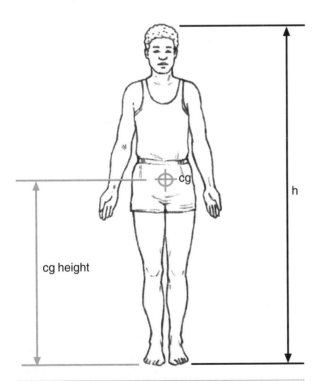

| Figure 8.13 | In anatomical position, the height of your center of gravity is 55 to 57% of your standing height. |

a

© Claus Andersen

b

© Claus Andersen

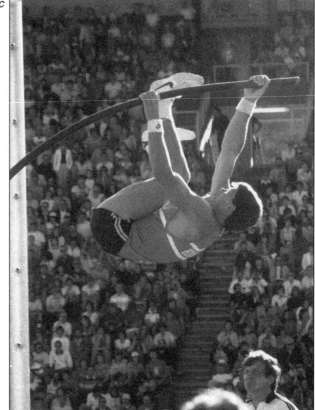

c

Figure 8.14   Where is the athlete's center of gravity located in each of these body positions?

The flexibility and complexity of the human body allows you to assume positions where your center of gravity may actually lie outside of your body.

Center of Gravity and Performance

Now let's consider how your limb movements may actually affect your performance of a skill by affecting your center-of-gravity position. In a vertical jump-and-reach test, the objective is to jump up and reach as high as possible with one hand. Try the different jump-and-reach test techniques described in the self-experiment below.

Using the first technique, you probably reached the highest. Using the second technique, your reach was probably 1 or 2 in. lower. And your reach with the third technique was probably 4 to 6 in. lower. Why?

With all three techniques, your center of gravity probably reached the same height if you jumped off the ground with the same effort each time. But wait—wouldn't your center of gravity be higher when you raised both arms over your head, and higher still when you lifted your legs up as well? Well, we did say that your center of gravity moved upward when any part of your body moved upward, but we should have said that center-of-gravity movement was a relative movement. In other words, raising your arms over your head causes your center of gravity to move higher in relation to

your other body parts, for instance, your head. When you raise both arms over your head, your center of gravity moves closer to your head. In absolute terms (i.e., movement measured in relation to a fixed point on the earth), this could mean that (1) your center of gravity actually did move upward and higher off the ground; (2) your center of gravity stayed the same height off the ground and your head and other parts moved downward or closer to the ground to compensate for other parts moving upward; or (3) some combination of (1) and (2).

The explanation for what happened in the jumping tests comes from our early discussion of the center-of-gravity concept. The application and interpretation of Newton's laws to the motion of a complex object are simplified by the center-of-gravity concept. It is the center of gravity whose motion these laws govern. When you jumped up in the air during the jump-and-reach tests, the only external force acting on you was the force of gravity (your weight). This force causes your center of gravity to accelerate downward at a constant rate. You were a projectile once your feet left the ground. The motion of your limbs when you were in the air could not alter the motion of your center of gravity because they were not pushing or pulling against anything external to your body. The net force acting on you was still only the downward pull of gravity. The path that your center of gravity followed was not changed by your limb actions, but these actions caused the motion of your other limbs and

SELF-EXPERIMENT

Find a room with a very high ceiling and a clear space next to a wall. You may have to go outside to perform this experiment. First, from a standing position, jump up and, reaching as high as you can, touch the wall with one hand. Keep your other hand down at your side. See how high you can touch on the wall. This might be difficult to do while you are jumping, so have a friend watch to see how high you reach. Try this three times to get a good idea of how high you can jump and reach using this technique.

Next, jump from a standing position again and reach as high as you can with one hand, but this time reach overhead with your other hand as well. Try this technique three times also. Did you reach as high?

Now jump from a standing position again and reach as high as you can with both hands, as you did in the previous case, but this time, when you are in the air, lift your knees and legs so your heels touch your buttocks. Try this technique three times also. Did you reach as high?

trunk to change. Lifting your arms caused something else to move downward in order for your center of gravity to continue moving along its parabolic path. When you raised your arms and legs, your head and trunk got lower to compensate for this movement.

Here's another way to interpret or explain why you were able to reach the highest with only one hand above your head versus with both arms, or with both arms and both legs raised. First, let's assume the center of gravity reached the same peak height. If this is the case, then to maximize the reach of one hand (its height above the ground), you want to maximize the vertical distance between the center of gravity and the outstretched arm. By raising both arms, or both arms and both legs, the center of gravity is closer to the head and thus closer to the reaching hand. By keeping all of the limbs and body parts (with the exception of the reaching arm and hand) as low as possible relative to the center of gravity, the distance from the reaching hand to the center of gravity is maximized. Figure 8.15 demonstrates this graphically.

A basketball player jumping up to block a shot can reach higher if only one hand reaches up and the other arm and the legs are not lifted in relation to the trunk. This is similarly true for a volleyball player trying to block a spike, although the volleyball player may be more effective reaching up with both hands. The player's hands won't reach as high, but they will block a greater area. In basketball, the direction of the shot is known to the defender—it's toward the goal, so only one hand is needed to block a shot. In volleyball, the direction of the spike is not known, so two hands are used to cover a greater area.

Let's consider one final jumping activity. How do basketball players, dancers, figure skaters, and gymnasts appear to "hang in the air"? During some of their jumps, it appears that their bodies are suspended in the air for a short time, rather than rising and falling in a parabolic path, as a projectile should. Can you explain how these athletes "defy gravity"? Perhaps figure 8.16 will help.

If we follow the path of the jumper's center of gravity, it does indeed rise and fall through a parabolic curve. But the jumper's head and trunk appear to be suspended at the same height during the middle stage of the leap. During this time, the jumper's legs and arms rise and then fall. These

Figure 8.15   Three different vertical jump techniques result in three different reach heights, but the height of the center of gravity (cg) above the ground may be similar.

Path of head

Path of cg

| Figure 8.16 | During a leap, the jumper's head stays at the same level but the center of gravity follows a parabolic path. |

movements account for the rise and fall of the center of gravity, so the head and trunk do not rise appreciably.

Center of Gravity and Stability

Our final topic concerning the center of gravity and performance in sports and human movement is stability. What is stability? When you say something or someone is stable, what characteristic are you speaking of? There are a variety of definitions for the word *stable*. *Webster's New World Dictionary* lists seven. The first defines stable as "not easily moved or thrown off balance; firm; steady" (p. 1302). This is probably similar to your own definition of the word stable. The fifth definition is "capable of returning to equilibrium or original position after having been displaced" (p. 1303). This is the mechanical definition of stable. Stability is thus "the capacity of an object to return to equilibrium or to its original position after being displaced" (p. 1302).

> **Stability is the capacity of an object to return to equilibrium or to its original position after being displaced.**

In many sports and human movement activities, the athletes or performers do not want to be moved from a certain stance or position. They want to be in a very stable position. Wrestlers, football linemen, even basketball players are more successful at certain skills if they adopt stable positions. In other sports, success may be determined by how quickly an athlete is able to move out of a position. The receiver of a serve in tennis or racquetball, a sprinter, a swimmer, a downhill skier, or a goalie in soccer are more successful if their stance is less stable.

Factors Affecting Stability

What factors affect stability? How can you make yourself more or less stable? For some insights, try balancing this book on the table. In what position is it most easily balanced (i.e., which is its most stable position)? The book is most stable when it is lying flat on the table. This is the orientation that puts its center of gravity at the lowest position. This is also the position that offers a base of support with the largest area. If the book weighed more, would it be more or less stable? A heavier book would be more stable. Apparently the stability of an object is affected by the height of the center of gravity, the size of the base of support, and the weight of the object. The base of support is the area within the lines connecting the outer perimeter of each of the points of support. Figure 8.17 shows examples of various stances and their bases of support. Which of these

Figure 8.17 Stances and their corresponding bases of support.

stances are most stable? Which of these stances are least stable?

> **The stability of an object is affected by the height of the center of gravity, the size of the base of support, and the weight of the object.**

What is the mechanical reason for why center-of-gravity height, base of support size, and weight affect stability? Let's examine the forces and moments of force that act on a book when a toppling force is exerted on it.

Stand a book on its edge and exert a horizontal force against it to tip it over. If the book remains in static equilibrium, the net force and torque acting on the book must be zero. The external forces acting on the book include the book's weight, W, acting through its center of gravity, cg; the toppling force, P; and the reaction force, R. If we examine the situation where the toppling force is just large enough that the book is almost starting to move, the free-body diagram shown in figure 8.18 is appropriate.

If moments of force are measured about an axis, a, through the lower left corner of the book, the sum of moments about this point is zero:

$$\Sigma T_a = 0$$
$$0 = (P \times h) - (W \times b)$$
$$P \times h = W \times b \qquad (8.4)$$

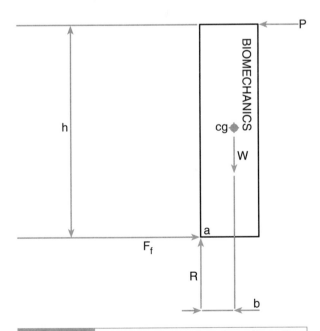

where

P = toppling force

h = moment arm of the toppling force

W = weight of the book (object)

b = moment arm of the book (object)

The terms on the left side of equation 8.4 are ones that are minimized to increase stability, and the terms on the right are ones that are maximized to increase stability. On the right side of equation 8.4, two variables appear: W, the weight of the object, and b, the moment arm of this weight. Increasing the weight will increase the stability because the moment of force keeping the object upright would be larger. Likewise, increasing the moment arm of the object's weight will increase stability. This dimension, b, is related to the size of the base of support, but it may be smaller or larger, depending on which direction the toppling force comes from. Figure 8.19 demonstrates this for objects of various shapes.

The triangular block is less stable and more likely to be tipped over if the toppling force is directed to the left (figure 8.19a) rather than to the right (figure 8.19b). Stability is directional. An object can be more stable in one direction than in another. It is not the size of the base of support that affects stability, but the horizontal distance between the line of gravity and the edge of the base of support in the direction that the toppling force is pushing or pulling.

On the left side of equation 8.4, the toppling force, P, is a factor that is not related to any characteristic of the object. The moment arm of this toppling force, h, is a characteristic of the object. This distance is related to the height of the object, which is also related to the height of the center of gravity. So a lower center of gravity, which implies a lower height and a shorter moment arm for the toppling force, increases stability.

Stability and Potential Energy

This explanation of why height or center of gravity affect stability is rather weak. Instead, it explains why the height of the object might affect stability. A better way of explaining why center-of-gravity height affects stability uses the concepts of work and potential energy. Consider the block shown in figure 8.20.

As long as the center of gravity of the block is to the left of the lower right corner, the weight creates a righting moment of force in opposition to the

Figure 8.18   Free-body diagram of a book standing on its edge with a toppling force, P, exerted on it.

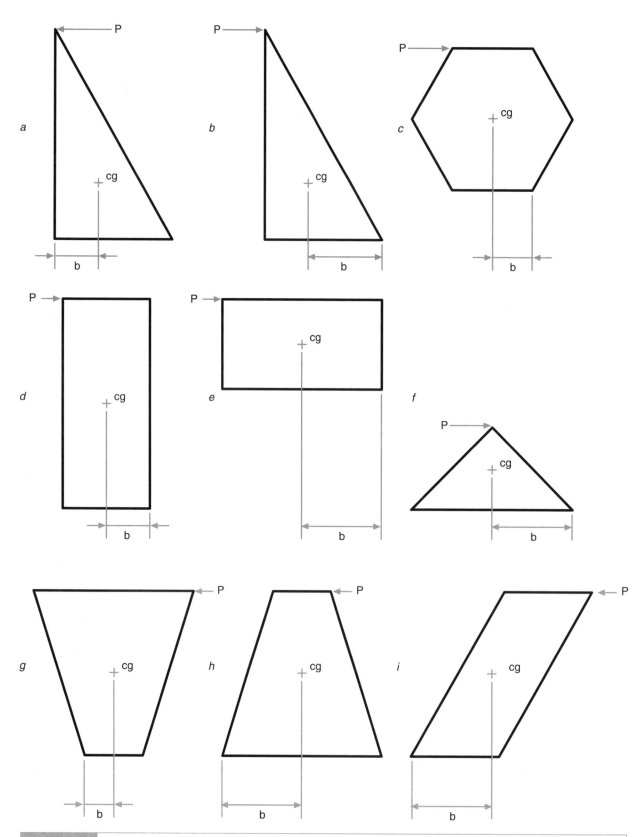

Figure 8.19   Stability is affected by the moment arm (b) of an object's weight around the edge of the base of support toward which the toppling force (P) pushes or pulls.

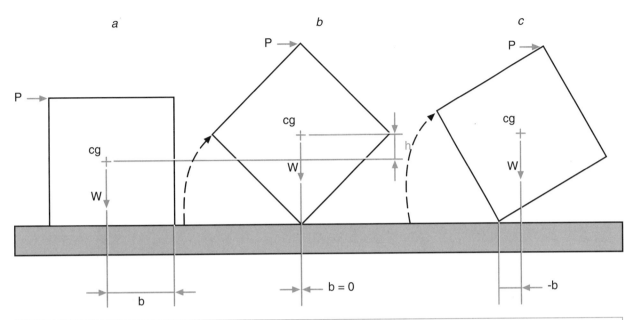

Figure 8.20 As an object is toppled, the center of gravity is raised and the righting moment of the weight decreases. In (*a*) the weight creates a righting moment (W × b); in (*b*) the weight creates no moment (W × 0); and in (*c*) the weight creates a toppling moment [W × (-b)].

toppling moment of force created by the force P. But when the block is tipped past the configuration shown in figure 8.20b, where the center of gravity lies directly over the supporting corner, the moment of force created by the weight changes direction and becomes a toppling moment that causes the block to topple, as shown in figure 8.20c.

To move the block from its stable position (figure 8.20a) to the brink of instability (figure 8.20b), the center of gravity had to be raised a distance, Δh. Work was required to do this, and the potential energy of the block increased.

Now let's examine what happens if the center of gravity is higher or lower. Figure 8.21 shows three blocks of the same shape and weight but with differing center-of-gravity heights.

The vertical displacement, Δh, each block undergoes before it topples is shown. The higher the center of gravity, the smaller this displacement, thus the smaller the change in potential energy and the smaller the amount of work done. So a block with a lower center of gravity is more stable because more work is required to topple it.

If the distance from the line of gravity to the edge of the base of support about which toppling will occur (the moment arm of the weight) is increased, the vertical displacement the center of gravity goes through before the object topples also increases, so the object is more stable. Figure 8.22 shows two blocks with identical center-of-gravity heights but

different horizontal distances from the line of gravity to the edge of the base of support.

The most stable stance or position an object or person can be in is the one that minimizes potential energy. Moving to any other position increases potential energy and requires work to be done on the object or person. Positions that place the center of gravity below the points of support are more stable than positions that place the center of gravity above the base of support.

> **The most stable stance or position an object or person can be in is the one that minimizes potential energy.**

If the center of gravity is below the base of support, after any displacement of the object, the object will return to its original equilibrium position. This is an example of stable equilibrium. A gymnast hanging from the horizontal bar is an example of stable equilibrium. When the center of gravity lies above the base of support, stability is maintained only as long as the line of gravity falls within the base of support (see figure 8.23).

Center of Gravity, Stability, and Human Movement

Our discussion of stability has focused primarily on rigid objects with fixed centers of gravity and

$$\Delta h_a > \Delta h_b > \Delta h_c$$
$$h_a < h_b < h_c$$

Figure 8.21   A block with a lower center of gravity must undergo a greater increase in its potential energy before it topples.

bases of support. The human body is not rigid and its center-of-gravity position and base of support can change with limb movements. Humans can thus control their stability by changing their stance and body position. Before examining how athletes manipulate their centers of gravity and bases of support to affect stability, let's try a few activities.

How do you initiate a walking step? You don't just lift your foot and place it in front of you. You lean forward until your line of gravity falls in front of your feet and you lose your stability. You begin to fall forward, and you step with one foot to catch your fall and reestablish your stability. So walking could be described as a series of falls and catches!

In athletic activities, athletes may want to maximize their stability in general or in a specific direction, or they may want to minimize stability (increase their mobility). During the first period of a wrestling match, the two wrestlers are standing and each is trying to take the other down. The direction of the toppling force is unknown; the other wrestler may be pulling or pushing forward, backward, or to the left or right. To maximize stability (while still maintaining the ability to move), the wrestler crouches to lower his center of gravity and widens

his base of support by placing his feet slightly wider than shoulder-width apart in a square stance (figure 8.24a) or by placing one foot in front of the other, again slightly farther than shoulder-width apart, in a staggered stance (figure 8.24b).

When the wrestler is in a defensive position on his stomach and trying not to be turned over onto his back, he maximizes his stability by sprawling his limbs to the sides to maximize the size of his base of support and to lower his center of gravity as much as possible (figure 8.25).

When force is expected from a specific direction, the base of support is widened in that direction. If a heavy medicine ball were thrown to you, the most stable stance for you to be in to catch the ball would be a staggered stance, with one foot in front of the other in line with the direction of the throw and leaning toward the front foot (figure 8.26).

This type of stance might be adopted by competitors at the start of a tug-of-war, except that they would shift their weight over the rear foot. Boxers also use staggered stances, as do tennis players, baseball batters, and so on. This is a popular stance in many sports, not only because it allows momentum to be reduced or increased by force applica-

$$h_a = h_b$$
$$b_a < b_b$$
$$\Delta h_a < \Delta h_b$$

a Narrow base

b Wide base

Figure 8.22 The greater the moment arm of the weight, the greater the vertical displacement of the center of gravity (Δh) before toppling occurs.

a

b

Figure 8.23 Line of gravity (*a*) and base of support (*b*) for normal stance.

tion over a long time, but also because it is a more stable position.

Unless some other part of our body or an implement we use touches the ground, the size of our base of support is limited by our shoe size and the stances being adopted. In sport, skis increase stability forward and backward. In physical rehabilitation and medicine, crutches, canes, walkers, and so on, are used to increase the base of support and stability of the injured, sick, or infirm.

In some activities, stability is minimized to enhance quick movement. For instance, in a track sprint start, in the set position, the sprinter raises her center of gravity and moves it forward to the edge of the base of support over her hands. At the starter's signal, lifting her hands off the track puts her line of gravity well in front of her base of support, and the sprinter falls forward. A similar strategy is used in swimming starts.

Summary

The turning effect created by a force is a moment of force, also called a torque. A torque is equal to the magnitude of the force times the perpendicular distance between the line of action of the force and the axis of rotation about which the torque is being measured. This perpendicular distance is called the moment arm of the force. Torque is a vector quantity. Its direction is defined by the orientation of the axis of rotation and the sense of the torque (clockwise or counterclockwise) about that axis of rotation. The units for torque are units of force times units of distance, or Nm in the SI system.

For an object to be in balance (in a state of equilibrium), the external forces acting on it must sum to zero and the moments of those forces must also sum to zero.

Figure 8.24   Two wrestling stances, square (*a*) and staggered (*b*), that represent compromises between stability and mobility.

Figure 8.25   The defensive sprawl of a wrestler maximizes his stability by minimizing the height of his center of gravity and maximizing his base of support.

Figure 8.26   Staggered stance provides directional stability for catching.

The center of gravity of an object is its balance point or the point about which the moments of force created by the weights of each of the parts of the object sum to zero. To locate an object's center of gravity, you can balance it, suspend it, or spin it. In the human body, the center of gravity lies on the midline left to right and front to back and between 55% and 57% of a person's height above the ground when standing in anatomical position.

Stability is affected by the position of an object's center of gravity above the ground and relative to the edges of the support base. Stability increases as the center of gravity is lowered and moved farther from any edge of the base of support. Increasing weight also increases stability.

Key Terms

base of support *(p. 147)*

center of gravity *(p. 140)*

centric force *(p. 128)*

eccentric force *(p. 128)*

force couple *(p. 128)*

moment arm *(p. 131)*

moment of force *(p. 128)*

stability *(p. 147)*

torque *(p. 128)*

REVIEW QUESTIONS

1. The Achilles tendon inserts on the calcaneus (the heel bone) at a distance of 8 cm from the axis of the ankle joint. If the force generated by the muscles attached to the Achilles tendon is 3000 N and the moment arm of this force about the ankle joint axis is 5 cm, what torque is created by these muscles about the ankle joint?

2. An athlete is doing a knee extension exercise using a 100-N dumbbell strapped to her ankle 40 cm from her knee joint. She holds her leg so that the horizontal distance from her knee joint to the dumbbell is 30 cm.

   a. For this position, what torque is created by the dumbbell about the axis through her knee joint?

   b. If the moment arm of the knee extensor muscles is 4 cm about the knee joint axis, what amount of force must these muscles produce to hold the leg in the position described? Ignore the weight of the leg.

3. A weight lifter is attempting to lift 150 N. The moment arm of this weight about her elbow joint is 25 cm. The force created by the elbow flexor muscles is 2000 N. The moment arm of the elbow flexor muscles is 2 cm. Is the athlete able to lift the weight with this amount of force in her flexor muscles?

4. A 75-kg athlete is standing with his back and heels against the wall. He is doing a shoulder flexor exercise with a barbell. His center of gravity is located 15 cm anteriorly away from the wall when his arm is flexed 90° at the shoulder joint. The farthest the barbell gets from the wall is 1.0 m, when the shoulders are flexed 90°. The length of the athlete's foot is 30 cm.

   a. What is the maximum amount of weight the athlete could hold in this position and still maintain static equilibrium? Assume that his shoulder flexor muscles are strong enough to lift this load.

   b. What torque does the barbell create about the axis of the shoulder joint when the barbell is held in the position described? (Assume the barbell is 1.0 m in front of the shoulder joint.)

c. If the moment arm of the shoulder flexor muscles is 2.5 cm, how much force must the shoulder flexor muscles in each shoulder produce to maintain static equilibrium? Assume the torque is distributed equally between both shoulders and ignore the weight of the arms.

5. Why is it recommended that you flex your knees and hips (squat down) when lifting a heavy load off the floor, rather than just flexing your hips and back (leaning over)?

6. A pole vaulter is running down the runway with the pole. The pole is 5 m long and is tilted upward at an angle of 30° from horizontal. The vaulter's right hand grips the pole in its top end and his left hand grips the pole 1 m below the top end. Although the pole is quite light (its mass is only 2.5 kg), the forces the vaulter must exert on the pole to maintain it in this position are quite large. How large are they?

Angular Kinematics: Describing Objects in Angular Motion

A hammer thrower steps into the throwing circle. After a few windups, the thrower begins rotating as he swings the hammer around and moves across the circle. The turning rate increases as he approaches the front of the circle. Suddenly he lets go of the hammer, and it becomes a fast-moving projectile. The 16-lb steel ball seems to fly forever before it buries itself into the ground with a thud 250 ft away. Wow! How did the thrower's rotary motion cause the hammer to move so fast and go so far linearly when he let go of it? To answer this question, you need to know something about angular kinematics and their relationship to linear kinematics. Both of these topics are discussed in this chapter.

OBJECTIVES

When you finish this chapter, you should be able to do the following:

■ Define relative and absolute angular position and distinguish between the two

■ Define angular displacement

■ Define average angular velocity

■ Define instantaneous angular velocity

■ Define average angular acceleration

■ Define instantaneous angular acceleration

■ Name the units of measurement for angular position, angular displacement, angular velocity, and angular acceleration

■ Explain the relationship between average linear speed and average angular velocity

■ Explain the relationship between instantaneous linear velocity and instantaneous angular velocity

■ Define tangential acceleration and explain its relationship to angular acceleration

■ Define centripetal acceleration and explain its relationship to angular velocity and tangential velocity

This chapter is concerned with angular kinematics, the description of angular motion. Recall that in chapter 5, we described kinematics as a branch of dynamics, which is a branch of mechanics. In chapter 5, we also learned about linear kinematics. Angular kinematics is the other branch of kinematics. Angular motion occurs when all points on an object move in circular paths about the same fixed axis. Angular motion is important because most human movements are the result of angular motions of limbs about joints. An understanding of how angular motion is measured and described is important.

Angular Position and Displacement

Before further discussion, a definition of *angle* is needed. What is an angle? An angle is formed by the intersection of two lines, two planes, or a line and a plane. The term angle refers to the orientation of these lines or planes to each other (figure 9.1).

An angle is formed by the intersection of two lines, two planes, or a line and a plane.

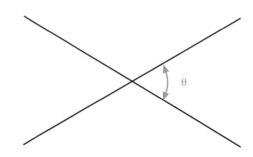

Figure 9.1	An angle (θ) formed by the intersection of two lines.

In figure 9.1, the Greek letter θ (theta) is used to represent the angle formed by the intersections of the lines and planes. Greek letters are used to represent many of the terms used in angular kinematics.

Angular Position

Angular position refers to the orientation of a line with another line or plane. If the other line or plane is capable of moving, the angular position is a relative angular position. If the other line or plane is fixed and immovable relative to the earth, the angular position is an absolute angular position. The angle your forearm makes with your upper arm describes a relative angular position of your forearm, or your elbow joint angle. Angles formed by limbs at joints describe relative angular positions of limbs.

The angle your forearm makes with a horizontal plane describes the absolute angular position of your forearm, because a horizontal plane is a fixed reference. These absolute angular positions will be our primary concern in this and the next chapter.

What units of measurement are used for angles? You probably are most familiar with measuring angles in degrees, but there are other units of measurement for angles besides degrees, just as there are other units of measurement for linear distance besides feet. If we wished to measure the absolute angle of line AB, as shown in figure 9.2a, with a horizontal plane, we might imagine a horizontal line BC with the same length as AB. We could then draw a circle using the length of AB as the radius and the point B as the center (figure 9.2b). The angle of ABC could then be described as the fraction of the circle created by the pie ABC. An angle of 1° represents 1/360 of a circle because there are 360° in a circle.

Another way to describe an angle is to measure how many radii are in the arc length AC if one radius is equal to the length of line segment BC or AB. In other words, if the length BC represents 1 radius, how long is the arc from A to C measured using this radius as the unit of measure? Mathematically, an angle measured in radius units (we'll call these radians) is:

$$\theta = \frac{\text{arc length}}{r} = \frac{\ell}{r} \qquad (9.1)$$

where

θ = angular measurement in radians

ℓ = arc length

r = radius

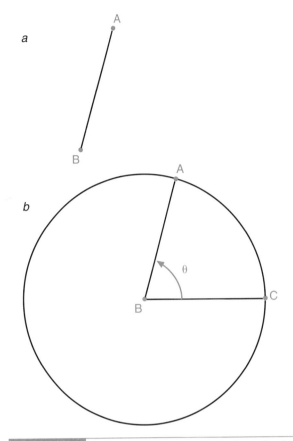

| Figure 9.2 | A circle is useful in describing the angle of a line (*a*) if the center of the circle coincides with the intersection of the lines defining the angle (*b*). |

The radian (abbreviated rad) unit of measure for an angle is really a ratio of arc length divided by the radius.

Figure 9.3 graphically shows the definition of an angle of 1 rad, π (pi) rad, and 2π rad. If you recall from geometry, the circumference of a circle is 2πr, so there are 2π rad in a circle or 2π rad in 360°. These conversions are shown below:

$$\frac{360°}{\text{circle}} = \frac{2\pi \text{ rad}}{\text{circle}} = \frac{6.28 \text{ rad}}{\text{circle}}$$

$$\frac{360°}{2\pi \text{ rad}} = 57.3° / \text{rad}$$

Angular Displacement

Angular displacement is the angular analog of linear displacement. It is the change in absolute angular position experienced by a rotating line. Angular displacement is thus the angle formed between the final position and the initial position of a rotating

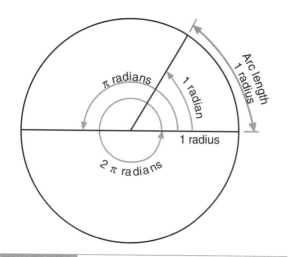

Figure 9.3   The relationships among the radius of a circle, the arc length along a circle, and an angle measured in radians.

line. (We often speak of the angular displacement of an object when the object is not a line. To measure such a displacement, choose any two points on the object. Imagine a line connecting these two points. If the object is rigid, the angular displacement of this line segment is identical to the angular displacement of the object.)

Angular displacement is the change in absolute angular position experienced by a rotating line.

As with linear displacement, angular displacement has direction associated with it. How is angular direction described? *Clockwise* and *counterclockwise* are common terms used to describe the direction of rotation. The hands of a clock rotate in a clockwise direction when you view a clock face from the front. If you could view a clock face from the back, would you still see the hands rotating in a clockwise direction? Because your viewpoint changed, the clock hands would now appear to be rotating counterclockwise! They didn't change the direction of rotation; you just changed your viewing position. If you describe an angular displacement as clockwise, a person must know your viewing position to be certain of the direction of the angular displacement.

One way of overcoming this possible source of confusion is to first identify the axis of rotation and the plane in which the part rotates. The axis of rotation is always perpendicular to the plane in which the motion occurs. This axis of rotation is like the axle of a bicycle wheel, and the spokes of the wheel lie in the plane of motion. Along the axis of rota-

tion, establish a positive direction. If you situate the thumb of your right hand so that it points in the positive direction along the axis of rotation, the direction in which your fingers curl is the positive direction of rotation. This is called the right-hand thumb rule.

Now consider the direction of rotation of the hands of a clock. The plane of their motion is the clock face, and the axis of rotation is a line through the plane of the clock face. If the positive direction along this axis is established as pointing out of the clock face toward you, the positive direction of rotation is counterclockwise as you face the clock. Check this by pointing your right thumb out away from the clock in the positive direction of the axis of rotation. Then observe the direction in which your fingers curl—counterclockwise (see figure 9.4).

Most screws, nuts, and bolts have right-handed threads. They follow the right-hand thumb rule. If you point your right thumb in the direction you want the screw or nut to move, your right fingers curl in the direction the screw or nut must be turned. This sign convention also applies to torques and measures of angular position.

Now let's consider how angular displacement is measured. A pitcher is being evaluated for shoulder joint range of motion. The measurement begins with the pitcher's arm at his side, as shown in figure 9.5. The pitcher then raises it away from his side as far as he can (he abducts his shoulder). What is the angular displacement of his arm?

The axis of rotation is the anteroposterior axis, a line through the shoulder joint with the positive direction pointing out of the page toward us. The

Figure 9.4   The right-hand thumb rule. The direction the fingers curl indicates the positive angular direction if the right thumb points in the positive linear direction along the axis of rotation.

Figure 9.5 Angular displacement of a pitcher's arm at the shoulder joint around the antero-posterior axis.

plane of motion is the frontal plane—the plane formed by his arms, legs, and trunk. If the initial position of the arm is 5° from the vertical and its final position is 170° from the vertical, the angular displacement is

$$\Delta\theta = \theta_f - \theta_i \qquad (9.2)$$
$$\Delta\theta = 170° - 5°$$
$$\Delta\theta = +165°$$

where

$\Delta\theta$ = angular displacement

θ_f = final angular position

θ_i = initial angular position

The displacement is positive because the rotation was in the direction in which the fingers of your right hand curl if your thumb is pointed in the posi-

tive direction of the axis of rotation (anteriorly and out of the shoulder).

As a coach or teacher, you will rarely measure angular displacements precisely, but in some sports, angular displacement is an important part of the skill. The number of twists or somersaults done in diving, gymnastics, or figure skating is a measure of angular displacement and plays an important role in how many points a judge awards. The angular displacement of a swing (range of motion) in golf or tennis affects the manner in which the ball is hit in these sports.

Angular and Linear Displacement

In chapter 8, we discovered that our muscles must produce very large forces to lift modest loads. The reason for this is that most muscles attach to bones close to the joint; thus they have small moment arms about the joint. Because of the small moment arm, large forces must be produced by the muscles to produce modest torques about the joints. The muscles are at a mechanical disadvantage for producing torque. Is there any advantage to this arrangement?

How far any point on the arm moves when you flex your arm, as in the self-experiment on page 162, is dependent on how far that point is from the elbow. This relationship is actually apparent in the definition of an angle measured in radians, as shown in equation 9.1:

$$\theta = \frac{\text{arc length}}{r} = \frac{\ell}{r} \qquad (9.1)$$

For our example, then, let's use an angular displacement measured in radians, and

$$\Delta\theta = \frac{\ell}{r} = \frac{\ell_a}{r_a} = \frac{\ell_b}{r_b}$$
$$\frac{r_a}{r_b} = \frac{\ell_a}{\ell_b} \qquad (9.3)$$

If the angular displacement of the arm was 1 rad, r_a was 10 in., and r_b was 1 in., then

$$\Delta\theta = \frac{\ell}{r}$$
$$\ell_a = \Delta\theta r_a = (1 \text{ rad})(10 \text{ in.}) = 10 \text{ in.}$$
$$\frac{r_a}{r_b} = \frac{\ell_a}{\ell_b}$$
$$\ell_b = \ell_a\left(\frac{r_b}{r_a}\right) = 10 \text{ in.}\left(\frac{1 \text{ in.}}{10 \text{ in.}}\right) = 1 \text{ in.}$$

SELF-EXPERIMENT

Let's look at the forearm and elbow again. Place your forearm on the desk or table in front of you. Now flex at the elbow and lift your hand off the table. Move it toward your shoulder as far as you can while keeping your elbow on the desktop, as shown in figure 9.6.

All the parts of the arm underwent the same angular displacement, but which moved farther, your hand (point A in figure 9.6) or the point of attachment of your biceps muscle (point B in figure 9.6)? Obviously, your hand moved farther. The linear distance it traveled [arc length AA (ℓ_a)] and its linear displacement [chord AA (d_a)] are larger than the linear distance traveled [arc length BB (ℓ_b)] and the linear displacement [chord BB (d_b)] of the biceps insertion.

In this example, the hand moves 10 times as far as the insertion of the biceps tendon. The linear distance (arc length) traveled by a point on a rotating object is directly proportional to the angular displacement of the object and the radius, the distance that point is from the axis of rotation of the object. If the angular displacement is measured in radians, the linear distance traveled (the arc length) is equal to the product of the angular displacement and the radius. This is only true if the angular displacement is measured in radians. This relationship is expressed mathematically in equation 9.4:

$$\ell = \Delta\theta r \qquad (9.4)$$

where

ℓ = arc length

$\Delta\theta$ = angle measured in radians

r = radius

The linear displacement of a point on a rotating object is also directly proportional to the distance that point is from the axis of rotation (the radius). This linear displacement is also related to the angular displacement, but it is not directly proportional to the angular displacement. The relationship is more complex.

The relationship between linear displacement and radius is easily seen in figure 9.7. The linear displacement and radii form sides of similar triangles, as shown in figure 9.7. From this, the following relationship can be established:

$$\frac{r_a}{r_b} = \frac{d_a}{d_b} \qquad (9.5)$$

| Figure 9.6 | The distance the hand or wrist (A) moves (ℓ_a or d_a) when your elbow (B) flexes is greater than the distance the insertion point of the biceps moves (ℓ_b or d_b). The ratio of these distances to each other is the same as the ratio of the radius r_a to the radius r_b. |

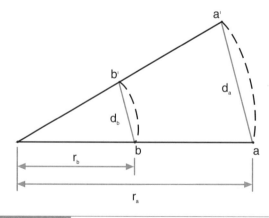

| Figure 9.7 | The ratio of the linear displacements of two points on a rotating object ($d_a \div d_b$) is equal to the ratio of the radii of these two points from the axis of rotation ($r_a \div r_b$). |

where

r = radius

d = linear displacement (or chord length)

One advantage of muscle insertions close to joints should now be clear. The muscle only has to contract and shorten a short distance to produce a large movement (linear displacement) at the end of the limb. Because the distance a muscle can shorten is limited to approximately 50% of its resting length, the range of motion at a joint would also be further limited if muscles inserted farther away from the joints.

This concept is useful when we use sports implements. Small movements of the hands produce large linear displacements at the end of a putter, a badminton racquet, a fishing pole, or a hockey stick, and so on.

Angular Velocity

Angular velocity is defined as the rate of change of angular displacement. Its units of measurement are radians per second (rad/s), degrees per second (°/s), revolutions per minute (rpm), and so on. Angular velocity is abbreviated with the Greek letter omega (ω). Angular velocity is a vector quantity, just like linear velocity, so it has direction associated with it. The direction of an angular velocity is determined using the right-hand thumb rule, as with angular displacement. Because angular velocity is a vector, a change in the size of the angular velocity or in the direction of its axis of rotation results in a change in angular velocity.

> **Angular velocity is defined as the rate of change of angular displacement.**

Average angular velocity is computed as the change in angular position (angular displacement) divided by time. Mathematically,

$$\overline{\omega} = \frac{\Delta\theta}{\Delta t} = \frac{\theta_f - \theta_i}{\Delta t} \tag{9.6}$$

where

$\overline{\omega}$ = average angular velocity

$\Delta\theta$ = angular displacement

Δt = time

θ_f = final angular position

θ_i = initial angular position

If one is concerned with how long it takes for something to rotate through a certain angular displacement, average angular velocity is the important measure. If one is concerned with how fast something is rotating at a specific instant in time, instantaneous angular velocity is the important measure. Instantaneous angular velocity is an indicator of how fast something is spinning at a specific instant in time. The tachometer on a car's instrument panel gives a measure of the engine's instantaneous angular velocity in revolutions per minute.

The average angular velocity of a batter's swing may determine whether or not he contacts the ball, but it is the bat's instantaneous velocity at ball contact that determines how fast and how far the ball will go. Similar situations exist in all racket sports and striking activities. To gymnasts, divers, and figure skaters, average angular velocity is the more important measure. It determines whether or not they will complete a certain number of twists or somersaults before landing or entering the water.

Angular and Linear Velocity

In several sports, especially ball sports, implements are used as extensions to the athletes' limbs. Golf, tennis, squash, lacrosse, racquetball, badminton, field hockey, and ice hockey are examples of these sports. One advantage of using these implements was described earlier—they amplify the motion (displacement) of our limbs. Now compare the velocities of the ball (or shuttlecock or puck) in each of these sports if it were thrown by hand versus if it were struck (or thrown) with the respective stick (or racket or club). Which is faster? The implements enable us to impart faster linear velocities to the ball (or shuttlecock or puck) in each of these sports. This is another advantage of using a stick, racket, or club. How is this accomplished?

Deriving the Relationship Between Linear and Angular Velocity

The relationship between angular displacement and linear distance traveled provides the answer. Consider a swinging golf club. All points on the club undergo the same angular displacement, and thus the same average angular velocity, because they all take the same time to undergo that displacement. But a point on the club closer to the club head (and farther from the axis of rotation) moves through a longer arc length than a point farther from the club head (and closer to the axis of rotation). Both points travel their respective arc lengths in the same time. The point farther from the axis of rotation must have a faster linear speed because it moves a longer distance but in the same time. Mathematically, this relationship can be derived from the relationship between angular displacement and linear distance traveled (arc length).

$$\ell = \Delta\theta r \qquad (9.4)$$

Dividing both sides by the time it takes to rotate through the displacement gives:

$$\frac{\ell}{t} = \frac{\Delta\theta r}{t}$$
$$\bar{s} = \bar{\omega}r \qquad (9.7)$$

where

ℓ = arc length

r = radius

$\Delta\theta$ = angular displacement (measured in radians)

t = time

$\bar{s}$ = average linear speed

$\bar{\omega}$ = average angular velocity (measured in radians per second)

The average linear speed of a point on a rotating object is equal to the average angular velocity of the object times the radius (the distance from the point on the object to the axis of rotation of the object).

At an instant in time, this relationship becomes:

$$v_T = \omega r \qquad (9.8)$$

where

v_T = instantaneous linear velocity tangent to the circular path of the point

ω = instantaneous angular velocity (measured in radians per second)

r = radius

> **The average linear speed of a point on a rotating object is equal to the average angular velocity of the object times the radius.**

The instantaneous linear velocity of a point on a rotating object is equal to the instantaneous angular velocity of the rotating object times the radius. The direction of this instantaneous linear velocity

is perpendicular to the radius and tangent to the circular path of the point. The instantaneous linear velocities for the two points on the golf club are shown in figure 9.8.

Applying the Relationship Between Linear and Angular Velocity

The sticks, clubs, and rackets used in the sports mentioned at the beginning of this section all utilize this relationship between linear and angular velocity and radius. The linear velocity of a point farther from the axis of rotation is faster if the angular velocity is the same. The woods, the longest clubs in a golf bag, are thus used for imparting a faster velocity to the ball to make it travel farther, whereas the shorter clubs are used for closer shots. If a batter can swing a longer bat with the same angular velocity as a shorter one, he'll be able to hit the ball farther with the longer bat. Gripping a tennis (or racquetball or squash) racket farther down on the grip (farther from the racket face) increases the radius of the swing and the linear velocity of the racket's striking surface (if the player is still able to swing the racket with the same angular velocity).

The implements used in these sports and other striking activities effectively lengthen the limbs of

Figure 9.8 The linear velocity of the clubhead (V_b) is faster than the linear velocity of a point on the shaft (V_a) because the clubhead is farther from the axis of rotation.

SELF-EXPERIMENT

Try this experiment to demonstrate the relationship between linear velocity, angular velocity, and radius. Take out the ruler you used previously. Line up five pennies in a row on a table alongside the edge of the ruler. Fix one end of the ruler by pushing down on it with your finger. This point will be the pivot or axis of rotation of the ruler. Now draw back the other end of the ruler and then strike it so that it swings forward and hits the pennies. Which penny went farthest and fastest? The penny farthest from the pivot end of the ruler went farthest and fastest. The average angular velocity of all points on the ruler was the same, but the radius, r, for the part of the ruler that hit this penny was largest, so this penny had the greatest linear velocity.

Let's try another experiment. Place the ruler on a pencil so that the pencil is at the 2-in. mark. Put a penny on the ruler at the 11-in. mark and another penny at the 4-in. mark. Now strike the ruler with your hand at the 0-in. end. Which penny flew higher and moved faster? The penny farthest from the axis of rotation (the pencil) went higher and faster. Now do some experimenting on your own to see how different locations of the axis of rotation affect the flight of a penny.

the performer, but the techniques used by the performer may increase the effective radius of the striking implements even further. Consider a golf swing. Where is the axis of rotation for the club during the swing? Actually, the club and arms act together as one long lever, and the axis of rotation at ball contact lies along the vertebral column in the trunk, as shown in figure 9.9. The effective radius of the club head is thus much longer than just the length of the club itself.

Consider other sports and human movement activities. Where is the axis of rotation and how long is the effective radius while serving in tennis? While chopping wood with a long-handled ax? While swinging a baseball bat? While bowling? While throwing a ball? While throwing a punch in boxing? While kicking a ball? While spiking a volleyball? While executing a badminton smash? While hitting a kill shot in racquetball? In many of these examples, the axis of rotation lies within your body, often between the shoulders.

The relationship among linear and angular velocities and radius of rotation also explains another advantage of the insertion of our muscles close to the joints. Consider the insertion of the biceps on the forearm. If it inserts only 1 in. from the elbow joint, and the hand is 10 in. from the joint, when the arm rotates, the linear velocity of the hand will be 10 times as fast as the linear velocity of the biceps insertion. As the velocity of muscular contraction

is limited, the insertion of muscles close to the joint allows a relatively slow contraction velocity of the muscle to be amplified at the end of a limb. Our hands and feet can move at linear velocities much

Figure 9.9 The axis of rotation and effective radius of the club head during a golf swing.

faster than the maximum velocities of the muscles that control the movements of the arm and leg.

Angular Acceleration

Angular acceleration is defined as the rate of change of angular velocity. Its units of measurement are radians per second per second (rad/s/s), degrees per second per second (°/s/s), or some unit of angular velocity per unit of time. Angular acceleration is abbreviated with the Greek letter α (alpha). Just like linear acceleration, angular acceleration is a vector quantity; it has direction and size. The right-hand thumb rule is used to describe directions of angular acceleration vectors.

> **Angular acceleration is defined as the rate of change of angular velocity.**

Average angular acceleration is computed as the change in angular velocity divided by time. Mathematically,

$$\bar{\alpha} = \frac{\Delta\omega}{\Delta t} = \frac{\omega_f - \omega_i}{\Delta t} \qquad (9.9)$$

where

$\bar{\alpha}$ = average angular acceleration

$\Delta\omega$ = change in angular velocity

Δt = time

ω_f = final angular velocity

ω_i = initial angular velocity

Angular acceleration occurs when something spins faster and faster or slower and slower, or when the spinning object's axis of spin changes direction.

> **Angular acceleration occurs when something spins faster and faster or slower and slower, or when the spinning object's axis of spin changes direction.**

Angular and Linear Acceleration

When the angular velocity of a spinning object increases, the linear velocity of a point on the object increases as well. The angular and linear acceleration of a point on a spinning object are related.

Tangential Acceleration

The component of linear acceleration tangent to the circular path of a point on a rotating object is called the tangential acceleration of that point. This linear acceleration is related to the angular acceleration of the object in the following way:

$$a_T = \alpha r \qquad (9.10)$$

where

a_T = instantaneous tangential acceleration

α = instantaneous angular acceleration (measured in rad/s/s)

r = radius

A point on a rotating object undergoes a linear acceleration tangent to its rotational path and equal to the angular acceleration of the object times the radius.

Centripetal Acceleration

Equation 9.10 indicates that a linear acceleration tangent to the path of rotation of a point occurs if the rotating object is being accelerated angularly. What if no angular acceleration occurs? Does a point on a rotating object experience any linear acceleration if the object spins at a constant angular velocity with no angular acceleration? Yes. Remember that linear acceleration occurs if something speeds up, slows down, or changes direction. A point on an object spinning at constant angular velocity doesn't speed up or slow down, but because the point follows a circular path, it is constantly changing direction and is thus experiencing a constant linear acceleration.

In which direction does this linear acceleration occur? Tie a weight tightly onto a sturdy string and spin it around so that it circles with a constant angular velocity, as shown in figure 9.10.

Which way must you pull on the string to keep the weight traveling in a circle? You pull on the string with a force directed toward the center of the circle, toward the axis of rotation. This must be the force that causes the linear acceleration of the weight, so the linear acceleration must be directed toward the center of the circle as well. This acceleration is called centripetal acceleration (or radial acceleration), and the force causing it is called centripetal force.

Mathematically, centripetal acceleration can be defined using two different equations:

Figure 9.10 Tie a weight on a string and swing it around in a circle. Which way do you pull on the string?

$$a_r = \frac{v_T^2}{r} \qquad (9.11)$$

$$a_r = \omega^2 r \qquad (9.12)$$

where

a_r = centripetal acceleration

v_T = tangential linear velocity

r = radius

ω = angular velocity

Centripetal acceleration is the linear acceleration directed toward the axis of rotation. It is directly proportional to the square of the tangential linear velocity and the square of the angular velocity. If angular velocity is held constant, centripetal acceleration is directly proportional to the radius of rotation. If the tangential linear velocity is held constant, centripetal acceleration is inversely proportional to the radius of rotation.

> **Centripetal acceleration is the linear acceleration directed toward the axis of rotation.**

When you run around a curve on the inside lane of a running track, you experience more centripetal acceleration and thus require more friction on your shoes to exert centripetal force on yourself than you would if you ran at the same linear speed in the outside lane (see figure 9.11). In these two cases, tangential linear velocity would be the same, but the radius of rotation would be larger in the outside lane. Equation 9.11 ($a_r = v_T^2/r$) would be the

appropriate equation to use to evaluate this situation because the linear velocities are the same.

On the other hand, a hammer thrower swinging a hammer with a 1.0-m chain (see figure 9.12a) would have to pull on the chain with greater force than if he swung a hammer with a 0.75-m chain at the same angular velocity (see figure 9.12b). The centripetal acceleration is greater (and thus the centripetal force provided by the thrower must be greater) for the hammer with the 1.0-m chain than for the one with the shorter chain because the radius of rotation is larger for the hammer with the 1.0-m chain. Equation 9.12 ($a_r = \omega^2 r$) would be the appropriate equation to use to evaluate this situation because the angular velocities are the same.

You can try this for yourself by turning in a circle while holding this book at arm's length and then pulling it closer to you while still turning at the same angular velocity. You have to exert a greater force on the book when it is farther from the axis of rotation.

Summary

Angular kinematics is concerned with the description of angular motion. Angles describe the orientation of two lines. Absolute angular position refers to the orientation of an object relative to a fixed reference line or plane, such as horizontal or vertical. Relative angular position refers to the orientation of an object relative to a nonfixed reference line or plane. Joint angles are relative, whereas limb positions may be relative or absolute.

When an object rotates, it undergoes an angular displacement. To define the angular displacement,

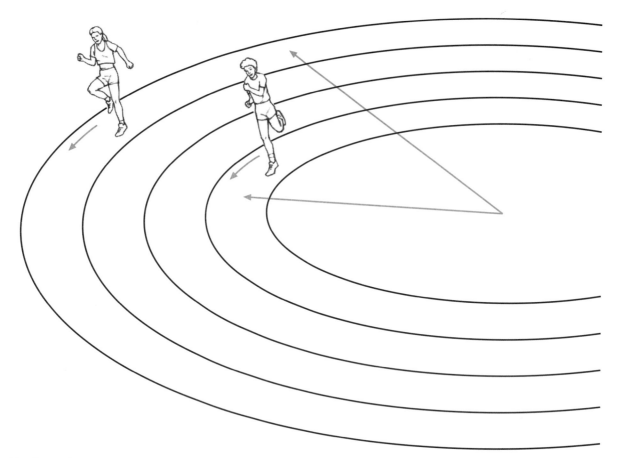

Figure 9.11   A runner in the inside lane must exert more centripetal force via friction than a runner in the outside lane if both runners have the same linear velocity ($a_{r_1} < a_{r_2}$ if $v_1 = v_2$). The centripetal acceleration of the runner in the inside lane is greater due to the smaller radius ($r_1 > r_2$).

Figure 9.12   A hammer thrower using a hammer with a 1.0 m chain (*a*) must exert a greater centripetal force than a thrower using a 0.75 m chain (*b*) if both hammers rotate with the same angular velocity ($a_{r_1} > a_{r_2}$ since $\omega_1 = \omega_2$ and $r_1 > r_2$). The centripetal acceleration of the hammer with the 1.0 m chain is greater due to the larger radius.

the axis and plane of rotation must be known. The direction of the angular displacement (and all other angular motion and torque vectors) is then established using the right-hand thumb rule. The definitions of angular displacement, angular velocity, and angular acceleration are similar to their linear counterparts.

The linear displacement and distance traveled by a point on a rotating object are directly proportional to the radius of rotation. The linear distance traveled equals the product of the angular displacement measured in radians times the radius of rotation.

Tangential linear velocity and acceleration of a point on a rotating object are directly proportional to the radius as well. The tangential linear velocity is equal to the product of the angular velocity times the radius of rotation. Lengthening the radius while maintaining the angular velocity is an important principle in a variety of striking skills. Tangential linear acceleration is equal to the product of angular acceleration times the radius of rotation.

Centripetal acceleration (also called radial acceleration) of an object rotating in a circular path is the component of linear acceleration directed toward the axis of rotation. It is directly proportional to the square of the tangential linear velocity or the square of the angular velocity. Centripetal force is the force exerted on the rotating object to cause the centripetal acceleration.

Key Terms

absolute angular position (p. 159)
angular acceleration (p. 166)
angular displacement (p. 159)
angular position (p. 159)
angular velocity (p. 163)
average angular velocity (p. 163)
centripetal acceleration (p. 166)
centripetal force (p. 166)
instantaneous angular velocity (p. 163)
radian (p. 159)
relative angular position (p. 159)
tangential acceleration (p. 166)

REVIEW QUESTIONS

1. A therapist examines the range of motion of an athlete's knee joint during rehabilitation from a knee injury. At full extension, the angle between the shank and thigh is 178°. At full flexion, the angle between the shank and thigh is 82°. During the test, the thigh was held in a fixed position and only the shank moved. What was the angular displacement of the shank from full extension to full flexion? Express your answer in degrees and in radians.

2. Figure skater Michelle Kwan performs a triple twisting jump. She rotates around her longitudinal axis three times while in the air. The time it takes to complete the jump from takeoff to landing is 0.8 s. What was Michelle's average angular speed in twisting for this jump?

3. When discus thrower Anthony Washington begins his throwing motion, he spins with an angular velocity of 5 rad/s. Just before he releases the discus, Anthony's angular velocity is 25 rad/s. If the time from the beginning of the throw to just before release is 1 s, what is Anthony's average angular acceleration?

4. The tendons from my knee extensor muscles attach to my tibia 1.5 in. away from the center of my knee joint, and my foot is 15 in. away from my knee joint. What

arc length does my foot move through when my knee extensor muscles contract and their point of insertion on my tibia moves through an arc length of 2 in.?

5. Hammer thrower Lance Deal spins with an angular velocity of 1700°/s. The distance from his axis of rotation to the hammerhead is 1.2 m.

   a. What is the linear velocity of the hammerhead?

   b. What is the centripetal acceleration of the hammerhead?

6. In golf, the longest club is the driver and the shortest club is the pitching wedge. Why is it easier to hit the ball farther with a driver than with a pitching wedge?

7. A hook in boxing primarily involves horizontal flexion of the shoulder while maintaining a constant angle at the elbow. During this punch, the horizontal flexor muscles of the shoulder contract and shorten at an average speed of 75 cm/s. They move through an arc length of 5 cm during the hook, whereas the fist moves through an arc length of 100 cm. What is the average speed of the fist during the hook?

8. What advantages does a longer-limbed individual have in throwing and striking activities?

9. Explain how step length (a linear kinematic variable) might increase if stretching exercises increase the range of motion at the hip joint, thus increasing the angular displacement (an angular kinematic variable) of the hip joint during a step.

10. A baseball pitcher pitches a fastball with a horizontal velocity of 40 m/s. The horizontal distance from the point of release to home plate is 17.50 m. The batter decides to swing the bat 0.30 s after the ball was released by the pitcher. The average angular velocity of the bat is 12 rad/s. The angular displacement of the bat from the batter's shoulder to hitting positions above the plate is between 1.5 and 1.8 rad.

    a. Will the bat be in a hitting position above the plate when the ball is above the plate? Assume that the pitch is in the strike zone.

    b. Assume that the batter does hit the ball. If the bat's instantaneous angular velocity is 30 rad/s at the instant of contact, and the distance from the sweet spot on the bat to the axis of rotation is 1.25 m, what is the instantaneous linear velocity of the sweet spot at the instant of ball contact?

Angular Kinetics: Explaining the Causes of Angular Motion

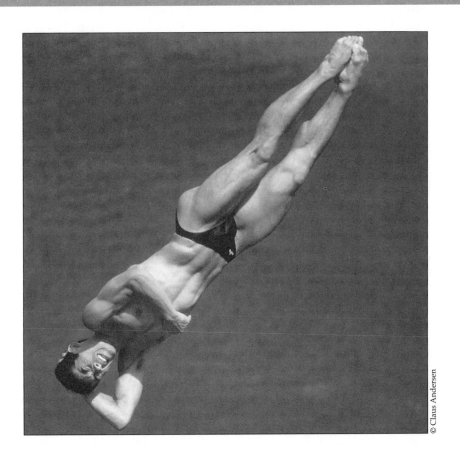

© Claus Andersen

A diver jumps off the 3-m platform. He is in a layout (a stretched-out position) and barely rotating at first. Then he flexes at the hips and he pikes (folds himself in half), and his rotation speeds up as if by magic. He spins three and a half times in this position as he falls toward the water. When he is just above the water, his spinning seems to stop as he opens up into a layout and enters the water with barely a splash. It seemed as if the diver could make himself spin faster or slower at will during the dive. How did he control his angular velocity (his rate of spin)? The information contained in this chapter should help us answer this question.

When you finish this chapter, you should be able to the following:

▌ Define moment of inertia

▌ Explain how the human body's moment of inertia may be manipulated

▌ Explain Newton's three laws of motion as they apply to angular motion

▌ Define angular impulse

▌ Define angular momentum

▌ Explain the relationship between angular impulse and angular momentum

This chapter is about the subbranch of mechanics called kinetics. Specifically, it is about angular kinetics, or the causes of angular motion. Many of the concepts developed in linear kinetics have similar counterparts in angular kinetics. The concepts learned in chapter 6 will thus be important for a good understanding of the concepts introduced in this chapter. An understanding of angular kinetics will help explain why a discus thrower spins a discus; why a long jumper "runs" in the air; how figure skaters, divers, and gymnasts can speed up or slow down their rate of spin when they're not in contact with the ground; and some aspects of the techniques used in a variety of other skills in sports and human movement.

Angular Inertia

In chapter 3, inertia was defined as the property of an object that resists changes in motion. Linear inertia was quantified as an object's mass. It is more difficult to speed up, slow down, or change the direction of a more massive object because it has more linear inertia. The property of an object to resist changes in its angular motion is angular inertia or rotary inertia. It is more difficult to speed up or slow down the rotation or change the axis of rotation of an object with more angular inertia. What factors affect the angular inertia of an object? How is angular inertia quantified?

The property of an object to resist changes in its angular motion is angular inertia or rotary inertia.

To learn some things about angular inertia, try the self-experiment on page 173. What characteristic of the book changed to cause this difference in angular inertia? The mass of the book stayed the same, but where the mass was located relative to the axis of rotation changed. When the book was opened to page 10, most of the mass of the book was on one side and close to the center of gravity and the axis of rotation. When the book was opened to the middle, about half of the book's mass was on each side of the spine. The center of gravity and axis of rotation was then through the spine of the book, and more mass was located farther from the axis of rotation. Angular inertia is affected by mass and how the mass is distributed relative to the axis of rotation. Again, consider the various sports implements and tools that you swing; the longer implements are more difficult to start or stop swinging. A longer bat is harder to swing than a shorter one.

Mathematical Definition of Angular Inertia

The quantity that describes angular inertia is called moment of inertia. Moment of inertia is abbrevi-

SELF-EXPERIMENT

Try the following experiments. Take this book and throw it up in the air so that it turns one revolution about an axis perpendicular to its cover (see figure 10.1). Now find a heavier book (or a lighter one) and do the same trick. Which book was easier to flip? Which book has less angular inertia? The lighter book was easier to flip because its angular inertia is smaller. Angular inertia is affected by the mass of an object. In fact, it is directly proportional to the mass of the object. If you consider the various sports implements and tools that you swing, the heavier implements are more difficult to start or stop swinging. A heavier bat is harder to swing than a lighter one.

Let's try another experiment. Open this book to page 10 (or some page close to the beginning of the book) and place it on a table with the open pages facing down. Apply a force couple to the book with your fingers and spin it on the table. Try this several times to see how large a torque you must apply to rotate the book through an angular displacement of 180°, or one-half revolution.

Now open the book to the middle pages and repeat the experiment. Try spinning it several times. In which configuration did the book turn more easily—when it was opened to the front or when it was opened to the middle? When the book was opened to page 10, it was easier to rotate. It took more torque to cause the same rotation of the book when it was opened to the middle. In this configuration, the book had more angular inertia.

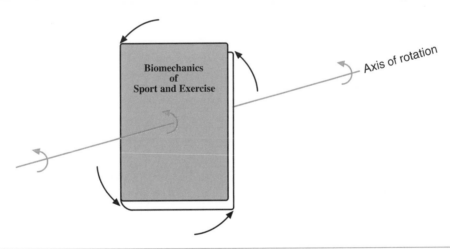

Figure 10.1   Spin a book about an axis of rotation perpendicular to its cover. Is a heavier or lighter book easier to spin?

ated with the letter I. Theoretically, an object may be considered to be composed of many particles of mass. The moment of inertia of such an object can be defined mathematically as follows:

$$I_a = \Sigma m_i r_i^2 \qquad (10.1)$$

where

I_a = moment of inertia about axis a

Σ = summation symbol

m_i = mass of particle i

r_i = radius (distance) from particle i to axis of rotation a

Each particle provides some resistance to change in angular motion. This change is equal to the mass of the particle times the square of the radius from the particle to the axis of rotation. The sum of all the particles' resistances to rotation is the total moment of inertia of the object. The units of measurement for moment of inertia are units of mass times units of length squared, or $kg \cdot m^2$ in SI units.

Moment of inertia may also be represented mathematically as:

$$I_a = mk_a^2 \qquad (10.2)$$

where

I_a = moment of inertia about axis a

m = mass of the object

k_a = radius of gyration about axis a

The radius of gyration is a length measurement that represents how far from the axis of rotation all of the object's mass must be concentrated to create the same resistance to change in angular motion as the object had in its original shape.

Whereas linear inertia is dependent on only one variable (mass), angular inertia is dependent on two variables: mass and the distribution of the mass. These two variables do not have equal effects on the moment of inertia. The influence of mass on angular inertia is much less than the influence of the distribution of the mass, because in equation 10.2, the radius of gyration is squared and mass is not. Doubling the mass of an object would double its moment of inertia, but doubling its radius of gyration would quadruple its moment of inertia. A batter who switches to a longer bat will have a more difficult time swinging it than if he switched to a heavier one.

Moments of Inertia About Different Axes

Objects have only one linear inertia, and it is described by their mass. Objects may have many different moments of inertia, though, because an object may rotate about many different axes of rotation.

> **Objects may have many different moments of inertia, though, because an object may rotate about many different axes of rotation.**

SELF-EXPERIMENT

Try these experiments. Take this book and flip it so that it rotates about an axis perpendicular to the front and back covers of the book, as you did earlier (see figure 10.1). Now hold the book so that the covers are parallel to the floor and flip it about an axis parallel to the spine of the book, as shown in figure 10.2.

Did it take the same effort to flip the book in both cases? No. The book was harder to flip in the first case. Its moment of inertia is larger about an axis perpendicular to its covers than about an axis parallel to its spine.

The book (or any object) has an infinite number of possible axes of rotation, so it also has an infinite number of moments of inertia. If an object is not symmetrical about all planes through its center of gravity, there will be one axis of rotation about which the moment of inertia is largest and one axis of rotation about which the moment of inertia is smallest. These two axes will always be perpendicular to each other and are the principal axes of the object. The third principal axis of an object is the axis perpendicular to these first two axes.

For this book, the largest moment of inertia is about an axis perpendicular to the book cover, and the smallest moment of inertia is about an axis parallel to the spine of the book. These are two of the principal axes of the book. The third principal axis is perpendicular to these two and is parallel to the pages of the book and perpendicular to the spine. The three principal axes are shown in figure 10.3.

The principal axes of the human body depend on the position of the limbs. For a standing position with the arms at the sides, the principal axes are shown in figure 10.4. These axes correspond to the anteroposterior, transverse, and longitudinal axes.

Figure 10.2 Spin a book about an axis of rotation parallel to its spine. Is it easier or harder to spin about this axis than about an axis perpendicular to its cover?

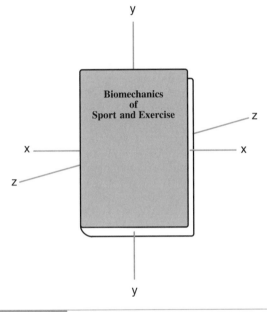

Figure 10.3 The three principal axes of rotation for a book.

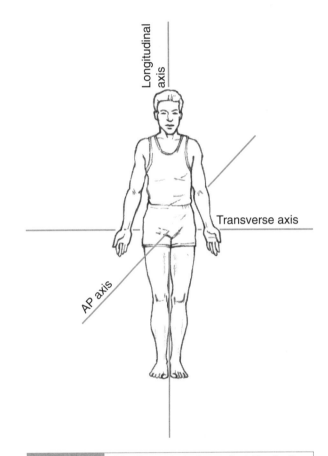

Figure 10.4 The three principal axes of rotation for the human body in anatomical position.

Manipulating the Moments of Inertia of the Human Body

A rigid object has many different moments of inertia because it may have many axes of rotation. But for any one axis of rotation, only one moment of inertia is associated with that axis. The human body is not a rigid object, though, because humans can move their limbs relative to each other. These movements may change the distribution of mass about an axis of rotation, thus changing the moment of inertia about that axis. A human's moment of inertia about any axis is variable. There is more than one value for the moment of inertia about an axis. This means that humans can manipulate their moments of inertia. A figure skater spinning about a longitudinal axis with her arms at her sides can more than double her moment of inertia about this longitudinal axis by abducting her arms to shoulder level.

A diver in a layout position and somersaulting about a transverse axis can reduce his moment of inertia to less than half that of the layout position by tucking as shown in figure 10.5.

A sprinter flexes her leg at the knee and hip when she accelerates angularly during the swing-through. This reduces the moment of inertia of the leg about the hip joint, where the axis of rotation is located (see figure 10.6).

In pole vaulting, the vaulter rotates about the top end of the pole where he grips it, and the pole itself (with the vaulter on it) rotates about the bottom end

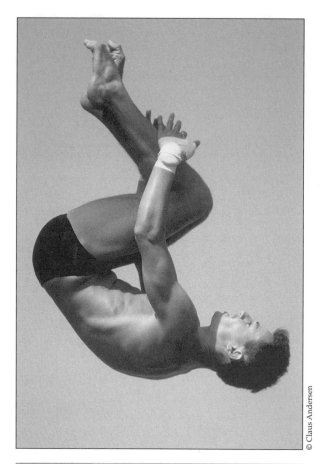

© Claus Andersen

Figure 10.5   Tucking reduces the diver's moment of inertia about his transverse axis.

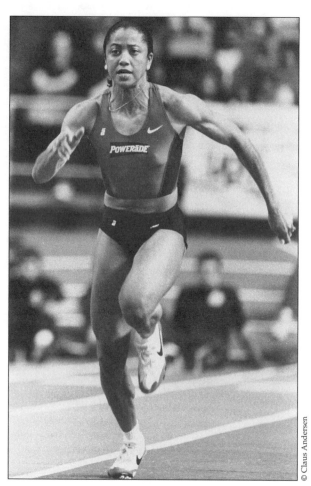

© Claus Andersen

Figure 10.6   Flexing the knee and hip of the recovery leg reduces the moment of inertia of the leg about the sprinter's hip joint.

of the pole. At the start of the vault, the vaulter is concerned with having the pole rotate about its bottom end, so that he can land safely in the pit. To make this rotation easier, the moment of inertia about the pole's axis of rotation (its bottom end) must be kept small. The vaulter stays in an extended position just after he leaves the ground, as shown in figure 10.7a. This keeps his center of gravity closer to the bottom end of the pole. The pole bends, which also shortens the distance between the vaulter's center of gravity and the bottom end of the pole. Both of these actions decrease the moment of inertia of the pole about its bottom end and facilitate the rotation of the pole. Once the vaulter is sure the pole is going to undergo an angular displacement that will be large enough to place him over the bar, he then wants to increase his rotation on the pole and decrease the rotation of the pole. He accomplishes this by tucking and moving his center of gravity toward his top handgrip, which is the axis of rotation for his body (see figure 10.7b). This decreases the moment of inertia of his body about his top

handgrip and facilitates his rotation about that axis. On the other hand, this action increases the moment of inertia of the pole about its axis of rotation, because the vaulter's center of gravity has moved farther from the bottom end of the pole. This slows down the rotation of the pole.

Dancers, divers, gymnasts, skaters, and other athletes regularly alter the moments of inertia of their entire bodies or the body parts they are swinging to perform stunts or skills more effectively.

Sports equipment designers are also aware of the effects an implement's moment of inertia may have on a skill. Downhill skiers use longer skis than slalom skiers use. The longer skis give the downhill skier greater stability—something desirable for a skier going 60 mi/hr down a mountain. Slalom skiers need more maneuverable skis—skis with a smaller moment of inertia about the turning axis. Slalom skis are thus shorter.

a

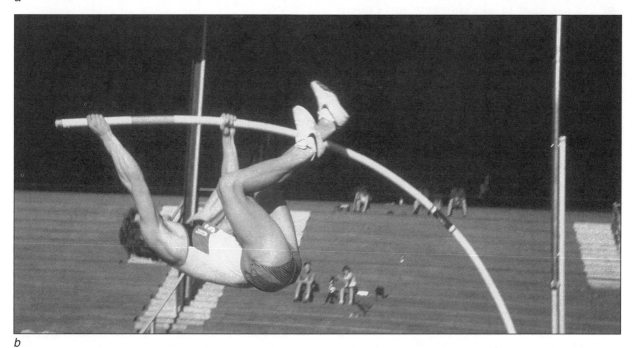

b

Figure 10.7 Vaulter in extended position (*a*) and tucked position (*b*). The moments of inertia of the vaulter about his top handgrip and the bottom end of the pole differ in these positions.

Moment of Inertia and Linear Velocity

In chapter 9, the advantage of using a longer striking implement was discussed and described mathematically. Recall that:

$$v_T = \omega r \qquad (10.3)$$

where

v_T = instantaneous linear velocity tangent to the circular path of the point

ω = instantaneous angular velocity (measured in radians per second)

r = radius

The longer radius of rotation of the striking surface of an implement causes a larger linear velocity at the striking end of the implement if the angular velocity is the same as the angular velocity of a shorter implement. The *if* at the end of the previous statement is a big qualifier. Unfortunately, as the length of an implement increases, its moment of inertia increases also. Due to the increase in angular inertia, it becomes more difficult to accelerate the implement angularly to achieve the same angular velocity. Increasing the length of a swinging implement to achieve a faster linear velocity at the end of the implement may not always yield the desired outcome unless the increase in moment of inertia caused by the increase in length is accounted for in some way.

Golf club designers have accommodated for the larger moment of inertia about the swing axis caused by the longer lengths of woods and drivers by making them lighter and less massive than the shorter irons. The smaller mass of a longer club causes a reduction in the club's moment of inertia that almost balances the increase in moment of inertia caused by the longer length (see figure 10.8).

Some golf club designers have also manipulated the moment of inertia about the twist axis of the clubs. During a long backswing and downswing, a golfer may inadvertently twist the club shaft so that the club face is not square to the ball at impact. The ball will not go where the golfer wants it to if this happens! An increase in the moment of inertia of the club about its longitudinal axis throughout the shaft would increase the club's resistance to twisting. Designers have increased the moment of inertia about the twist axis by distributing more of the mass of the club head toward the heel and toe of the club, as shown in figure 10.9.

By now, you should have a good understanding of angular inertia and moment of inertia. They are important elements in the angular analog of Newton's law of motion, but before we discuss them, another important element included in them must be understood.

Angular Momentum

In chapter 6, the concept of momentum was introduced. Linear momentum is the product of mass and velocity. Mathematically, linear momentum is

$$L = mv \qquad (10.4)$$

Figure 10.8   The longer club, the driver (*a*), has a greater moment of inertia than the nine-iron (*b*) does despite its smaller mass.

where

L = linear momentum

m = mass

v = linear velocity

Linear momentum quantifies the linear motion of an object. Angular momentum quantifies the angular motion of an object.

Angular Momentum of a Rigid Body

Angular momentum is the angular analog of linear momentum, so it is the product of the angular analog of mass (moment of inertia) times the angular analog of linear velocity (angular velocity). Mathematically, then, the angular momentum of a rigid body is

$$H_a = I_a \omega_a \qquad (10.5)$$

| Figure 10.9 | Traditional (*a*) and perimeter-weighted (*b*) club head designs have different moments of inertia about the twist axis (x-x) of the club. |

where

H_a = angular momentum about axis a

I_a = moment of inertia about axis a

ω_a = angular velocity about axis a

Angular momentum is abbreviated with the letter H. The units for angular momentum are kilogram meters per second (kg · m/s). Angular momentum is a vector quantity, just like linear momentum, so it has size and direction. The direction of angular momentum is the same as the direction of the angular velocity that defines it. The right-hand thumb rule is used to determine direction.

Linear momentum depends on two variables: mass and velocity. But the mass of most objects does not change; it is not variable. Changes in linear momentum thus depend on changes in only one variable: velocity. Angular momentum also depends on two variables: moment of inertia and angular velocity. For rigid objects, changes in angular momentum also depend on changes in only one variable—angular velocity—because the moment of inertia of a rigid object does not change. For nonrigid objects, however, changes in angular momentum may result from changes in angular velocity or changes in moment of inertia, or both, because

angular velocity and moment of inertia are both variable.

Angular Momentum of the Human Body

Equation 10.5 mathematically defines angular momentum and seems simple, but does it adequately explain how the angular momentum of the human body is determined? If the body acts as a rigid object, so that all the body segments rotate with the same angular velocity, this angular velocity is used as ω and the moment of inertia of the whole body is used as I_a in equation 10.5.

What if some limbs rotate at different angular velocities than other limbs? How is the angular momentum of the body determined then? Mathematically, the angular momentum about an axis through the center of gravity of a multisegment object such as the human body is defined by equation 10.6 (and illustrated in figure 10.10):

$$H_a = \Sigma(I_i\omega_i + m_i r_{i/cg}^2 \omega_{i/cg}) \qquad (10.6)$$

where

H_a = angular momentum about axis a through the center of gravity

Σ = summation symbol

I_i = moment of inertia of segment i about its own center of gravity

ω_i = angular velocity of segment i

m_i = mass of segment i

$r_{i/cg}$ = distance from center of gravity of segment i to center of gravity of entire body

$\omega_{i/cg}$ = angular velocity of $r_{i/cg}$ about center of gravity of entire body

Wow! Equation 10.6 is too complicated to understand easily or use in qualitative analyses in coaching or teaching. Instead, the following gross approximation is a more useful representation of the angular momentum of the entire body:

$$H_a \cong \Sigma(H_i) \cong \Sigma(I_{i/cg}\omega_i)$$
$$= (I\omega)_{r.arm} + (I\omega)_{l.arm} + (I\omega)_{r.leg} + (I\omega)_{l.leg} + (I\omega)_{trunk} \quad (10.7)$$

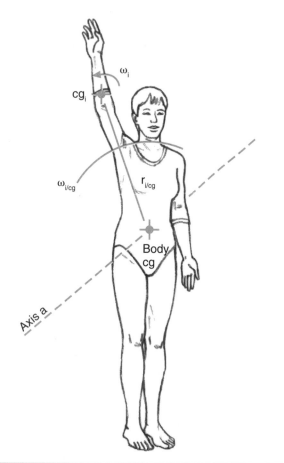

| Figure 10.10 | Terms used to compute the angular momentum of the human body (from only one segment, the right forearm). |

where

H_a = angular momentum about axis a through the center of gravity

Σ = summation symbol

H_i = angular momentum of segment i about the entire body's center of gravity

$I_{i/cg}$ = I = moment of inertia of segment i about the entire body's center of gravity

ω_i = angular velocity of segment i

In other words, the sum of the angular momenta of all the body segments gives an approximation of the angular momentum of the entire body.

Consider the running motion. The left arm swings backward as the right arm swings forward. The left leg swings forward as the right leg swings backward. What is the angular momentum of the runner about a transverse axis (left to right) through the runner's center of gravity? Using the right-hand thumb rule, with the positive direction of the axis to the left, the angular momentum of the left arm is positive, the angular momentum of the right arm is negative, the angular momentum of the right leg is positive, and the angular momentum of the left leg is negative. The trunk isn't rotating, so its angular momentum is zero. If we use equation 10.7 to approximate the total angular momentum of the body, it would appear that the angular momenta of the arms sum to zero (they cancel each other out) and the angular momenta of the legs sum to zero also. The total angular momentum of the body is zero.

Angular Interpretation of Newton's First Law of Motion

For linear motion, Newton's first law states that every body continues in its state of rest, or of uniform motion in a straight line, unless it is compelled to change that state by forces impressed upon it. This law is sometimes referred to as the law of inertia. It may also be referred to as the law of conservation of momentum. The momentum of a body is constant unless a net external force acts on it.

The angular analog of Newton's first law may be stated as: the angular momentum of an object remains constant unless a net external torque is exerted on it. For a rigid object whose moments of inertia are constant, this law implies that the angular velocity remains constant. Its rate of rotation and its axis of rotation do not change unless an external torque acts to change it.

> **The angular momentum of an object remains constant unless a net external torque is exerted on it.**

SELF-EXPERIMENT

Throw your pen up into the air and give it a flip as you release it so it rotates end over end. As the pen falls, does its angular velocity change? Does its rotation speed up or slow down after you've released it? Does the axis of rotation change direction? The answer to all of these questions is no. The angular momentum of the pen was constant because no external torque acted on it once it was in the air. What about the force of gravity? Doesn't it create an external torque on the pen? The force of gravity (the weight of the pen) acts through the center of gravity of the pen. Once it leaves your hand, the pen becomes a projectile, and the axis of rotation of a projectile is through its center of gravity. The force of gravity can't create a torque on the pen because the moment arm of the force of gravity is zero (see figure 10.11).

How does the angular version of Newton's first law apply to the human body? The angular momentum of the human body is constant unless external torques act on it. Mathematically, this can be represented as:

$$H_i = I_i\omega_i = I_f\omega_f = H_f \tag{10.8}$$

where

H_i = initial angular momentum

H_f = final angular momentum

I_i = initial moment of inertia

I_f = final moment of inertia

ω_i = initial angular velocity

ω_f = final angular velocity

Because the body's moment of inertia is variable and can be changed by altering limb positions, the body's angular velocity also changes to accommodate the changes in the moment of inertia. In this case, Newton's first law does not require that the angular velocity be constant, but rather that the product of the moment of inertia times the angular velocity be constant if no external torques act. For this to occur, any increases in moment of inertia, created by the person moving limbs farther from the axis of rotation, cause decreases in angular velocity to keep angular momentum constant. Likewise, any decreases in moment of inertia, created by the person bringing limbs closer to the axis of rotation, require increases in angular velocity to keep angular momentum constant. Even though angular momentum remains constant in situations where no external torques act, athletes may control their angular velocity by changing their moment of inertia.

Figure 10.11   The force of gravity (W) acting on a pen flipped into the air does not create a torque on the pen because the moment arm of the weight around the axis of rotation (the center of gravity) is zero.

Newton's first law does not require that the angular velocity be constant.

Conservation of angular momentum is well demonstrated by a figure skater doing a spin. The torque created by friction between the ice and the skates is minimal and may be ignored. As a skater begins a spin, one leg and both arms may be held up and away from the body, as shown in figure 10.12. The skater thus has a large moment of inertia about the longitudinal axis. As the spin progresses, though, the skater adducts the arms and legs, bringing them closer to the body, thus reducing the moment of inertia. Angular momentum remains the same, so the reduction in the moment of inertia must be accompanied by an increase in angular velocity, which is exactly what happens to the skater.

Gymnasts, figure skaters, dancers, divers, and other athletes use this principle of conservation of angular momentum to control their angular velocities when somersaulting and twisting. A gymnast tucks her body to speed rotation. A figure skater abducts his arms to slow down his spin. A dancer adducts her arms to speed up her spin. A diver extends from a pike to slow down somersaulting.

Another way athletes make use of this principle is not to control the angular velocity of their entire body, but to control the individual angular velocities of their limbs or trunk. This situation is best described by combining equation 10.7, which describes an approximation of the angular momentum of the entire body when its segments act independently, with equation 10.8, which explains the angular version of Newton's first law. Combining these gives:

$$
\begin{aligned}
H_i &= [(I\omega)_{r.arm} + (I\omega)_{l.arm} + (I\omega)_{r.leg} + (I\omega)_{l.leg} \\
&\quad + (I\omega)_{trunk}]_{initial} \\
&= [(I\omega)_{r.arm} + (I\omega)_{l.arm} + (I\omega)_{r.leg} + (I\omega)_{l.leg} \\
&\quad + (I\omega)_{trunk}]_{final} \\
&= H_f
\end{aligned}
\tag{10.9}
$$

where

H_i = initial angular momentum of entire body

H_f = final angular momentum of entire body

I = moment of inertia of segment about the entire body's center of gravity

ω = angular velocity of segment

Consider the hurdler shown in figure 10.13. When the hurdler leaves the ground during the hurdling stride, her angular momentum about a longitudinal axis through her center of gravity is zero. Because the hurdler becomes a projectile once her foot leaves the ground, no external torques act on her and her angular momentum remains constant, in this case at zero. During the flight over the hurdle, the hurdler must swing her trailing leg (her left leg in this case) over the hurdle very quickly. When she does this, the trailing leg has a large angular momentum because it has a large angular velocity and a large moment of inertia about the longitudinal axis through the hurdler's center of gravity. But the hurdler's total angular momentum is still zero because no external torque has acted on her. This means that the angular momentum of some other body part or parts must be in the opposite direction to keep the angular momentum of the entire body zero. The hurdler's right arm swings in the opposite direction, but its angular momentum is not large enough to counter the angular momentum of the trailing leg. The trunk also turns slightly in the opposite direction of the trailing leg. Because

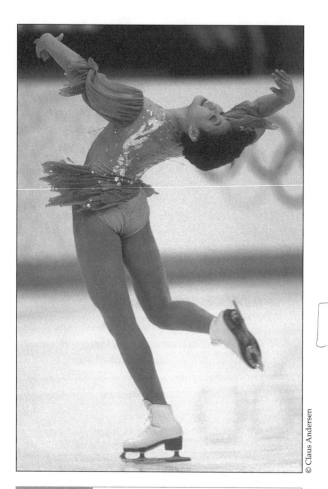

© Claus Andersen

Figure 10.12 A figure skater begins a spin with a relatively large moment of inertia.

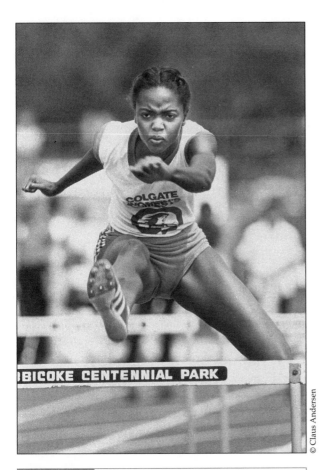

© Claus Andersen

| Figure 10.13 | The angular momentum of the hurdler's trail leg must be countered by the angular momentum of the arms and trunk. |

Angular Interpretation of Newton's Second Law of Motion

Newton's second law of motion describes what happens when external forces do act on an object. It states that the change in motion of an object is proportional to the force impressed and is made in the direction of the straight line in which the force is impressed. Stated more simply, this law says: if a net external force is exerted on an object, the object will accelerate in the direction of the net external force, and its acceleration will be proportional to the net external force and inversely proportional to its mass. Recall from chapter 6 that this law is stated mathematically as:

$$\Sigma F = ma \qquad (10.10)$$

where

ΣF = net external force

m = mass of the object

a = linear acceleration of the object

The angular analog of Newton's second law of motion may be stated as follows: the change in angular momentum of an object is proportional to the net external torque exerted on it, and this change is in the direction of the net external torque. The net external torque exerted on an object is proportional to the rate of change in angular momentum. For a rigid object with constant moments of inertia, this law can be stated more simply by substituting torque for force, angular acceleration for acceleration, and moment of inertia for mass in the linear version of this law: if a net external torque is exerted on an object, the object will accelerate angularly in the direction of the net external torque, and its angular acceleration will be directly proportional to the net external torque and inversely proportional to its moment of inertia. Mathematically, for a rigid object with constant moments of inertia, this law is stated as:

$$\Sigma T_a = I_a \alpha_a \qquad (10.11)$$

where

ΣT_a = net external torque about axis a

I_a = moment of inertia of the object about axis a

α_a = angular acceleration of the object about axis a

If the external torques acting on an object do not sum to zero, the object will experience an angular

the trunk is bent over, it has a large moment of inertia relative to the longitudinal axis. The angular momentum of the trunk along with the angular momentum of the right arm are enough to counter the angular momentum of the trailing leg.

If the hurdler swings the trailing leg too far to the left (i.e., if the trailing leg is too extended when the hurdler swings it over the hurdle), it will have a very large angular momentum due to its large moment of inertia. In such a case, the arms and trunk must swing too far in the opposite direction to counter this angular momentum, and the hurdler is likely to fall. The technique that better hurdlers use is to flex the trailing leg and keep it as close to the body as possible when clearing the hurdle. Another advantage is that a more tightly tucked trailing leg has a smaller moment of inertia about the hip and can thus be moved more quickly. Similar strategies are used by long jumpers, runners, skiers, gymnasts, bronco riders, and other athletes to prevent loss of balance when one body part has too much angular momentum relative to the other parts.

acceleration in the direction of the net torque. Its angular velocity will speed up or slow down, or its axis of rotation will change direction. If an object's angular velocity or axis of rotation changes, a net external torque must be acting on the object to cause the angular acceleration.

> The change in angular momentum of an object is proportional to the net external torque exerted on it, and this change is in the direction of the net external torque. The net external torque exerted on an object is proportional to the rate of change in angular momentum.

For a nonrigid object with a variable moment of inertia, equation 10.11 does not apply. In this case, the net external torque equals the rate of change of momentum. Mathematically, this can be expressed for average net torque as follows:

$$\Sigma \overline{T}_a = \frac{\Delta H_a}{\Delta t} = \frac{\left(H_f - H_i\right)_a}{\Delta t} \qquad (10.12)$$

where

$\Sigma \overline{T}_a$ = average net external torque about axis a

ΔH_a = change in angular momentum about axis a

H_f = final angular momentum about axis a

H_i = initial angular momentum about axis a

Δt = change in time

A net external torque acting on a nonrigid object with variable moments of inertia will cause a large and quick change in angular momentum if the net torque is large and a small and slow change in angular momentum if the net torque is small, provided the torques act for equal time intervals. The change in angular momentum may be seen as (1) a speeding up or slowing down of the object's angular velocity, (2) a change in the direction of the axis of rotation, or (3) a change in the moment of inertia. The angular acceleration of the object or a change in its moment of inertia does not necessarily indicate the presence of a net external torque, because the total angular momentum of a nonrigid object may remain constant even if it angularly accelerates or if its moment of inertia changes.

As with the linear version of Newton's second law, the angular version only indicates what happens at an instant in time when a net torque acts. In most sports and human movements, we are more concerned with the final outcome of external torques acting on an athlete or implement over some duration of time. The impulse-momentum relationship provides us with this information.

Angular Impulse and Angular Momentum

Linear impulse is the product of force times the duration of that force application. The impulse produced by a net force causes a change in momentum of the object the net force acts on, as indicated by equation 10.13:

$$\Sigma \overline{F} \Delta t = m(v_f - v_i) \qquad (10.13)$$
$$\text{impulse} = \text{change in momentum}$$

where

$\Sigma \overline{F}$ = average net force acting on an object

Δt = interval of time during which this force acts

m = mass of object being accelerated

v_f = final velocity of the object at the end of the time interval

v_i = initial velocity of the object at the beginning of the time interval

The angular analog of the impulse momentum relationship may be derived from equations 10.12 and 10.13:

$$\Sigma \overline{T}_a = \frac{\Delta H_a}{\Delta t} = \frac{\left(H_f - H_i\right)_a}{\Delta t} \qquad (10.12)$$

$$\Sigma \overline{T}_a \Delta t = (H_f - H_i)_a \qquad (10.14)$$

angular impulse = change in angular momentum

where

$\Sigma \overline{T}_a$ = average net external torque about axis a

Δt = time interval during which this force acts

ΔH_a = change in angular momentum about axis a

H_f = final angular momentum about axis a

H_i = initial angular momentum about axis a

In many sport skills, the athlete must cause a change in the angular momentum of the entire body or an implement or individual body part. The angular impulse-angular momentum relationship shown in equation 10.14 indicates how this is accomplished. A larger external torque acting over a longer duration will create a larger change in angular momentum. Larger torques can be created by using longer moment arms, as discussed in chapter 8. Increasing the time of application of a torque seems straightforward at first, but it may be more difficult than it seems.

Consider how a dancer or gymnast initiates a spin or twist on the floor. The axis of rotation for the twisting action is a longitudinal axis that passes through one foot on the floor (see figure 10.14). The other foot pushes on the floor to create a frictional force that creates a torque about the longitudinal axis. This foot must be some distance away from the twisting foot to maximize the moment arm and the torque created. If the moment of inertia of the dancer is small as the torque is created, the dancer begins to turn faster and faster and the foot must stop pushing on the floor because it begins to rotate with the rest of the body. How can the dancer prolong the pushing action to create a larger impulse and thus a larger change in angular momentum? If the dancer has a larger moment of inertia as the torque is created, she begins to turn as a result

of the torque, but she doesn't turn as fast, so she has more time to push with her foot before her body has turned far enough that the foot must come off the floor. This prolonging of the torque application by increasing the moment of inertia to slow down the spin results in a larger impulse and thus a larger change in momentum. After her foot comes off the floor, the dancer can reduce her moment of inertia and increase her angular velocity.

Discus throwers use a similar technique. At the start of their throwing action, they have a large moment of inertia, but at the release of the discus, the moment of inertia is smaller. In activities where the goal is to spin very fast, the athlete should start with a large moment of inertia during the torque-production period so that the duration of torque application is maximized. This increases the impulse and thus the change in angular momentum that occurs. Once the torque-production phase is complete, the performer then reduces the moment of inertia, and the angular velocity increases accordingly.

Novosti/Corbis-Bettmann

Figure 10.14   The size of a dancer's moment of inertia affects how much angular momentum she can generate at the start of a spin.

Angular Interpretation of Newton's Third Law of Motion

Newton's third law of motion states that for every action there is an equal but opposite reaction. A clearer statement of this law is that for every force exerted by one body on another, the other body exerts an equal force back on the first body but in the opposite direction. The angular version of this law states that for every torque exerted by one object on another, the other object exerts an equal torque back on the first body but in the opposite direction.

A common confusion regarding the linear version of this law is that it is the forces that are equal but opposite in direction, not the effects of the forces. The effects of forces are often accelerations, and the accelerations of the objects these forces act on depend on the masses of the objects as well as any other forces that act on the objects.

In interpreting the angular version of Newton's third law, it must be remembered that the torques acting on the two objects have the same axis of rotation. Also, the effects of these torques are different because they act on different objects. The effects of the torques depend on the moments of inertia of the objects and whether or not any other torques act on the objects.

One angular example of Newton's third law involves muscles. As discussed in chapter 8, muscles produce torques about joints by creating forces on the limbs on either side of the joint. The vastus group of the quadriceps femoris are knee extensor muscles. When these muscles contract, they create a torque on the leg that causes it to rotate (or tend to rotate) in one direction, and they create an equal but opposite torque on the thigh that causes it to rotate (or tend to rotate) in the opposite direction. These two opposite rotations produce extension at the knee joint.

Let's look at another example of how Newton's third law is used. Tightrope walkers use long poles to help them stay on the tightrope, as shown in figure 10.15. How does a long pole help an acrobat stay balanced on the tightrope? One reason is that it can lower the center of gravity of the acrobat and pole system as well as increase its moment of inertia. These effects both increase the stability of the acrobat. The most important use of the pole can be explained using Newton's third law. Suppose the tightrope walker begins to fall clockwise to his left. What should he do with his pole to help him regain his balance?

If the acrobat exerted a clockwise torque on the pole and moved it clockwise in the same direction of his fall, the pole would exert an equal but oppositely directed torque on the acrobat. This torque would act counterclockwise on the acrobat, opposite his direction of falling, and possibly move him into a position where he could regain his balance. If the acrobat moved the pole counterclockwise in the opposite direction of the fall, the reaction torque exerted on the acrobat would be in the clockwise direction and would move him faster in the direction he was falling!

Summary

The basics of angular kinetics, the causes of angular motion, are explained by angular interpretations of Newton's laws of motion. Angular analogs of inertia and momentum must be understood to make these interpretations. Angular inertia, called moment of inertia, is an object's resistance to change in its angular motion. Mathematically, it is defined as the product of mass times radius of gyration squared. Radius of gyration is a length dimension representing, on average, how far an object's mass is located from an axis of rotation. Objects have many and different moments of inertia, one for each of their possible axes of rotation. Angular momentum, like linear momentum, is a measure of an object's motion. Angular momentum is the product of moment of inertia and angular velocity. It is a vector quantity that is specific for an axis of rotation. Angular momentum

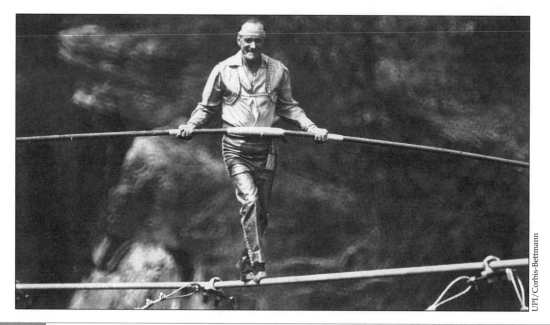

Figure 10.15   Long poles are used by tightrope walkers to maintain their balance. By creating a torque on the pole in one direction, the acrobat causes an equal but opposite torque to be created on him.

SELF-EXPERIMENT

Stand with your toes just behind a line on the floor. Imagine that the line is the edge of a cliff, and you are standing right on the edge when you lean too far forward and start to fall, so that your body rotates counterclockwise. Go ahead and try it. Which way did you swing your arms to try to catch yourself as you started to fall? You probably swung your arms back, upward, and then forward so that they circled in a counterclockwise direction (unless you thought too hard before trying this). Your shoulder muscles exerted a counterclockwise torque on your arms to cause this rotation. The reaction to this was a clockwise torque exerted by these same muscles on your trunk. This clockwise torque would tend to rotate your trunk and legs clockwise or slow down their counterclockwise rotation. This might have saved you from falling off the cliff. When a gymnast loses her balance on the balance beam, you will often see this wild arm swinging, which is an attempt to regain balance. Similar arm flailing can be observed in ski jumpers if their forward rotation is too great. Recreational skiers can often be seen doing this when a bump gives them some unexpected angular momentum.

may be constant even if angular velocity varies, so long as the variation in angular velocity is accompanied by an inverse variation in moment of inertia. The angular interpretation of Newton's first law says that objects do not change their angular momentum unless a net external torque acts on them. The angular interpretation of Newton's second law explains what happens if a net external torque does act on an object. A rigid object will accelerate angularly in the direction of the net external torque, and its angular acceleration will be inversely related to its moment of inertia. For objects with variable moments of inertia, the impulse-momentum relationship is a more applicable angular interpretation of Newton's second law. A net external torque acting over some duration of time causes a change in angular momentum in the same direction as the net external torque. An angular interpretation of Newton's third law explains that torques act in pairs. For every torque, there is an equal torque acting on another object but in the opposite direction.

Table 10.1 gives a mathematical comparison of the angular kinetic parameters and principles, along with their linear counterparts. This table is an excellent summary of the elements of this chapter.

TABLE 10.1   Comparison of Linear and Angular Kinetic Quantities

	Linear	
Quantity	Symbol and equation of definition	SI units
Inertia (mass)	m	kg
Force	F	N
Linear momentum	$L = mv$	kg · m/s
Impulse	$\Sigma \bar{F} \Delta t$	N · s

(continued)

	Angular	
Quantity	Symbol and equation of definition	SI units
Moment of inertia	$I = \Sigma mr^2 = mk^2$	$kg \cdot m^2$
Torque or moment of force	$T = F \times r$	Nm
Angular momentum	$H = I\omega$	$kg \cdot m^2/s$
Angular impulse	$\Sigma \bar{T} \Delta t$	$Nm \cdot s$

TABLE 10.1 *(continued)*

Key Terms

angular impulse *(p. 184)*

angular inertia *(p. 172)*

angular momentum *(p. 178)*

moment of inertia *(p. 172)*

principal axis *(p. 174)*

radius of gyration *(p. 174)*

rotary inertia *(p. 172)*

REVIEW QUESTIONS

1. At the instant of takeoff, a 60-kg diver's angular momentum about his transverse axis is 20 kg · m²/s. His radius of gyration about the transverse axis is 1.0 m at this instant. During the dive, the diver tucks and reduces his radius of gyration about the transverse axis to 0.5 m.

   a. At takeoff, what is the diver's angular velocity about the transverse axis?

   b. After the diver tucks, what is his angular velocity about the transverse axis?

   Questions 2 through 5 refer to the four dancers described below. Assume that each dancer is in anatomical position unless stated otherwise.

   - Ann is standing on her left foot. Her right leg is abducted approximately 20° at the hip. Her left arm is abducted 90° at the shoulder.
   - Barbara is standing on her right foot. Her left leg is flexed 90° at the knee. Each arm is slightly abducted at the shoulder and fully flexed at the elbow.
   - Carol is standing on her right foot. Her left leg is abducted 60° at the hip. Each arm is abducted 100° at the shoulder.
   - Donna is standing on both her feet. Her left arm is flexed 180° at the shoulder so it is outstretched over her head. Her right arm is slightly adducted past anatomical position so it crosses in front of her body.

2. Which dancer has the largest moment of inertia about her AP axis?

3. Rank the dancers' positions described above from largest to smallest moment of inertia about the longitudinal axis.

4. If each dancer shown above is rotating with the same angular velocity about the longitudinal axis, which dancer has the greatest angular momentum about the longitudinal axis?

5. If each dancer shown is rotating with the same angular momentum about the longitudinal axis, which dancer is spinning with the fastest angular velocity about the longitudinal axis?

6. Which sprinting technique is more effective—flexing the knee of the recovery leg more during the swing-through or flexing the knee of the recovery leg less during the swing-through? Why?

7. A volleyball player jumps into the air to spike the ball. She leaves the ground without any angular momentum. As she swings her arm forward to spike the ball, what happens to her legs if her trunk and non-hitting arm do not rotate?

8. A skier has skied over a bump and become airborne. He has some angular momentum that is rotating him forward (clockwise). If he continues to rotate, he will probably fall on his face when he lands. In which direction should he rotate his arms to prevent a fall?

9. Three dives are performed such that the diver rotates through the same number of revolutions in the same amount of time in all three dives. In each dive the diver's position at takeoff and just prior to entry are identical layout positions. In dive 1, the diver is in a layout position in midflight. In dive 2, the diver is in a pike position in midflight. In dive 3, the diver is in a tuck position in midflight.

   a. In which dive does the diver have the largest angular momentum about the transverse axis?

   b. In which dive will the diver be rotating the slowest about the transverse axis just prior to entry?

10. An ice skater is spinning at 20 rad/s about her longitudinal axis when she abducts her arms and doubles her radius of gyration about her longitudinal axis from 30 to 60 cm. If her angular momentum is conserved, what is her angular velocity about her longitudinal axis after she increases her radius of gyration?

11. The mass of the discus is 2 kg, and its radius of gyration about the spin axis is 12 cm. The average net torque exerted on a discus about its axis of spin is 100 Nm during a throw. If the discus is not spinning at the start of the throwing action, and the throwing action lasts for 0.20 s, how fast is the discus spinning when it is released?

Fluid Mechanics:
The Effects of Water and Air

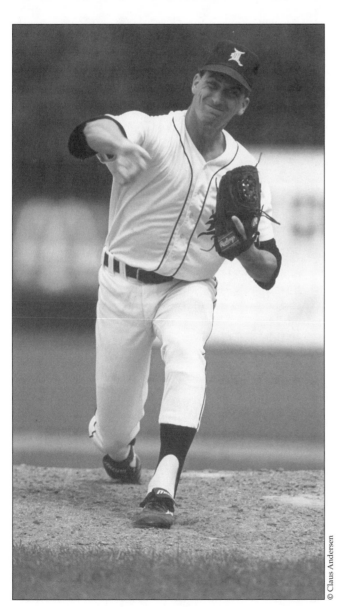

© Claus Andersen

You are standing in the batter's box, waiting for the next pitch. The pitcher winds up and throws. The ball is coming at you, and it looks like the pitcher has thrown the ball right in the middle of the strike zone. You start your swing, anticipating the crack of the bat hitting the ball. Whiff! Whomp! "Strike three, you're out!" Rather than feeling and hearing the crack of the bat hitting the ball, all you feel is the bat slicing through air, and all you hear is the thud of the ball hitting the catcher's mitt. Just as you were visualizing the ball sailing over the left-field fence, the ball seemed to swerve down and to the left. What happened? How could a ball that seemed to be on a perfect trajectory toward the strike zone (and your bat) swerve out of your bat's path? The answer to this question comes from fluid mechanics.

OBJECTIVES

When you finish this chapter, you should be able to do the following:

▮ Define relative and absolute motion

▮ Define buoyancy and buoyant force

▮ Define pressure

▮ Define fluid

▮ Explain how a fluid exerts forces on an object moving through it

▮ Identify the components of fluid forces

▮ Define drag force

▮ Distinguish between surface drag and form drag

▮ Define lift force

▮ Describe Bernoulli's principle

▮ Describe the Magnus effect

▮ Identify the various factors that determine the effects of fluid forces on an object

This chapter is about the branch of mechanics called fluid mechanics. Specifically, it is about the forces fluids exert on objects in them or moving through them. Unlike solids, liquids and gases can flow and change shape quickly and easily without separating, so they are classified as fluids. The fluids we are most concerned about in sport biomechanics are air and water. Air is the medium we move through in all land-based sports and human activities, and water is the medium we move through in all aquatic sports and activities.

In swimming and other aquatic activities, fluid forces are large, and their importance to success in these activities is obvious. In many land-based activities, fluid forces (air resistance) may be so small that they can be ignored. But in other land-based activities, fluid forces may be large enough to affect the movements of bodies or implements or so large that they determine the outcome of a movement skill. Consider the importance of air resistance in the following activities: sprint running, baseball pitching, cycling, sailboarding, discus throwing, sailing, speedskating, downhill ski racing, hang gliding, and skydiving. In the last two activities, one's life depends on air resistance!

As fluid forces are fundamental for success in aquatic sports and activities, as well as certain land-based activities, a basic understanding of fluid forces is desirable. Information presented in this chapter will assist you in gaining a basic understanding of fluid forces.

Buoyant Force: Force Due to Immersion

Two types of forces are exerted on an object by a fluid environment: a buoyant force due to its immersion in the fluid and a dynamic force due to its relative motion in the fluid. The dynamic force is usually resolved into two components: drag and lift forces. The buoyant force, on the other hand, always acts vertically. A buoyant force acts upward on an object immersed in a fluid. You are probably most familiar with this principle regarding objects in water.

> Two types of forces are exerted on an object by a fluid environment: a buoyant force due to its immersion in the fluid and a dynamic force due to its relative motion in the fluid.

SELF-EXPERIMENT

Fill a bathtub with water. Take a basketball (or any other large inflatable ball) and try to push it into the water. Now try it again with a tennis ball (or racquetball). Which is easier to push underwater? It's easier to push the tennis ball underwater. You may also notice that the farther you push the ball into the water, the greater the force you have to push with. The force that pushed upward on the tennis ball or basketball was buoyant force. Buoyant force seems to be related to the size of the object immersed in water and how much of the object is immersed.

Pressure

Let's see if we can explain the cause of this buoyant force. Suppose you were in a pool of still water. As you dive deeper and deeper into the pool, the water exerts greater and greater pressure on you. The pressure the water exerts is due to the weight (force) of the water above you. The water pressure doesn't just act downward on you; the water below you pushes upward on you, and the water to either side pushes laterally on you. Water pressure acts in all directions with the same magnitude, as long as you stay at the same level. The deeper you go, the greater the pressure. Pressure is defined as force per unit area. One cubic meter (m^3) of water weighs about 9800 N, so at a depth of 1 m the water pressure is 9800 N/m^2. At a depth of 2 m, the 2 m^3 of water above a 1-m square weigh 19,600 N, so the pressure is 19,600 N/m^2. Figure 11.1a illustrates how water pressure increases linearly with depth.

Imagine removing a cube of this water from 1 m below the surface, as shown in figure 11.1b. Draw a free-body diagram of the cube of water, as shown in

figure 11.1c. What vertical forces act on the cube of water? The water above the cube exerts 9800 N/m^2 pressure on top of the cube. This pressure acting on the square-meter top face of the cube creates a 9800-N resultant force acting downward on the cube. Let's represent this force with R_u, for upper force, and with an arrow acting downward on the cube. The water below the cube exerts 19,600 N/m^2 pressure on the bottom of the cube. This pressure acting on the square-meter bottom face of the cube creates a 19,600-N resultant force acting upward on the cube. Let's represent this force with R_l, for lower force, and with an arrow acting upward on the cube. Some forces also act on the sides of the cube, as shown, but because we are only concerned with the vertical forces, we won't try to quantify these lateral forces. Do any other vertical forces act on the cube of water? The cube of water weighs something, about 9800 N, because gravity exerts a force on it. Let's represent this force with W_w, for weight, and with an arrow acting downward on the cube. If the cube is in equilibrium and not accelerating, then according to Newton's second law:

$$\Sigma F = R_l + (-R_u) + (-W_w) = 0 \qquad (11.1)$$
$$\Sigma F = 19,600 \text{ N} - 9800 \text{ N} - 9800 \text{ N} = 0$$

Sure enough, the net vertical force acting on the cube of water is zero, so the water is in equilibrium. What happens if we take out a cube at a depth of 4 m? At 4 m, the pressure is 39,200 N/m^2 (4 m $\times$ 9800 N/m^2), so the force R_u is 39,200 N. At 5 m, the pressure is 49,000 N/m^2 (5 m $\times$ 9800 N/m^2), so the force R_l is 49,000 N. The cube still weighs the same, so equation 11.1 becomes:

$$\Sigma F = R_l + (-R_u) + (-W_w) = 0 \qquad (11.1)$$
$$\Sigma F = 49,000 \text{ N} - 39,200 \text{ N} - 9800 \text{ N} = 0$$

No matter what depth we choose, the difference in pressures at the top and bottom of the cube accounts for the weight of the cube of water, and equation 11.1 is valid.

Buoyant force is the difference between the force acting upward on the cube, R_l, and the force acting downward on the cube, R_u. If we let F_b, for buoyant force, represent this difference, equation 11.1 becomes:

$$\Sigma F = R_l + (-R_u) + (-W_w) = 0 \qquad (11.1)$$

$$\Sigma F = F_b + (-W_w) = 0$$
$$F_b = W_w \qquad (11.2)$$

The size of the buoyant force is equal to the weight of the volume of fluid displaced by the

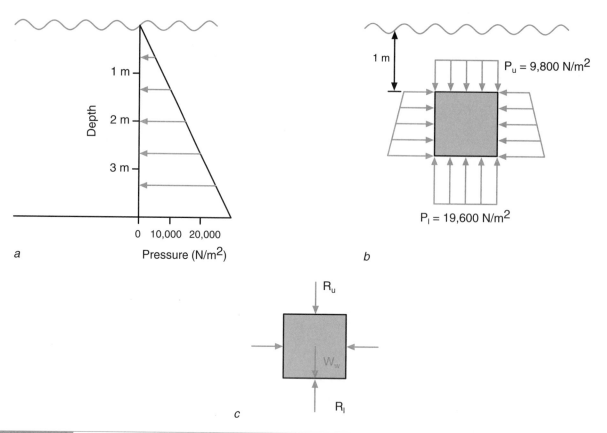

Figure 11.1   Graphical representations of the relationship between water pressure and depth (*a*), a 1.0 m³ cube of water 1.0 m below the water (*b*), and a free-body diagram of the forces acting on the cube of water (*c*).

object. This principle was discovered by the Greek mathematician Archimedes more than 2000 years ago. He lived from 287-212 B.C. It is called Archimedes' principle. For simplicity, you may think of buoyant force as similar to a reaction force from the water, but the reaction force depends on how much water is pushed out of the way by the object.

> **The size of the buoyant force is equal to the weight of the volume of fluid displaced by the object.**

What happens if we fill the hole made by the cube of water we took out with a cube made of some material? The pressure on top of the cube will still be the same and will result in the same force, and the pressure on the bottom of the cube will still be the same and will result in the same force. The only difference will be in the weight of the cube, represented by W_c. Applying Newton's second law, equation 11.2 becomes:

$$\Sigma F = F_b + (-W_c) = ma$$

If the cube of material weighs more than the cube of water, the buoyant force acting upward is less than the weight acting downward, and the cube will accelerate downward. If the cube of the material weighs less than the cube of water, the buoyant force acting upward is more than the weight acting downward, and the cube will accelerate upward. If you let go of the basketball after submerging it, this would be demonstrated clearly. If the cube of material weighs the same as the cube of water, the buoyant force acting upward equals the weight acting downward, and the cube is in equilibrium. For something to float, then, the buoyant force must equal the weight of the object.

Specific Gravity and Density

Whether or not something floats is determined by the volume of the object immersed and the weight of the object compared to the weight of the same volume of water. **Specific gravity** is the ratio of the weight of an object to the weight of an equal volume of water. Something with a specific gravity of 1.0 or less will float. Another measure that can be used to determine if a material will float is density. **Density** is the ratio of mass to volume. The density of water is about 1000 kg/m³. The density of air is only about 1.2 kg/m³.

$$\rho = \frac{m}{V} \qquad (11.3)$$

where

ρ = density

m = mass

V = volume

Buoyancy of the Human Body

Muscle and bone have densities greater than 1000 kg/m³ (specific gravities greater than 1.0), whereas fat has a density less than 1000 kg/m³ (specific gravity less than 1.0). These differences in density are the basis for the underwater weighing techniques used to determine body composition.

Someone who has low body fat can still float, because the lungs and other body cavities may be filled with air or other gases that are much less dense than water. A forced exhalation of the air out of the lungs may cause a lean person to sink, however. The volume of your chest increases when you inhale and decreases when you exhale, so you actually have some control over the total density of your body. To increase your buoyancy, increase your volume by inhaling forcibly. To decrease your buoyancy, decrease your volume by exhaling forcibly.

Most people are able to float if they take a large breath and hold it. But why is it so difficult to float with your legs horizontal? If you try to float on your back with your legs horizontal, you will notice that your legs will drop or your entire body will rotate toward a more vertical orientation, as shown in figure 11.2.

The reason the body rotates to a more vertical position or the legs drop is that the buoyant force and the force of gravity are not colinear. The line of action of the buoyant force is through the center of volume of the body rather than through the center of gravity. For many objects (the basketball, for instance), these two points will coincide, but for objects with parts of different densities, this may not be the case. Because your legs (mostly muscle and bone) are more dense than your abdomen and chest (lots of cavities filled with air or other gases), the center of volume of your body is farther toward your head than your center of gravity is. The conditions for equilibrium are not met when you try to hold your legs horizontal, because the buoyant force and the gravitational force create a torque. The net torque acting on the body is not zero, so rotation occurs until these forces align and the torque is zero.

Buoyant force is a significant force for activities in the water, but what about land-based activities? Does the air exert a buoyant force? To answer this question, consider a helium balloon, a hot-air

a

b

| Figure 11.2 | When the body is floating horizontally in the water (*a*), the weight and buoyant force are not aligned, thus creating torque. When the body is partially vertical (*b*), the weight and buoyant force are aligned, creating a more stable floating position. |

balloon, or a blimp. What force keeps these things from falling? Air does exert a buoyant force, but the size of this force is very small unless the volume of air displaced is very large. The density of air (at sea level) is approximately 1.2 kg/m³ (a weight density of 11.8 N/m³). A normal-sized human being takes up approximately 1/12 of a cubic meter, so the buoyant force exerted by air on a human body is approximately 1 N = (1/12 m³) (11.8 N/m³). This is about 0.2 lb, an amount so small compared to our weight that we can ignore it.

Dynamic Fluid Force: Force Due to Relative Motion

Buoyant force is the vertical force exerted on an object immersed in a fluid. It is present whether the object is at rest or is moving relative to the fluid. When an object moves within a fluid (or when a fluid moves past an object immersed in it), dynamic fluid forces are exerted on the object by the fluid. The dynamic fluid force is proportional to the density of the fluid, the surface area of the object immersed in the fluid, and the square of the relative velocity of the object to the fluid. Equation 11.4 summarizes this relationship:

$$F \propto \rho A v^2 \qquad (11.4)$$

where

$\propto$ = proportional to

F = dynamic fluid force

ρ = fluid density

A = surface area of the object

v = relative velocity of the object with respect to the fluid

In equation 11.4, fluid density and object surface area are linear terms; an increase in either will cause a proportional increase in the dynamic fluid force. If the area doubles, the dynamic fluid force doubles. If the fluid becomes three times as dense, the dynamic fluid force becomes three times larger. The relative velocity term in equation 11.4 is not linear, it is squared, so if the relative velocity doubles, the dynamic fluid force quadruples (2²). If the relative velocity becomes three times as fast, the dynamic fluid force becomes nine (3²) times larger. Figure 11.3 demonstrates these relationships. Relative velocity is obviously the most important factor in determining dynamic fluid forces, so before going further, an explanation of relative velocity is in order.

> **The dynamic fluid force is proportional to the density of the fluid, the surface area of the object immersed in the fluid, and the square of the relative velocity of the object to the fluid.**

Relative Velocity

When considering the dynamic fluid forces acting on an object, the velocity of the object as well as the velocity of the fluid itself must be considered. Relative velocity is used to represent the effects of these two absolute velocities. Relative velocity is the difference between the object's velocity and the fluid's velocity. Suppose you were standing still on a running track and the wind was blowing in your face at 5 m/s (see figure 11.4a). The relative velocity is the difference between your velocity and the wind's velocity: 0 m/s – (–5 m/s) = 5 m/s. (Because the wind was blowing opposite to your direction, we indicate its velocity with a negative sign.) Now suppose there was no wind and you were running down the track at 5 m/s. The relative velocity is the difference between your velocity and the wind's ve-

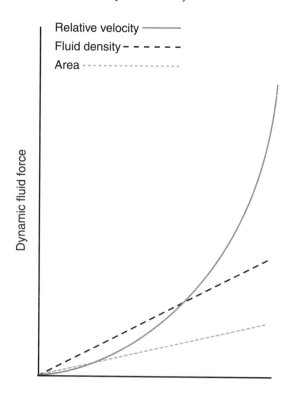

Figure 11.3 The relationship between dynamic fluid force and object area, fluid density, and relative velocity.

locity: 5 m/s – 0 m/s = 5 m/s (see figure 11.4b). The wind that you feel on your body would be the same in both cases. In one case, the air was moving at 5 m/s, and in the other case, you were moving at 5 m/s, but the fluid forces produced by the air were identical.

Suppose the wind was blowing in your face at 5 m/s and you were running at 5 m/s (see figure 11.4c). What is the relative velocity? The relative velocity is again the difference between your velocity and the wind's velocity: 5 m/s – (–5 m/s) = 10 m/s. What if the wind was blowing at your back at 5 m/s and you were running at 5 m/s (see figure 11.4d)? The relative velocity would be 5 m/s – 5 m/s = 0 m/s.

Quantitatively, to determine relative velocity, we take the difference between the object's absolute velocity and the fluid's absolute velocity. Qualitatively, you can consider relative velocity as follows: If you are sitting on the object that is moving through the fluid, how fast does the fluid move past you? The velocity of the fluid moving past you is the relative velocity of the fluid with respect to the object.

Drag Force

For practical reasons, the dynamic fluid force that results from motion within a fluid is commonly resolved into two components: the drag force and the lift force. Figure 11.5 illustrates the resolution of the dynamic fluid force into lift and drag components.

> **The dynamic fluid force that results from motion within a fluid is commonly resolved into two components: the drag force and the lift force.**

Drag force, or drag, is the component of the resultant dynamic fluid force that acts in opposition to the relative motion of the object with respect to the fluid. A drag force will tend to slow down the relative velocity of an object through a fluid if it is the only force acting on the object. Drag force is defined by equation 11.5:

a

b

c

d

Figure 11.4   Relative velocity between a runner and the wind.

a

b

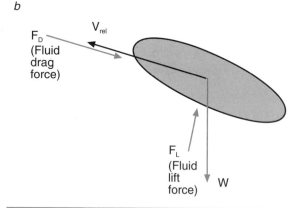

Figure 11.5 The resultant dynamic fluid force (F_R) acting on an object (*a*) and the drag (F_D) and lift (F_L) components of this force (*b*).

$$F_D = C_D \rho A v^2 \qquad (11.5)$$

where

F_D = drag force

C_D = coefficient of drag

ρ = fluid density

A = cross-sectional area of the object perpendicular to the flow direction

v = relative velocity of the object with respect to the fluid

Qualitatively, drag force can be considered in the following manner. Drag is the backward force the molecules of the fluid exert on the object moving relative to the fluid. According to Newton's third law, an equal but opposite force is exerted by the object on the molecules of the fluid. The size of the drag force is thus proportional to the acceleration (slowing down) of the fluid molecules as they pass the object as well as to the mass of the molecules that are slowed down. The greater the decrease in velocity of the molecules and the

faster the rate of this decrease, the greater the total drag.

Drag forces are produced by two different means: surface drag and form drag. Surface drag may be thought of as equivalent to the sum of the friction forces acting between the fluid molecules and the surface of the object (or between the fluid molecules themselves). Form drag may be thought of as equivalent to the sum of the impact forces resulting from the collisions between the fluid molecules and the object.

> **Drag forces are produced by two different means: surface drag and form drag.**

Surface Drag

Surface drag is also called skin friction or viscous drag. As a fluid molecule slides past the surface of an object, the friction between the surface and the molecule slows down the molecule. On the opposite side of this molecule are fluid molecules that are now moving faster than this molecule, so these molecules are also slowed down as they slide past the molecules closest to the object. These molecules, in turn, slow down the molecules next to them. So the surface drag is proportional to the total mass of the molecules slowed down by the friction force and the average rate of change of velocity of these molecules.

The size of the surface drag is affected by the factors listed in equation 11.5: the coefficient of drag, the density of the fluid, the cross-sectional area of the object, and the square of the relative velocity. The coefficient of drag is influenced by several other factors associated with surface drag. The roughness of the surface is one of these factors. Rougher surfaces will create larger friction forces between the fluid molecules and the object. Thus the equipment used and clothing worn by athletes in many activities are smooth so as to reduce surface drag. Examples of surface drag reduction in sports are shown in figure 11.6 and include the sleek, skintight suits worn by speedskaters, cyclists, downhill skiers, lugers, and even sprint runners. Swimmers shave their bodies to reduce surface drag.

The coefficient of drag is also influenced by the viscosity of the fluid associated with surface drag. Viscosity is a measure of the internal friction between the layers of molecules of a fluid or the resistance of a fluid to shear forces. Slower-flowing fluids are more viscous than faster-flowing fluids. Motor oil is more viscous than water, which is more vis-

Figure 11.6 Athletes in certain sports wear skintight apparel to reduce surface drag.

cous than air. Larger friction forces between the fluid and the surface of an object moving through it are created by fluids with greater viscosity, so surface drag increases with increased viscosity. As most athletes have no control over the properties of the fluid they move through, examples of making changes in fluid viscosity in sports are rare.

Form Drag

Besides surface drag, form drag is the other means by which drag is produced. Form drag is also called shape drag, profile drag, or pressure drag. As a fluid molecule first strikes the surface of an object moving through it, it bounces off, but then the molecule strikes another fluid molecule and is pushed back toward the surface of the object. Thus the molecule will tend to follow the curvature of the object's surface as the object moves past it. Because a change in direction is an acceleration, the object must exert a force on the molecule to change its direction. The molecule, in turn, exerts an equal but opposite force on the object as it changes direction. The larger the changes in direction, the larger the forces exerted.

On the leading surfaces of the object, the forces exerted by the fluid molecules have components directed toward the rear of the object. These forces contribute to the form drag. On the trailing surfaces of the object, the forces exerted by the fluid molecules have components directed toward the front of the object. These forces reduce the form drag; however, these forces are only present if the fluid molecules stay close to the surface of the object by being deflected by other fluid molecules. This occurs during what is known as laminar flow, as shown in figure 11.7a. But if the change in the surface curvature is too large or the relative velocity is too fast, the impact forces between fluid molecules are not large enough to deflect the molecules back toward the surface, and the molecules separate from the surface. Turbulent flow results, and the molecules no longer press against the surface of the object. If the fluid molecules are not pressing against the trailing surface of the object, then nothing is, and a vacuum may be created behind the object. Less force (or no force at all) is thus exerted against the trailing surfaces of the object. The difference between these forward- and backward-directed forces is form

a　　　Laminar flow

Laminar
b　flow

Turbulent
flow

Figure 11.7	Examples of objects experiencing laminar flow (*a*) and laminar and turbulent flow (*b*).

drag. Form drag thus increases as the amount of turbulent flow increases. Figure 11.7b illustrates an example of an object experiencing laminar and turbulent flow.

Like surface drag, the size of the form drag is affected by the factors listed in equation 11.5: drag coefficient, density of the fluid, cross-sectional area of the object (frontal area), and the square of the relative velocity. The drag coefficient is influenced by several other factors associated with form drag as well. The shape of the object has the greatest influence on the form drag contribution to the drag coefficient. To minimize turbulent flow, the fluid molecules have to follow the shape of the object. If the molecules must change direction quickly, large forces are required to cause this acceleration, so the reaction forces the molecules exert against the object are large, or turbulence occurs. To reduce form drag, then, the surface of an object should curve gently or be "streamlined," without any protrusions or rough spots. An aerodynamic or streamlined shape is one that is long in the direction of flow, so that the surface curvature of the object is gentle and not sudden. The tail of the streamlined object essentially fills up the vacuum or empty space created by the turbulent flow. An illustration of non-streamlined and streamlined shapes is shown in figure 11.8, along with examples of how athletes adopt streamlined positions or use equipment with streamlined shapes.

Like surface drag, form drag is also influenced by surface texture. A rough surface will cause turbulent flow at lower velocities than will a smooth surface. Because turbulence increases form drag, in most cases, an object with a rougher surface will experience greater form drag than will an object with a smoother surface.

Sometimes, however, a rougher surface will actually decrease form drag. This paradox is due to the turbulent flow induced by the rough surface. When turbulent and laminar flow are present, form drag is large due to the difference in pressure (and thus forces) pushing against the leading and trailing surfaces of the object. When turbulent flow occurs at the leading surface of the object and all the flow around the object is turbulent, form drag actually decreases. How does this happen? When the fluid flow around the object is completely turbulent, a layer of turbulent fluid surrounds the object and is carried along with it. The flow of the rest of the fluid is laminar. This laminar-flowing fluid doesn't exert much drag force on the object and its layer of turbulent fluid. Figure 11.9 illustrates this concept. The dimples on a golf ball reduce the form drag the ball experiences by creating a layer of turbulent air around it. The fuzz on a tennis ball and the stitches on a baseball may produce the same effect.

Strategies for Reducing Drag Force

In many sports, athletes want to minimize drag forces to maximize their performance. Equipped with our knowledge of the sources of drag forces, let's summarize the ways athletes can reduce drag force by examining the variables in equation 11.5, which is repeated here:

$$F_D = C_D \rho A v^2 \qquad (11.5)$$

where

F_D = drag force

C_D = coefficient of drag

ρ = fluid density

A = cross-sectional area of the object perpendicular to the flow direction

v = relative velocity of the object with respect to the fluid

Four variables occur in equation 11.5: drag coefficient, fluid density, area of the object, and relative velocity. Fluid density is usually not considered to be a variable under the athlete's control. However, record attempts in cycling and track-and-field

Figure 11.8 Examples of nonstreamlined and streamlined shapes that affect the magnitude of form drag.

events have been specifically planned at high-altitude locations where air density is lower. An athlete may also attempt to control this variable by choosing warm-weather competitions, as warmer air is less dense than cooler air and humid air is less dense than dry air.

Athletes do have several means of reducing the drag coefficient of their bodies and equipment. One way is to make the body surfaces and clothing (or equipment) smoother. Another way is to streamline the shape of the body or equipment. For the athlete, positions that put the body or parts of the body in an elongated orientation relative to the fluid flow will reduce form drag. Cyclists, swimmers, speed-skaters, downhill skiers, and lugers all manipulate their positions to reduce form drag. A third way is

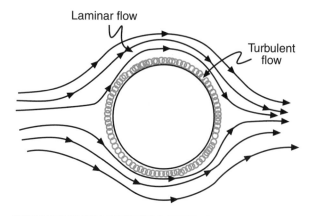

Laminar flow

Turbulent flow

Figure 11.9 A rough surface might decrease form drag by creating a layer of turbulent flow completely surrounding the object on the leading and trailing edges. Outside of this turbulent layer is laminar flow.

to reduce the total surface area exposed to the flow, which reduces surface drag. Swimmers (and other aquatic athletes) may also have one other means of reducing the drag coefficient: by choosing pools with warmer water. Warmer water is less viscous (and thus creates less surface drag) than cooler water.

> **Form drag accounts for most of the drag force at faster velocities, whereas surface drag accounts for most of the drag force at slower velocities.**

This difference must be considered when deciding on strategies to use in reducing drag. Exposing a smaller surface area to the fluid flow will reduce surface drag, yet streamlining a shape to reduce form drag usually involves an increase in surface area. If you or your equipment are moving through the fluid at fast speeds (greater than 10 m/s or 20 mi/hr), go for the streamlined shape. Otherwise, try to reduce the surface area.

The cross-sectional (or frontal) area of the athlete or equipment may also be controlled to some extent. Basically, the athlete wants to reduce the frontal area exposed to the flow. Most of the streamlined positions taken to reduce form drag in the above examples also reduce frontal area. Besides making the surface smoother, skintight clothing also reduces frontal area. Compare a baggy T-shirt and shorts to a Lycra bodysuit. Athletes orient the implements they use in various sports to minimize frontal area. A football is thrown so that the airflow is along its long axis, and similar orientations are used with a javelin, discus, or Frisbee. These orientations minimize drag.

Relative velocity is the last variable in equation 11.5. Because this term is squared, it has the greatest effect on drag force, so it is the most important variable that the athlete can control. But how can athletes control this variable to minimize drag—when it's their velocity that they want to maximize? One way is by timing a performance. Running velocity is an important determinant of success for long jumpers, triple jumpers, and pole vaulters. They have 1 or 1.5 min to make a jump, and most will use this time to wait for the tailwind to pick up (or for the headwind to slow down). For the same reason, track-and-field meet promoters will often choose to run the straightaway races in the direction of the prevailing winds. However, records in the long jump, the triple jump, and the short sprints and hurdles do not count if the wind velocity in the direction of the jump or sprint is greater than 2.0 m/s.

Runners, cyclists, and swimmers use another way to reduce the relative velocity. Runners and cyclists rely on a teammate or opponent to lead the race, and they draft the leader. Swimmers may ride the wake of another swimmer who is leading them. In these cases, the fluid has been disturbed by the person leading the athlete. The relative velocity of the fluid passing the trailing athlete is thus much lower, and the drag forces exerted on this athlete are smaller. An athlete must run or ride very close to the athlete in front of them to maximize this effect, however. An athlete who leads a cycling or running race is doubly disadvantaged by this tactic. First, the leading athlete must work harder because the drag force is larger due to the faster relative velocity of the fluid. And second, the athletes behind the leader (if they can stay close enough) don't have to work as hard because the drag force is lower due to the slower relative velocity of the fluid. These athletes are thus more likely to sprint past the leader at the end of the race!

Lift Force

Lift force is the dynamic fluid force component that acts perpendicular to the relative motion of the object with respect to the fluid. Rather than opposing the relative motion of the object through the fluid, the effect of lift force is to change the direction of the relative motion of the object through the fluid. The word *lift* implies that the lift force is directed upward, but this is not necessarily the case. A lift force can be directed upward, downward, or in any direction. The possible directions of the lift force are determined by the direction of flow of the fluid. The lift force must be perpendicular to this flow. Lift force is defined by equation 11.6:

$$F_L = C_L \rho A v^2 \qquad (11.6)$$

where

F_L = lift force

C_L = coefficient of lift

ρ = fluid density

A = cross-sectional area (frontal area) of the object perpendicular to the flow direction

v = relative velocity of the object with respect to the fluid

Qualitatively, lift force can be considered in the following manner. Lift is caused by the lateral deflection of fluid molecules as they pass the object. The object exerts a force on the molecules that causes this lateral deflection (an acceleration, because the molecules change direction). According to Newton's third law, an equal but opposite lateral force is exerted by the molecules on the object. This is the lift force. Lift force is thus proportional to the lateral acceleration of the fluid molecules and the mass of the molecules that are deflected. Figure 11.10 graphically illustrates lift force.

Lift force is the dynamic fluid force component that acts perpendicular to the relative motion of the object with respect to the fluid.

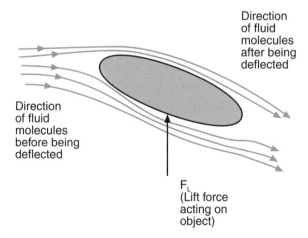

F_L
(Lift force acting on object)

Direction of fluid molecules after being deflected

Direction of fluid molecules before being deflected

Figure 11.10	The fluid molecules passing by an object are deflected laterally. The change in direction is a lateral acceleration caused by the force exerted by the object. The reaction to this force is the lift force acting on the object.

SELF-EXPERIMENT

You can demonstrate lift force to yourself by reaching your hand out the window of a moving car. With your fingers extended and your palm facing the ground, slowly supinate your forearm so that your palm faces forward. As you do this, you'll notice forces that try to push your hand backward (drag force) or upward (lift force). These forces will change as the orientation of your hand changes relative to the airflow or as the velocity of the airflow changes. Now clench your fingers into a fist and try the same thing. Are the lift forces the same? The shape of the object in the fluid flow also affects lift force.

The example of your hand in the airflow beside the car demonstrates several factors that influence lift force. Again, relative velocity is the most important. Increasing the car's velocity (and thus the relative velocity of the air) greatly increases the lift force, but the shape and orientation of your hand greatly affect lift force as well. These two factors are partly accounted for in equation 11.6 by the frontal area of the object. The other variable that takes these factors into account is the coefficient of lift.

Some common examples of the use of lift forces, shown in figure 11.11, may further clarify this force. Wings (airfoils) on planes use lift force to keep the aircraft from falling. Likewise, submerged wings (hydrofoils) on some boats (hydroplanes) use lift force to lift the hull of the boat out of the water. A rudder on a boat or plane uses lift force to change the direction of the craft. Propellers on boats and airplanes use lift forces to actually propel the craft. The spoilers on the rear of some race cars use lift force to push the rear-drive wheels against the ground for better traction. Ski jumpers use long, wide skis and position themselves so that the lift forces generated keep them in the air longer so they jump farther. When you tread water, you scull your hands in a horizontal plane, and the lift forces generated by your hands help keep your head above water. The propulsion techniques used in various swimming strokes are a resultant of the lift and drag

Figure 11.11   Examples of objects that create lift forces.

forces acting on the hands. Sails on sailboats and sailboards use lift forces to propel them forward when they are tacking.

In most of these examples, the object that generates the lift force is longer in the dimension parallel to the flow and shorter in the dimension perpendicular to the flow. Also, lift is generated in most of these examples when the longer dimension of the object is not aligned parallel to the fluid flow. Some objects will generate lift even though they are not longer in one dimension than another (e.g., a spinning ball), or even if their longest dimension appears to align with the fluid flow (e.g., an airfoil). How do these objects deflect the fluid laterally to generate lift forces?

Bernoulli's Principle

In 1738, Daniel Bernoulli (1700-1782), a Swiss mathematician, discovered that faster-moving fluids exert less pressure laterally than do slower-moving fluids. This principle, called Bernoulli's principle, may explain why some objects are able to generate lift forces without being longer in one dimension than another or if their longest dimension is aligned with the flow.

Faster-moving fluids exert less pressure laterally than do slower-moving fluids.

Let's examine an airfoil to see how Bernoulli's principle works. An airfoil is an example of an object whose longest dimension appears to align with the fluid flow, even while it creates lift. Figure 11.12a shows an airfoil in cross section. One surface (the upper surface of an airplane wing) of the airfoil is more curved than the other. The curvature is very gentle, and the airfoil is streamlined. When oriented such that its long dimension aligns with the fluid flow, the airfoil's streamlined shape produces laminar flow (and minimal drag). Imagine two rows of four air molecules each as they approach the airfoil (figure 11.12b). When they strike the leading edge of the airfoil, one row travels along the upper, curved surface of the airfoil and the other row travels along the lower, flat surface. If the flow is laminar, each molecule on top reaches the trailing edge of the airfoil at the same time as the corresponding molecule on the bottom. The molecules on the upper surface travel farther to get to the trailing edge of the airfoil, but they get there at the same time as the molecules on the lower surface. The

a

b

c

d

| Figure 11.12 | A possible explanation for Bernoulli's principle, using an airfoil as an example. |

molecules on top are moving faster, so according to Bernoulli, the lateral pressure exerted by the faster-moving molecules is less than that exerted by the slower-moving molecules, and an upward lift force

is generated. The faster-moving upper molecules have a downward component to their velocity as they slide off the trailing edge of the airfoil, so some downward deflection of the air occurs.

Now let's try to determine the basis for Bernoulli's principle. Back to the two rows of molecules on the airfoil. When the first molecules reach the trailing edge of the airfoil, the molecules will be distributed along the surfaces of the airfoil, as shown in figure 11.12c. All of these molecules will exert some force against the surface of the airfoil due to the pressure from adjacent molecules. The forces exerted by the molecules on the lower surface of the airfoil will be directed normal to this surface, or upward. The forces exerted by the molecules on the upper surface of the airfoil will be directed perpendicular to this surface. But the upper surface is curved, so these forces will be directed downward as well as backward or forward, as shown in figure 11.12d. If the resultant force each molecule exerts is the same from the upper to the lower surfaces, the net upward force exerted by the molecules on the lower surface is greater than the net downward force exerted by the molecules on the upper surface, because the forces from the molecules on the upper surface have forward or backward components as well. The difference in these net forces is the lift force that is generated.

Spin and the Magnus Effect

In 1852, the German scientist, Gustav Magnus, noticed that lift forces are also generated by spinning balls. This effect is called the **Magnus effect**, and a lift force caused by a spin is called a Magnus force. But how can an object that doesn't have any broad, relatively flat surface deflect the air laterally to cause a lift force? Let's examine what happens to the air molecules as they approach and pass a ball with topspin. Figure 11.13 shows an illustration of a ball (moving left to right across the page) with topspin.

> **A lift force caused by a spin is called a Magnus force.**

Notice that on a ball with topspin, the upper surface of the ball has a forward velocity (to the right in figure 11.13) relative to the center of the ball, and the lower surface has a backward velocity (to the left) relative to the center of the ball. The air molecules, on the other hand, all have a backward velocity (to the left) relative to the center of the ball. When molecules strike the lower surface of the ball, they don't slow down as much because this surface

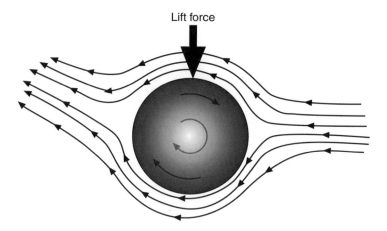

Figure 11.13   A ball with topspin has a lift force that acts downward on the ball.

is moving in the same direction as the molecules (backward or to the left) relative to the center of the ball. When the molecules strike the top surface of the ball, they are slowed down more because this surface is moving in the opposite direction (forward or to the right) relative to the center of the ball. The velocity of the air over the top surface of the ball is lower than the velocity of the air over its bottom surface. According to Bernoulli's principle, then, less pressure will be exerted by the faster-moving molecules on the bottom surface of the ball. This difference in pressure results in a lift force acting downward on the ball, as shown in figure 11.13.

The lift force can also be explained by imagining how the slower-moving molecules get bunched up on the ball's upper surface, whereas the faster-moving molecules spread out on the ball's lower surface. The bunched-up molecules exert more pressure (and thus more force) on the ball than do the spread-out molecules. Imagine a crowd of people exiting a theater or arena through only one or two doors. As they approach the doors, they bunch together more and slow down, and more bumping occurs. This is analogous to what happens to the molecules on top of the ball. More bumping translates to more pressure.

You may also notice that the faster molecules on the bottom of the ball have an easier time staying on the surface of the ball. They don't break away from the ball until they are past the most rearward point on the ball. Thus the molecules are deflected upward (and the ball is deflected downward).

Lift force due to the Magnus effect is responsible for the curved flights of balls observed in a variety of sports. Backspin will keep a ball aloft longer, whereas topspin will cause a ball to drop to the ground sooner. Sidespin will cause a ball to swerve right or left. Golf clubs are designed to impart back-spin on the ball so that it stays in the air longer and thus carries farther. Many golfers unintentionally impart sidespin to the ball that causes it to hook or slice to the left or right. Professional golfers may use this sidespin intentionally. Baseball pitchers use sidespin, topspin, or backspin in their repertoire of pitches to make the path of the pitched ball less predictable to the batter. Soccer players use sidespin to cause their corner kicks or penalty kicks to swerve around opposing players. Tennis players use top-spin to force their shots to drop into the court sooner. Perhaps the most striking example of the Magnus effect in a ball sport is in table tennis, where the spin of the ball causes wide swerves in its flight.

Center of Pressure

The resultant of the lift and drag forces that act on an object is the dynamic fluid force. This force is actually the result of pressures exerted on the surfaces of the object. The theoretical point of application of this force to the object is called the center of pressure. If the resultant force acting at the center of pressure is not on a line passing through the center of gravity of the object, a torque is produced that will cause the object to rotate. A Frisbee or discus or football thrown with no spin is more likely to wobble because of this torque. An airplane whose cargo or passengers are loaded too far to the rear will have a dangerous torque act on it that will tend to tip the nose of the plane upward. The IAAF specifications adopted in 1986 for men's javelins essentially force the center of pressure to be behind the grip (and the center of gravity) of the javelin. The torque produced by the dynamic forces acting on the javelin during flight cause it to rotate so that the tip of the javelin strikes the ground. Figure 11.14 illustrates this effect.

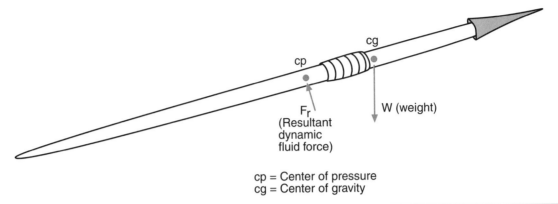

cp = Center of pressure
cg = Center of gravity

Figure 11.14 The forces acting on a javelin—its weight and the dynamic fluid force—act at different locations, thus creating a torque that causes the javelin to rotate so the tip drops.

> **The resultant of the lift and drag forces that act on an object is the dynamic fluid force. The theoretical point of application of this force to the object is called the center of pressure.**

Effects of Dynamic Fluid Forces

The dynamic fluid forces that act on an object moving through a fluid have been explained in the previous pages. But athletes are more concerned with the effects of these forces than with the forces themselves. Newton's second law is expressed as:

$$\Sigma F = ma \qquad (11.7)$$

where

ΣF = net force

m = mass of the object

a = acceleration of the object

The effect of a force is the acceleration that would be caused if this were the net force acting, or:

$$\Sigma F = ma \qquad (11.7)$$

$$\frac{\Sigma F}{m} = a$$

The effects of dynamic fluid forces can be determined from equation 11.4:

$$F \propto \rho A v^2 \qquad (11.4)$$

$$a = \frac{\Sigma F}{m} \propto \frac{\rho A v^2}{m}$$

$$a \propto \frac{\rho A v^2}{m}$$

Accelerations are the effects of dynamic fluid forces, and these accelerations are related to all of the factors that the dynamic fluid forces are related to; however, the acceleration is inversely proportional to the mass of the object. This means that two objects similar in size and shape will experience the same dynamic fluid forces, but the more massive object will experience less acceleration. The wind blowing in the face of a 50-kg distance runner will have a greater effect than the same wind blowing in the face of a 70-kg distance runner. It's easier to throw a curveball with a Nerf ball than with a real baseball. The Magnus force on a table tennis ball has a much greater effect than the same Magnus force on a racquetball. The lift force acting on a Frisbee is similar to the lift force acting on a discus, but the effects are much different. The relative effects of lift and drag forces on projectiles can be seen in figure 11.15. In this figure, the trajectories of a variety of balls are shown. The initial velocities, angles, and heights of projection of these objects are identical, but their masses are different.

> **Two objects similar in size and shape will experience the same dynamic fluid forces, but the more massive object will experience less acceleration.**

Summary

Fluid forces affect movements in a variety of sports. Buoyant force affects all aquatic sport participants, and dynamic fluid forces affect participants in aquatic as well as land-based sports where fast velocities are involved. Buoyant force is an upward force equal in magnitude to the weight of water

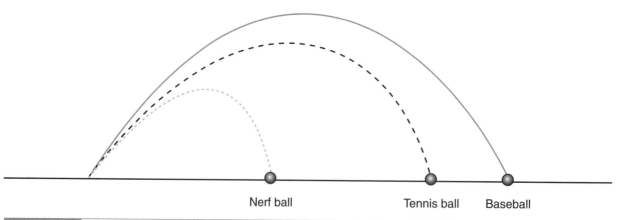

| **Figure 11.15** | The trajectories of three balls with different masses, demonstrating that the effect of fluid forces is influenced by the mass of the object. |

Nerf ball   Tennis ball   Baseball

displaced by the immersed object. The dynamic fluid force is resolved into drag and lift components and is proportional to the density of the fluid, the frontal area of the object, and the square of the relative velocity of the fluid. Relative velocity of the fluid thus has the greatest influence on drag and lift forces, because these forces are proportional to

the square of the velocity. Drag can be reduced by adopting a streamlined shape, keeping the surface of the object smooth (in most cases), and reducing the frontal area. Lift can be controlled by the shape and orientation of an object. Spin on a ball can also cause lift, thus making ball games much more interesting.

Key Terms

Bernoulli's principle *(p. 204)*

buoyant force *(p. 192)*

center of pressure *(p. 206)*

density *(p. 195)*

drag force *(p. 197)*

dynamic fluid force *(p. 206)*

form drag *(p. 198)*

laminar flow *(p. 199)*

lift force *(p. 202)*

Magnus effect *(p. 205)*

pressure *(p. 193)*

specific gravity *(p. 194)*

surface drag *(p. 198)*

turbulent flow *(p. 199)*

viscosity *(p. 198)*

REVIEW QUESTIONS

1. **What drag force is reduced by putting dimples on a golf ball?**

2. **In a human-powered vehicle such as a bicycle, the rider's body creates most of the drag force. If you were designing a human-powered vehicle, which rider position would minimize drag force: upright; crouched, as on a racing bicycle; reclining, as on a recumbent bicycle; or prone, with head and arms forward?**

3. Why are bugs (those that cannot fly) often able to walk away from falls that would be lethal to humans?

4. How do the stitches on baseballs affect their flight?

5. How does the spin of a tennis ball affect its flight?

6. In which sporting event does air resistance have the greatest effect: 100-m sprinting; 10,000-m running; or 500-m speedskating?

7. How is the optimal angle of projection in the discus throw affected by air resistance on the discus during flight?

8. Rank the following items according to the effect air resistance has on each of them: shot, discus, baseball, Nerf ball, basketball, beach ball, tennis ball, Frisbee.

9. Explain how it is possible for a sailboat to have a velocity component into the wind.

10. Why do swimmers shave their bodies?

Internal Biomechanics: Internal Forces and Their Effects on the Body and Its Movement

The movements of the body as a whole are determined by the external forces that act on it, but what about the internal forces that hold the body together? How do these forces affect the body and its movements? These questions are addressed in part III, most of which is devoted to the *bio-* part of biomechanics. Part III includes four chapters. Chapter 12 is concerned with the mechanics of biological materials. The next two chapters, chapters 13 and 14, are concerned with the structural mechanics of the musculoskeletal system. Chapter 15 presents a brief overview of the control of the neuromuscular system.

12

Mechanics of Biological Materials: Stresses and Strains on the Body

© Claus Andersen

Afootball running back is trying to gain enough yardage for a first down when he is hit from the side by a linebacker. A loud pop is heard as the running back's knee bends sideways. Ouch! The force from the linebacker's hit was directed medially against the lateral part of the knee, and the medial collateral ligament of the knee snapped. Why wasn't the knee able to sustain this load without injury? Bones break, ligaments tear, cartilage rips, and muscles pull. These failures occur when the loads imposed cause stresses that exceed the strength of the material. This chapter explains the mechanical behavior of biological materials and how they are able to withstand the loads they encounter.

OBJECTIVES

When you finish this chapter, you should be able to do the following:

▌ Define mechanical stress

▌ Define strain

▌ Define tension (tensile strain and tensile stress)

▌ Define compression (compressive strain and compressive stress)

▌ Define shear (shear strain and shear stress)

▌ Identify and describe bending loads

▌ Describe the stress patterns that develop under a bending load

▌ Identify and describe torsion loads

▌ Describe the general stress-strain relationship for an elastic material

▌ Define elastic modulus (Young's modulus)

▌ Describe elastic and plastic behavior

▌ Define the various descriptors of material strength: yield strength, ultimate strength, and failure strength

▌ Define stiff, ductile, brittle, and pliant

▌ Define toughness

▌ Understand the material properties of bone

▌ Understand the material properties of tendons and ligaments

▌ Understand the material properties of muscle

The external forces that act on the body affect the movements of the entire body. These forces also impose loads on the body that affect the internal structures of the body—cartilage, tendons, ligaments, bones, and muscle. An understanding of the mechanical properties of these internal structures is important for preventing injury and evaluating the causes of injury.

> **The external forces that act on the body impose loads that affect the internal structures of the body.**

Stress

In part II, we considered the body to be a system of linked rigid segments and thus used rigid-body mechanics to study the effects of the forces acting on the body. In part III, we will discard our assumption of rigidity and consider the segments to be deformable bodies. The external forces that act on the body are resisted by internal forces and cause deformation of the body. The amount of deformation produced is related to the stress caused by the forces and the material that is loaded.

Mechanical **stress** is the internal force divided by the cross-sectional area of the surface on which the

SELF-EXPERIMENT

Find a couple of rubber bands. Take one rubber band and pull it until it is twice as long as it was. The rubber band stretches or deforms because of the force exerted. Now take another rubber band and pull on it, as well as on the first one. If you don't have two rubber bands, double the first one by folding it in half. Pull with about the same force you needed to double the length of the single rubber band. Do the two rubber bands (or the doubled-over rubber band) stretch as much as the single rubber band did under the same force?

The deformation or stretch of the two rubber bands (or the one doubled-over rubber band) was not as large as that of the single rubber band unless you exerted twice as much force. The stress in the single rubber band was twice as large as the stress in the double rubber band.

internal force acts. Stress may vary within an object and is associated with a specific internal surface. The three principal stresses are tension, compression, and shear. Mathematically, stress is represented by the Greek letter σ (sigma) and is defined as:

$$\sigma = \frac{F}{A} \qquad (12.1)$$

where

σ = stress

F = internal force

A = cross-sectional area of the internal surface

As force is measured in newtons and area is measured in units of length squared (m^2), the unit for stress is N/m^2, which is a derived unit also called a pascal (for the French physicist, philosopher, and mathematician, Blaise Pascal).

> **Mechanical stress is the internal force divided by the cross-sectional area of the surface on which the internal force acts.**

Before going further, a clear explanation of the internal force, F, and the cross-sectional area, A, in equation 12.1 must be presented. Consider the single rubber band. If it is held still in a stretched position, it is in static equilibrium. A free-body diagram of the rubber band is shown in figure 12.1a.

If we ignore the weight of the single rubber band in the self-experiment above, the pulling force, P_1, at the left end must equal the pulling force, P_2, at

the right end. Let's call the magnitude of this force P. Now imagine a plane that cuts at right angles through the rubber band at A-A. A free-body diagram of the rubber band to the left of this plane is shown in figure 12.1b. A free-body diagram of the rubber band to the right of the imaginary cut plane is shown in figure 12.1c. The piece of rubber band shown in figure 12.1b is not in static equilibrium, as shown. The pulling force, P, that acts on the left end of the rubber band must be countered by some force, equal in magnitude to P, that acts to the right. That force is shown in figure 12.1d. This is the force that is used to define stress in equation 12.1. What agent is responsible for this force?

The bonds between the molecules of the rubber band to the right and left of the plane of the imaginary cut produce the pulling force, P, that acts at the cut. This force does not act at a single point, as depicted in figure 12.1d, however. This force actually represents the resultant of all the individual intermolecular bond forces that act across the surface of the imaginary cut and hold the rubber band together (see figure 12.1e). The cross-sectional area of the rubber band at this imaginary cut plane (let's call this the analysis plane) is the area, A, used to define the stress at this plane in equation 12.1.

Tension

Tension is one of the two axial stresses (also called normal or longitudinal stresses), and one of the three principal stresses. The stress described and defined above is an example of tensile stress. Tensile stress is the axial or normal stress that occurs at the

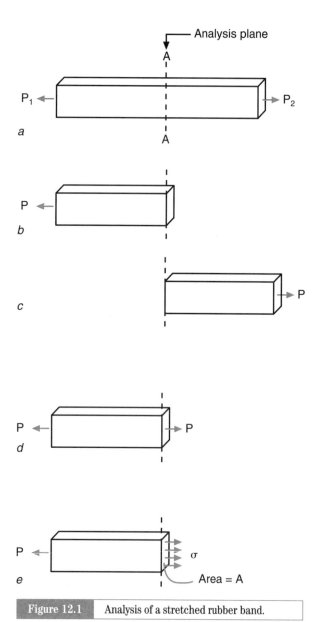

Figure 12.1   Analysis of a stretched rubber band.

mension than in the other two dimensions. Stress in such structures is typically analyzed by considering analysis planes that cut through the structure perpendicular to its longest dimension. The stresses that act perpendicular to the analysis planes (and thus along the long axis of the structure) are called axial, normal, or longitudinal stresses. Consider your humerus (the bone between your elbow and shoulder). If forces pulled on both ends of your humerus, which might be the case when you hang from a chin-up bar, the humerus would be loaded in axial tension. This is an example of axial loading, a loading situation in which the forces act in the direction of the long axis of the bone. A free-body diagram of this situation is shown in figure 12.2.

If an analysis plane were cut perpendicular to the long axis of the bone, as shown in figure 12.2b, we could determine the stress in the bone at that location, as we did with the rubber band. Divide the load acting across the plane by the cross-sectional area of the bone at the surface of the plane. Because the force tends to pull the bone apart, and the stress acts away from the analysis plane, the stress is a tensile stress.

The rubber band had the same cross-sectional area along its length, so no matter where you placed the analysis plane, the stress was the same. The cross-sectional area of the humerus changes as you move proximally to distally along its length. The stress varies with these changes as well. Where the cross-sectional area is larger, the stress is smaller and the bone is stronger. Where the cross-sectional area is smaller, the stress is larger and the bone is weaker.

When an object or material is axially loaded in tension with forces pulling on either end, tensile stress is produced within the object, and the object also tends to deform by stretching or elongating in the direction of the external loads. For most materials, this elongation is directly proportional to the magnitude of the stress. The increase in deformation of the rubber band was easily seen as it stretched farther with an increasing load (and increasing tensile stress). In the human body, very large tensile loads may sprain or rupture ligaments and tendons, tear muscles and cartilage, and fracture bones.

analysis plane as a result of a force or load that tends to pull apart the molecules that bond the object together at that plane. Tensile stress acts perpendicular or normal to the analysis plane and is thus called a normal stress or axial stress.

> **Tensile stress is the axial or normal stress that occurs at the analysis plane as a result of a force or load that tends to pull apart the molecules that bond the object together at that plane.**

Many anatomical structures (long bones, muscles, tendons, ligaments) are longer in one di-

> **When an object or material is axially loaded in tension, the object tends to deform by stretching or elongating in the direction of the external loads.**

| Figure 12.2 | The humerus is loaded axially in tension when you do a chin-up. |

Compression

Compression or compressive stress is the axial stress that results when a load tends to push or squash the molecules of a material more tightly together at the analysis plane. Your femur and tibia are under compression when you are standing, as a result of your body weight pushing down on the proximal end and the reaction force from below pushing up on the distal end.

Let's take a look at the humerus again, but this time when it is axially loaded in compression. During a push-up, forces push on each end of the humerus and it is axially loaded in compression (see figure 12.3).

> **Compression or compressive stress is the axial stress that results when a load tends to push or squash the molecules of a material more tightly together at the analysis plane.**

As before, an analysis plane is cut through the humerus. A free-body diagram is drawn of the humerus on either side of the analysis plane, and the stress is determined by dividing the internal force acting at the cut surface by the cross-sectional area of the humerus at the analysis plane. In this case, because the force acting at the analysis plane is pushing toward the surface of the plane, the resulting stress is compressive.

When an object is axially loaded in compression with forces pushing on either end, compressive stress is produced within the object, and the object tends to deform by shortening in the direction of these external forces. If you squeeze a rubber ball, the deformation of the ball can be easily seen. The ball is under compression. In the human body, large compressive loads may cause bruising of soft tissues and crushing fractures of bones.

> **When an object is axially loaded in compression, the object tends to deform by shortening in the direction of these external forces.**

Shear

The third of the three principal stresses is shear. Compression and tension are axial stresses that act perpendicular to the analysis plane as a result of forces acting perpendicular to this plane. Shear stress is a transverse stress that acts parallel to the analysis plane as a result of forces acting parallel to this plane. These forces tend to slide the molecules of the object past each other.

The humerus is loaded axially in compression when you do a push-up.

SELF-EXPERIMENT

Take a pencil (or pen) and hold it near the middle between the thumb and index finger of your right hand. Place your left index finger on top of the pencil, just to the left of your left thumb, as shown in figure 12.4a. Hold the pencil still while pushing down with your left index finger and pushing up with your right thumb. A free-body diagram of the pencil is shown in figure 12.4b.

Consider an analysis plane cut through the pencil just between the two forces created by your right thumb and left index finger. A free-body diagram of the left half of the pencil is shown in figure 12.4c. Because the left half of the pencil is in static equilibrium, the force pushing down on the pencil must be counteracted by a force pushing up to maintain force equilibrium. No other external forces act on the left half of the pencil, so the counteracting force must be an internal force acting at the analysis plane. Figure 12.4d shows this force, which arises from the molecular bonds between the two surfaces of the pencil at the analysis plane.

Figure 12.4   Analysis of a pencil withstanding a shear load.

> **Shear stress is a transverse stress that acts parallel to the analysis plane as a result of forces acting parallel to this plane.**

Shear stress at the analysis plane is equal to the force at the analysis plane (which acts parallel to the analysis plane) divided by the cross-sectional area of the pencil at the analysis plane, or:

$$\tau = \frac{P}{A} \qquad (12.2)$$

where

   τ = shear stress

   P = shear force

   A = cross-sectional area at the analysis plane

Whereas axial stresses (tension and compression) are represented by the symbol σ (sigma), transverse stress (shear) is represented by its own symbol, τ (tau).

Scissors are an example of a tool that creates large shear forces, thus creating large shear stresses in the material worked on. Scissors are also referred to as "shears." In the human body, shear loads cause blisters of the skin, and large shear loads acting on the extremities may cause joint dislocations or shear fractures of bones. A shear load causes a deformation of the object as well, but rather than stretching or squashing it, shear causes a change in the orientation of the sides of the object, or a skewing. The three different types of stress are illustrated in figure 12.5.

Mechanical Loads

The external forces acting on an object impose a mechanical load on the object. Specific arrangements of external forces will produce different types of loads. In other words, the number, direction, and location of the external forces acting on an object, as well as the shape of the object itself, define the type of load imposed. Simple tensile, compressive, or shear loads were described in the examples above. In these examples, the external forces acted along the same line (uniaxial load), the analysis plane was chosen perpendicular to the long axis of the loaded object, and the stresses present at the analysis plane were of only one type—tension alone, compression alone, or shear alone. When you stretched the rubber band, it was an example of a uniaxial tensile load, and the stress produced at the analysis plane was tension alone. This stress was also uniform across the analysis plane (i.e., at any point within the analysis plane, the tensile stress was the same as at any other point within the analysis plane).

a   Tensile stress

b   Compressive stress

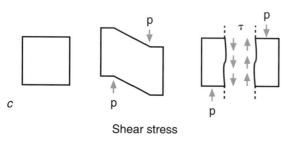

c   Shear stress

Figure 12.5 Illustration of simple loads that produce the three mechanical stresses: tension (*a*), compression (*b*), and shear (*c*).

Bending

Simple uniaxial loads produce only one type of stress, which is uniform across the analysis plane. Most loading situations are not so simple, and multiple stresses occur at the analysis plane, or the stress may vary across the analysis plane. A bending load is an example of a load that will produce different stresses at the analysis plane. Try the self-experiment on page 221. The self-experiment and figure 12.6 illustrate how bending loads produce tensile and compressive stresses.

To maintain force equilibrium in the pencil in the self-experiment, the net tensile force acting at the analysis plane must equal the net compressive force acting at the analysis plane. Because these net forces act some distance away from each other, they create the force couple that causes a torque counter to the

torque created by your thumb and index finger. The moment arm (the distance between these internal forces) of the internal force couple is obviously limited by the depth of the object withstanding the bending load. For the pencil, this moment arm is small, so the internal forces (and thus the stresses) must be large to create a large enough countering torque. Recall the equation for torque we learned in chapter 8:

$$T = F \times r \qquad (12.3)$$

where

T = torque (or moment of force)

F = force

r = moment arm (or perpendicular distance)

An object with greater depth (and more cross-sectional area farther from its neutral axis) is able to withstand greater bending loads because it has a larger moment arm. Under similar bending loads, the stresses in such an object will be smaller. Structural members used in construction are designed with this in mind. In houses, 2×6, 2×8, or 2×10 boards are used as beams. Which dimension is the depth of the beam? The longer dimension is aligned in the plane of the bending loads to maximize the moment arm. I beams used in steel construction are an even better example, because the beam depth is great and the area of the cross section is largest near the top and bottom. Figure 12.7 illustrates the effect of an object's depth on its ability to withstand bending loads.

> **An object with greater depth (and more cross-sectional area farther from its neutral axis) is able to withstand greater bending loads because it has a larger moment arm.**

The primary structural members of the human body are long bones. Whereas beams used in building construction are designed for specific loading situations with forces acting to bend them in a plane, long bones must contend with a wide variety of loading situations with forces acting to bend them in many different directions. Thus the cross sections of long bones do not have definite deeper and narrower dimensions. Instead, long bones are more like pipes, round in cross section but most dense near the perimeter, where the most stress occurs, and least dense (or hollow) near the central (or neutral) axis, where the least stress occurs.

An object, such as a beam, subjected to a bending load will deform by curving. The tension side

SELF-EXPERIMENT

Take a pencil (or pen) and grip it by the ends with your right and left hands so that your index fingers are closest to the pencil ends and your thumbs are closer to the middle. Bend the pencil by pushing away with your thumbs and pulling toward yourself with your fingers. The forces produced by the index finger and thumb of each hand produce equal but opposite force couples or torques on the pencil. A diagram of the pencil is shown in figure 12.6a.

With an analysis plane cutting through the pencil between the two thumbs, the free-body diagram of the left side of the pencil is shown in figure 12.6b. The forces acting on this pencil piece are in equilibrium (if the pulling force from the index finger is equal to the pushing force from the thumb), but the torques created by these forces are not in equilibrium. The two forces acting on the left piece of the pencil create a force couple, and the torque caused by this force couple would tend to rotate and angularly accelerate this pencil piece counterclockwise. Other forces must act on the pencil to create a torque to counter the force couple. The only place where forces could act to create a countering torque is at the analysis plane. To create a torque, though, any forces acting at this analysis plane must be paired as force couples to maintain force equilibrium while still creating a torque. Figure 12.6c shows how forces could act as a force couple at the analysis plane and create a torque to counter the unbalanced torque. Such a force couple would be created by the molecules near the top of the right pencil piece pulling on the molecules at the top of the left pencil piece and the molecules near the bottom of the right pencil piece pushing on the molecules near the bottom of the left pencil piece. The forces shown acting perpendicular to the analysis plane in figure 12.6c represent the net forces created by the pulling and pushing of the molecules at the analysis plane. These forces are distributed among the molecules across the area of the analysis plane as tensile stress on the upper half of the plane and compressive stress on the lower half of the plane, as shown in figure 12.6d. The stresses become greater as the distance from the center line (neutral axis) of the pencil increases.

Figure 12.6   A bending load creates both tensile and compressive stresses.

Figure 12.7   Cross sections of a variety of beam shapes used in construction. The depth of the beam (d) greatly affects its ability to withstand bending loads.

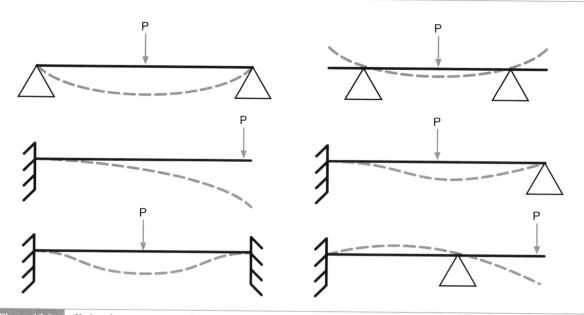

Figure 12.8   Various beam support configurations and the resulting deflections under a single concentrated load.

of the beam will elongate, whereas the compression side will compact and shorten, thus causing the beam to bend. Figure 12.8 shows a variety of other bending load configurations and the deformation of the beams loaded.

Most bones in the human body are constantly under some bending load as a result of the forces exerted on them by gravity, muscles, tendons, ligaments, and other bones. The neck of the femur is a good example of a cantilever beam; a force is exerted on the head of the femur, and the shaft of the femur provides the rigid support for the neck of the femur. Tensile stress develops in the upper part of the neck, whereas compressive stress de-

| Figure 12.9 | The neck of the femur acts as a cantilever beam and develops tensile stress in its upper part and compressive stress in its lower part. |

velops in the lower part of the neck. Figure 12.9 illustrates these stresses.

> **An object, such as a beam, subjected to a bending load will deform by curving. The tension side of the beam will elongate, whereas the compression side will compact and shorten, thus causing the beam to bend.**

When the foot is bearing weight, it is an example of an anatomical beam in which the load is distributed among several structures. Figure 12.10 illustrates the foot during weight bearing. It behaves as a simple beam. Supports are at the heel and metatarsal-phalangeal joints, and the load is applied from the tibia to the talus. In this situation, the bones

| Figure 12.10 | The foot as an example of a simple beam. |

of the foot carry the compressive stress, whereas the plantar fascia and dorsal muscles carry the tensile stress.

Torsion

A **torsion load** is another type of load that produces something other than uniaxial stresses. Torsion loading occurs when torques act about the long axis of the object at each end. Take a pencil and hold it with one hand at each end. Now try to twist the pencil in one direction with your right hand and in the opposite direction with your left hand. Figure 12.11a shows a three-dimensional free-body diagram of the whole pencil.

A free-body diagram of the left half of the pencil with an analysis plane cutting through it is shown in figure 12.11b. This pencil piece is not in equilibrium. A single torque acts on the pencil, which would cause it to spin about its longitudinal axis. Another torque must be present to counteract this torque. The countering torque is created by the shear force between the molecules of the pencil at the right and left sides of the analysis plane. Figure 12.11c shows a free-body diagram of the left half of the pencil with this internal torque included. Figure 12.11d shows the shear stresses that produce this internal torque at the analysis plane. The stresses become larger with increasing distance from the central axis of the pencil (the axis of rotation for the torques). The shear stress is thus not uniform across the analysis plane.

The moment arm between the internal shear forces on either side of the longitudinal axis is limited by the diameter of the object under torsion. For the pencil, this moment arm is small, so the internal forces (and stresses) must be large to create a large enough torque (refer to the equation for torque, equation 12.3). An object with a larger diameter is thus able to withstand greater torsional loads. Figure 12.12 illustrates this graphically.

This is similar to the situation with bending loads, except that increases in width and depth (increases in diameter) are desirable for increasing torsional strength. Thus structural members or tools designed to resist torsional loads are usually circular in cross section, such as driveshafts, axles (although axles carry bending loads as well), and screwdriver shafts. This is another reason for the circular cross section of bones. Torsion loading of bones is common in a variety of situations. For example, when you step onto your foot and then pivot, torque is produced around the longitudinal axis of your tibia and it becomes torsionally loaded.

Figure 12.11   A pencil withstanding a torsional load is subjected to shear stress.

Figure 12.12   The diameter (d) of an object's cross section greatly affects its ability to withstand torsional loads.

Combined Loads

Because of their construction, certain anatomical structures such as muscles, tendons, and ligaments behave like ropes or cables and effectively carry only one type of load: uniaxial tension. Bones and cartilage, on the other hand, may be loaded in a variety of ways, from uniaxial tension, compression, or simple shear loads that produce uniform stress to bending and torsion loads that produce more complex stress patterns. Bones and cartilage will often encounter a combination of these loading configurations. This type of load is called a combined load.

Analysis of the stresses produced by combined loads is complex, but basically involves simplify-ing the loading into several fundamental load configurations. The stresses produced under each of these loading configurations are then added to determine the actual stresses produced by the combined load. Consider the shaft of the femur during weight bearing, as shown in figure 12.13a. Ignoring the muscle and ligament forces that act on it, the femur basically acts as a column (to support the weight of the body) and as a beam (because the body's weight is eccentric to the shaft of the femur), so it carries a bending as well as a compressive load. The stresses in the shaft of the femur resulting from the axial compressive column load are shown in figure 12.13b, and the stresses resulting from the bending load are shown in figure 12.13c. The combined stresses are shown in figure 12.13d.

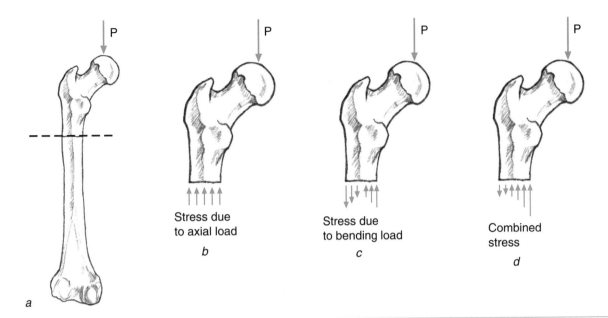

P   P   P   P

Stress due
to axial load
b

Stress due
to bending load
c

Combined
stress
d

a

Figure 12.13 The combined stresses due to the compressive load and bending load on the femur.

Strain

Objects deform when they are subjected to external forces. These deformations may be large or small, depending on the nature of the material and the stresses involved. Strain is the quantification of the deformation of a material. Linear strain occurs as a result of a change in the object's length. Linear strain is produced by compressive or tensile stresses. Shear strain occurs as a result of a change in the orientation of the object's molecules.

**Strain is the quantification
of the deformation of a material.**

Linear Strain

When the loading of an object causes tensile or compressive stress within the object, some change in length accompanies this stress. This deformation can be measured in absolute terms that describe the change in length of the object as a result of the loading; for example, the rubber band stretched 10 cm or the intervertebral disk compressed 2 mm. The deformation may also be described in relative terms as a proportion of the change in length (the difference between the undeformed and the deformed lengths) divided by the original length (the undeformed length). This relative deformation measure is called linear strain.

SELF-EXPERIMENT

Take the rubber band you used at the beginning of this chapter and lay it, unstretched, alongside a ruler. With a pen, make a mark on the rubber band about 1 cm from the left end. Align this mark with zero on the ruler, and then mark the rubber band at 5-mm (1/4-in.) intervals along its length. Now pull on the right end of the rubber band while keeping the leftmost mark aligned with the zero mark on the ruler. Stretch the rubber band until it is twice its original length.

**Linear strain occurs as a result
of a change in the object's length.**

Suppose the distance between the marks closest to either end of the rubber band was 5 cm to begin with and 10 cm after stretching; the absolute deformation of the rubber band would then be:

10 cm – 5 cm = 5 cm

The relative deformation or strain would be:

$$\varepsilon = \frac{\text{change in length}}{\text{original length}}$$

$$\varepsilon = \frac{1 - 1_o}{1_o} \qquad (12.4)$$

$$\varepsilon = \frac{10\ \text{cm} - 5\ \text{cm}}{5\ \text{cm}}$$

$$\varepsilon = 1.0\ \text{cm} / \text{cm, or } 100\%$$

where

ε = linear strain

l = stretched length

l_o = original, undeformed length

$l - l_o$ = change in length

Look at any 5-mm section of the rubber band. Each of them has stretched to 10 mm long, so the strain in every 5-mm section is:

$$\varepsilon = \frac{1 - 1_o}{1_o} \qquad (12.4)$$

$$\varepsilon = \frac{10\ \text{mm} - 5\ \text{mm}}{5\ \text{mm}}$$

$$\varepsilon = 1.0\ \text{mm} / \text{mm, or } 100\%$$

The strain between any two adjacent marks on the rubber band is also 100%. This makes sense; because the tensile stress is uniform throughout the rubber band, it follows that the strain would be also.

You may have noticed that we have reported strain in centimeters per centimeter, millimeters per millimeter, and as a percentage. Actually, strain is a dimensionless quantity because it is a ratio of length to length. So centimeters per centimeter is the same as millimeters per millimeter is the same as inches per inch, because the units cancel out. Usually, strain is reported as a percentage, which seems to make more sense. In that case, the ratio is multiplied by 100% to get the percentage of strain. In the above example, 1 cm/cm was the same as 1 mm/mm because each equals 100%. Most biological materials are not as elastic as a rubber band, and rupture or failure will result at strains much less than 100%.

Shear Strain

Linear strain occurs with a change in length as a result of molecules being pulled apart or pushed together. Shear strain occurs with a change in ori-

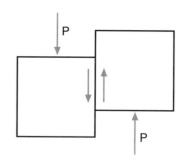

Two unconnected blocks slide relative to each other

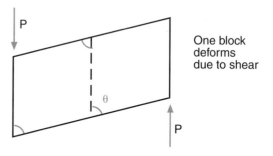

One block deforms due to shear

Figure 12.14 Illustration of deformation caused by shear. The change in the angle (θ) indicates the shear strain.

entation of adjacent molecules as a result of these molecules slipping past each other. Figure 12.14 graphically illustrates shear strain.

Shear strain occurs with a change in orientation of adjacent molecules as a result of these molecules slipping past each other.

Shear strain is measured as follows. Imagine a line perpendicular to the analysis plane through the object at the location of interest and another line within the analysis plane in the direction of interest. These two lines are perpendicular and form a right angle. When a shear load is applied to the object, this angle deforms. The change in the angle (θ) is the measure of shear strain in the direction of interest. Shear strain is abbreviated with the Greek letter λ (lambda) and is measured in radians.

Poisson's Ratio

Consider the rubber band again. As you stretched it, what happened to the width of the rubber band? Its width became narrower as it was stretched in length. Consider a rubber ball loaded in compression. The diameter of the ball shortens in the direction of the compressive load, but what about the diameter in the lateral direction? When the rubber ball is compressed, it becomes wider in the lateral direction while shortening in the direction of the compressive load. This is called the Poisson effect, after S.D. Poisson, a French scientist who investigated this phenomenon in the 1820s. A specific ratio of strain in the axial direction to strain in the transverse direction exists for each different type of material. This ratio is called Poisson's ratio. Values of Poisson's ratio can be as low as 0.1 and as high as 0.5, but for most materials are between 0.25 and 0.35.

The intervertebral disks are a good example of the Poisson effect. During the day, your intervertebral disks are loaded in compression (unless you are lying down all day). This compressive load shortens the vertical dimension of the disks, but laterally, the disks bulge out. Under extremely large compressive loads, a disk may bulge out too much and rupture.

Mechanical Properties of Materials: The Stress-Strain Relationship

How much stress can an intervertebral disk withstand before it ruptures? How far can a bone bend? How much energy can a ligament absorb before it breaks? How far can a tendon stretch before the stretch becomes permanent? These questions all relate to the manner in which materials respond to forces. The relationship between the stress and strain in a material may help explain the behavior of a material under load.

Elastic Behavior

Let's consider the rubber band once more. The more you stretch it, the greater the force you have to pull with. If we could measure this force and the corresponding elongation of the rubber band, we would probably get a plot similar to that shown in figure 12.15.

Because the rubber band was stretched by a uniaxial load, the stress and strain in the rubber band are uniform. The plot shown in figure 12.15 could be represented by a similar plot of stress versus strain, as shown in figure 12.16. The load-deformation plot shown in figure 12.15 was specific to the rubber band being stretched, but the stress-strain curve shown in figure 12.16 is a characteristic of the material the rubber band is made of.

The stress-strain curve shown in figure 12.16 is an example of the stress-strain curve for a linearly elastic material. Rubber is elastic (versus plastic). It stretches under a tensile load but returns to its original shape when the load is removed. This property is called elasticity. Rubber is linearly elastic because, as the stress increases, the strain increases a proportional amount. The ratio of stress to strain (which is shown graphically as the slope of the stress-strain

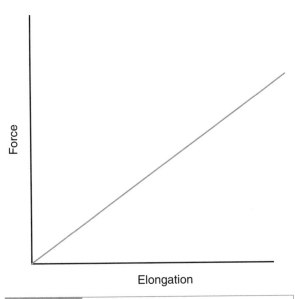

Figure 12.15 A plot of tensile force applied and the corresponding elongation of a rubber band.

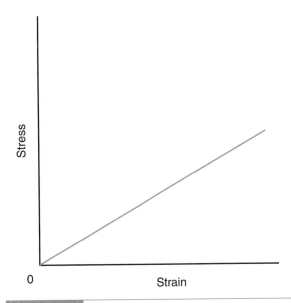

Figure 12.16 A plot of stress and strain for a rubber band's material.

curve) is called the elastic modulus of a material. Mathematically, the elastic modulus, which is also called Young's modulus or the modulus of elasticity (bulk modulus for compression or shear modulus for shear), is defined as:

$$E = \frac{\Delta\sigma}{\Delta\varepsilon} \qquad (12.5)$$

where

E = elastic modulus

$\Delta\sigma$ = change in stress

$\Delta\varepsilon$ = change in strain

A material that is more stiff has a steeper slope of the stress-strain curve, and thus a larger elastic modulus, than a material that is more pliant. Steel is stiffer than rubber. Bones are stiffer than ligaments or tendons. Figure 12.17 shows the stress-strain curves for materials with different stiffnesses.

> **The ratio of stress to strain (which is shown graphically as the slope of the stress-strain curve) is called the elastic modulus of a material.**

Plastic Behavior

The behavior of a material under load may change as the load increases. Under small loads, the object

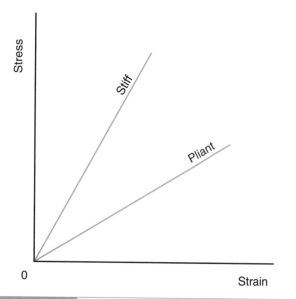

Figure 12.17   Stress-strain curves for stiff versus pliant materials.

SELF-EXPERIMENT

Consider a paper clip. Pull lightly on one of its wire ends so that it tends to unbend one of the curves in the wire. If you pull with only a small force and then let go, the wire springs back to its original shape. This loading occurs in the elastic region of the metal's stress-strain curve. But if you pull with a larger force, the wire does not spring back to its original paper clip shape. This loading occurs in the plastic region of the metal's stress-strain curve. In this region, the stress is no longer proportional to the strain, and the material will not return to its original shape when the stress is removed. An example of the stress-strain curve for the metal of the paper clip is shown in figure 12.18.

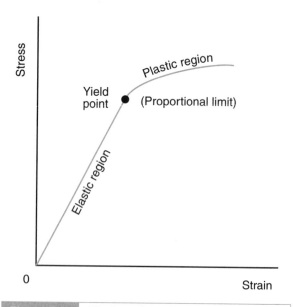

Figure 12.18   Elastic region, plastic region, and yield point on a stress-strain curve for the metal of a paper clip.

may be elastic; if the load is removed, the object returns to its original shape and dimensions. But if the load exceeds a certain size, some permanent deformation of the object may occur. This is called plastic behavior.

The point on the curve where further stress will cause permanent deformation is called the yield point or the elastic limit. It often coincides with the proportional limit, the end of the linear elastic range of the curve. Below this load, the material behaves elastically, and above this load it behaves plastically. Most materials exhibit some elastic and some plastic behavior.

Material Strength and Mechanical Failure

Physiologically, muscle strength is defined as the ability of a muscle to produce force. Mechanically, the strength of a material has to do with the maximum stress (or strain) the material is able to withstand before failure. The key word here is *failure*. Failure may be defined as the inability to perform a function. Thus there are several quantifications for the strength of a material, depending on which function of the material is of interest. These are illustrated in the stress-strain curve shown in figure 12.19.

> **Mechanically, the strength of a material has to do with the maximum stress (or strain) the material is able to withstand before failure.**

The stress at the elastic limit of a material's stress-strain curve is the yield strength of the material. Although no breakage or rupture of the material occurs, stresses beyond the yield strength will cause permanent changes in the dimensions of the mate-

rial. Beyond this point, the material fails in the sense that it is unable to regain its shape. A severe strain of the ankle ligaments may cause a permanent lengthening, and thus loosening, of these ligaments.

The ultimate strength of a material is the maximum stress the material is capable of withstanding. This is represented by stress corresponding to the highest point on the stress-strain curve shown in figure 12.19. If the function of interest is the ability of the material to withstand large loads, the ultimate stress determines how much load may be carried. The load is the measure of the material's strength in this case.

The rupture or failure strength of a material is the stress where failure actually occurs. Failure in this sense means breakage or rupture. Failure strength is represented by stress corresponding to the endpoint of the stress-strain curve in figure 12.19. Failure strength (rupture strength or fracture strength) often has the same value as ultimate strength.

Material strength may also be expressed as the failure strain. Failure strain is the strain exhibited by a material when breakage occurs. It is the strain corresponding to the endpoint of the stress-strain curve in figure 12.19. A material able to withstand a large plastic strain is called a ductile material, whereas a material with a small failure strain is called a brittle material. Glass is brittle, whereas rubber is ductile. Bones become more brittle with age (primarily because of disuse). Figure 12.20 illustrates these properties.

Figure 12.19   Material measures of strength shown on a stress-strain curve.

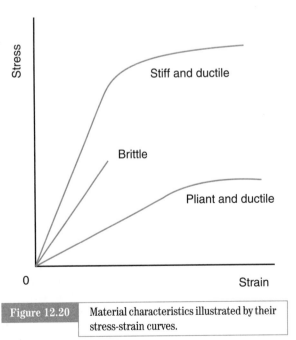

Figure 12.20   Material characteristics illustrated by their stress-strain curves.

Perhaps the best way to characterize the strength of a material is by its toughness. Mechanically, toughness is the ability of a material to absorb energy. Put another way, a material is tougher if it takes more energy to break it. An estimation of the toughness of a material is given by the area under the stress-strain curve, as shown in figure 12.21.

> **Mechanically, toughness is the ability of a material to absorb energy.**

Recall the work-energy relationship we discussed in chapter 7. It is the basis behind the measurement of material toughness. Something that is able to exert great force, but only over a very short displacement, will not do as much work or cause as great a change in energy as something that is able to exert less force over a much longer displacement. A hard, brittle material is able to withstand great stress but only small strain, so its energy-absorbing ability is less than a softer but more ductile material. Dry bone is an example of a hard, brittle material. Although it may withstand large stresses, it is relatively easy to break, because it cannot deform easily. Moist, live bone, on the other hand, deforms more easily and is less brittle. It is much more difficult to break than dry bone because it is tougher; more work must be done to fracture it.

Mechanical Properties of the Musculoskeletal System

Muscle tissue and connective tissue form the structural units of the musculoskeletal system. The specific connective tissues important to the structure of the musculoskeletal system are bone, cartilage, ligament, and tendon. Muscles may be thought of as the active elements of the musculoskeletal system, whereas connective tissues are the passive elements.

All connective tissues are composed of living cells as well as extracellular components consisting of collagen, elastin, ground substance, minerals, and water. The mechanical properties of the various tissues are determined in part by the proportions of these components and their arrangement. Collagen, which is a fibrous protein, is the most abundant substance in all connective tissue. Molecules of collagen align together to form collagen fibrils that bind together to form collagen fibers. Collagen is thus very stiff (failure strain of 8% to 10%) and has high

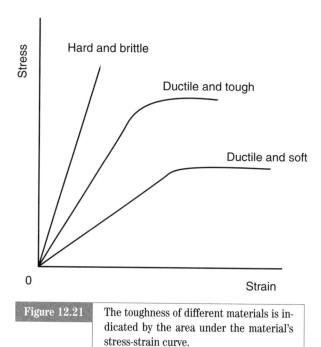

Figure 12.21 The toughness of different materials is indicated by the area under the material's stress-strain curve.

tensile strength. On the other hand, collagen is unable to resist compression because of its length. It collapses or buckles like a rope. Elastin is also fibrous, but unlike collagen, it is pliant and very extensible (failure strain as high as 160%). The ground substance consists of carbohydrates and proteins that, combined with water, form a gel-like matrix for the collagen and elastin fibers.

The composition of connective tissues (and most biological tissues) causes them to be anisotropic, unlike many synthetic materials, which are isotropic. Isotropic materials have the same mechanical properties in every direction. Anisotropic materials have different mechanical properties depending on the direction of the load. For example, the ultimate tensile strength of a tendon is very high if the tensile load is in the direction of the tendon fibers, but its ultimate tensile strength is low if the tensile load is perpendicular to these fibers (see figure 12.22).

Age and activity affect the mechanical properties of all connective tissues. Bone, cartilage, tendon, and ligament strengths increase with regular cycles of loading and unloading. Usually this strength increase is due to an increase in size of the tissue cross section, but the stiffness and ultimate strength of these tissues may also increase, an indication that size alone is not responsible for the strength increase. The exact threshold of the loads and number of loading cycles required to stimulate these strength gains is unclear. Inactivity and im-

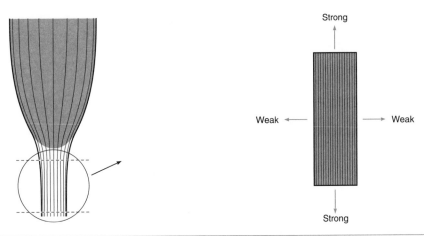

Figure 12.22 Anisotropic behavior of a tendon. The tendon is stronger if the tensile load is aligned with the tendon fibers rather than transverse to the fibers.

mobilization result in decreased strength of these tissues and a shortening of the ligaments and tendons.

Age and activity affect the mechanical properties of all connective tissues.

All of these connective tissues show an increase in ultimate strength with age until the third decade of life, after which strength decreases. Bones become more brittle and less tough with increasing age. Tendons and ligaments, on the other hand, become less stiff.

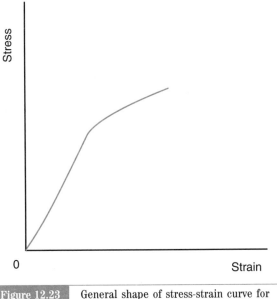

Figure 12.23 General shape of stress-strain curve for cortical bone.

Bone

Bones carry almost all of the compressive loads experienced by the body, and they are capable of resisting large shear and tensile loads as well. By weight, 30% to 35% of bone is collagen, 1% to 2% is ground substance, 45% is mineral, and about 20% is water. Bone is the strongest and stiffest material of the musculoskeletal system. The stress-strain curve for a piece of cortical bone removed from a long bone and loaded axially is shown in figure 12.23. Cortical or compact bone is found in the dense and hard outer layers of bone. Cancellous (trabecular or spongy) bone is the less dense, porous bone that is spongy in appearance and found deep to cortical bone near the ends of long bones.

The elastic region of the stress-strain curve is not quite linear but slightly curved. Values for the elastic modulus for bone thus have some inherent error associated with them. In the plastic region of the curve, the loading is still able to increase past the elastic limit. Values for the elastic modulus, yield strength, ultimate strength, failure strength, and failure strain of cortical bone vary from bone to bone as well as from location to location on an individual bone. In addition, values of these properties change with the age of the bone. The porosity of the bone primarily determines its strength and stiffness.

Another factor that affects the mechanical strength and stiffness of bone is the rate of loading. Bone is stronger and stiffer if a load is applied quickly but weaker and less stiff if a load is applied slowly. Figure 12.24 illustrates this rate dependency. Thus a load applied slowly across a joint may result in an avulsion fracture (the ligament pulls bone out of its attachment site), whereas the same load

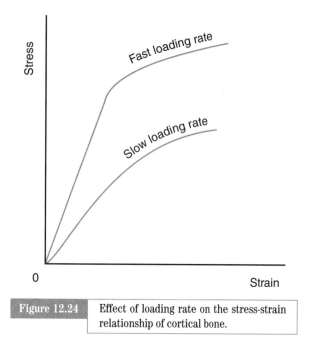

Stress

Fast loading rate

Slow loading rate

0   Strain

Figure 12.24 Effect of loading rate on the stress-strain relationship of cortical bone.

applied quickly across the joint will rupture the ligament.

The high mineral content of bone, primarily calcium and phosphate, accounts for its high compressive strength. The collagen fibers intertwined with these mineral salts give bone its high tensile strength.

> **Bones are strongest in compression and weakest in shear.**

Cartilage

There are three different types of cartilage: hyaline cartilage, fibrous cartilage, and elastic cartilage. Hyaline cartilage, also called articular cartilage, is the cartilage that covers the ends of long bones at joints. Fibrous cartilage (fibrocartilage) is found within some joint cavities (the menisci of the knee), in the intervertebral disks (the annulus fibrosus), at the edges of some joint cavities, and at the insertions of tendons and ligaments into bone. Elastic cartilage is found in the external ear and in several other organs that are not part of the musculoskeletal system. Cartilage is able to withstand compressive, tensile, and shear loads.

By weight, hyaline cartilage consists of 10% to 30% collagen, 3% to 10% ground substance, and 60% to 80% water. Superficially, collagen fibers are arranged in layers parallel to the articular surface. Deeper, the collagen fibers seem to be arranged ran-

domly. At the deepest layer, where the hyaline cartilage joins with bone, the collagen fibers are perpendicular to the articular surface and stick into the bone like fingers. This arrangement fuses the articular cartilage to the bone. Articular cartilage (as well as fibrous and elastic cartilage) has no nerve or blood supply, so it must be very thin to allow the diffusion of nutrients to cells for normal metabolism. Human articular cartilage is usually only 1 to 3 mm thick.

Articular cartilage transmits the compressive loads from bone to bone at joints. But how does a material composed primarily of collagen and water resist compression loads, when collagen can withstand only small compressive stresses? Imagine a balloon filled with water. As you push down on the balloon, its sides bulge out and stretch. Because water doesn't compress, it pushes out on the sides of the balloon. The sides of the balloon are thus under tension. If the balloon is made of a material capable of resisting large tensile stress, you can push on the balloon with more force. Articular cartilage is similar. Collagen's tensile strength holds the cartilage together under compressive loads due to the arrangement of the collagen fibers near the exterior surface of the cartilage.

Unlike a balloon, however, articular cartilage is not watertight, so some fluid is exuded (squeezed out) when it is loaded in compression. This behavior causes creep and stress relaxation effects. Articular cartilage loaded under a constant compressive stress will not experience a corresponding constant strain. The strain will increase with time as fluid is exuded from the cartilage until it reaches a point where no more fluid is exuded. This increase in strain under a constant stress is called creep. Creep rate is how quickly the cartilage reaches a constant strain. Creep rate depends on the magnitude of the compressive stress, the thickness of the articular cartilage, and the permeability of the cartilage. The time it takes to reach constant strain may range from 4 to 16 hr for human articular cartilage. Figure 12.25 illustrates creep.

What happens to the fluid exuded by articular cartilage? It may assist with lubricating the articular surfaces, and it is reabsorbed by the cartilage as the compressive stresses are reduced.

Stress relaxation is another effect caused by fluid squeezing out of articular cartilage. If articular cartilage is loaded so that it experiences a constant strain, it will not experience a corresponding constant stress. The initial strain will cause an increase in stress, but the stress will then decrease (or relax) to some lower value.

| Figure 12.25 | Creep in articular cartilage under constant compressive stress. |

SELF-EXPERIMENT

Again, take a water balloon and place it on a table. Initially, only a small area of the balloon makes contact and carries the load. This part of the balloon is thus under great pressure because of the small area of contact. But this part of the balloon gives so that more of the balloon comes in contact with the table and pressure is reduced. Because the balloon deforms quickly and conforms to the shape of the supporting surface, pressure on the balloon is reduced. Articular cartilage and a water balloon behave similarly when under stress, but unlike the balloon, cartilage deforms slowly. Thus large stresses may be present at first, and these stresses reduce as the rest of the cartilage deforms.

Imagine the surfaces of the articular cartilages. There may be only a small area of contact between these surfaces in the unloaded condition. As load is applied, this small area of contact is compressed and fluid is squeezed to other areas of the cartilage, which then expand and contact the other cartilage

surface to carry some of the load. This increase in contact area results in a decrease or relaxation of stress at the initial contact point. This process is illustrated with the plot of stress versus time in figure 12.26.

The elastic modulus of articular cartilage is much smaller than the elastic modulus of bone, but articular cartilage transmits compressive loads from one bone to another. The difference in stiffness of these materials accounts for a slight damping effect of articular cartilage. But because articular cartilage is so thin, this shock-absorbing effect is limited.

Tendons and Ligaments

Tendons and ligaments are similar in composition and structure, so we will consider them together. By weight, tendons and ligaments consist of approximately 70% water, 25% collagen, and 5% ground substance and elastin. Ligaments have more elastin than tendons. Other than water, collagen is the primary component of tendons and ligaments, but this is also true of cartilage. The major difference between cartilage and tendons or ligaments is in the arrangement of their collagen fibers. The groups of collagen fibers in tendons are bound together in parallel (tendon) or nearly parallel (ligament) bundles that lie along the functional axis of the tendon. The parallel arrangement produces a structure that is very stiff and high in tensile strength but has little resistance to compression or shear. The slight difference in the arrangement of the bundles

| Figure 12.26 | Stress relaxation in articular cartilage under constant compressive strain. |

of collagen fibers and the slightly larger elastin component in ligaments make them less stiff and slightly weaker than tendons. But because the collagen fibers of ligaments are not as well aligned as those of tendon, ligaments can carry loads that are nonaxial. Figure 12.27 illustrates the fiber arrangements in tendons and ligaments.

Collagen fibers have a wavy or crimped appearance. These crimps or waves in the collagen fibers may account for the unusual behavior of ligaments and tendons under uniaxial tensile loads and the stress-strain curves that result. When a tendon is pulled with low tensile force, it stretches easily and behaves like a fairly elastic material. But beyond a certain load, the tendon is very stiff. The low stresses that accompany the initial strains are a result of the straightening of the crimps or waves in the collagen. Once these have been straightened, the actual stretching of the collagen begins, and larger stresses are required to produce the same change in strain. The stress-strain curve for tendon is shown in figure 12.28. The tensile stress-strain curves for ligament and cartilage have shapes similar to those of tendons.

The first part of this curve is called the toe region and corresponds to the straightening of the crimps in the collagen. Past the toe region, the behavior is fairly linear; an increase in stress results in a proportional increase in strain. The modulus of elasticity for tendons and ligaments is the modulus that corresponds to this linear region of the stress-strain curve.

Tendons and ligaments also exhibit creep and stress relaxation behavior similar to cartilage, as was shown in figures 12.25 and 12.26. The creep response of these tissues is relevant to stretching exercises. Stretching exercises to increase flexibility are more effective if you move into the stretch position slowly and then hold this position by keeping the tissues involved under constant stress. The strain in the tendons and ligaments, and thus their lengths, will slowly increase. If the stretch is done very quickly, using swinging or bouncing movements, the strain in the tendons and ligaments is not as large, and the change in their lengths is not significant.

Muscle

Unlike connective tissues, muscle tissue is capable of actively contracting to produce tension within itself and the structures to which it attaches. The active contractile components of muscle thus determine the stiffness of the muscle at any instant. Muscle stiffness varies as a function of the number of active contractile elements. We will examine the structure and function of muscle more thoroughly in chapter 14. Some things can be learned about muscle if we consider it in its relaxed or passive state.

If a passive muscle is slowly stretched, there will be some resistance to this stretching and thus some stress developed. This resistance to stretching originates from two different sources: the contractile elements and the connective tissue sheaths that surround the muscle fibers and are connected to the tendons. The stiffness of the passive contractile component caused by the filaments of the contractile elements sliding past each other is very low, so the resistance to this movement is small as well. Stretching of the muscle past the point where the contractile filaments no longer overlap is resisted by the connective tissue component of the muscle. This connective tissue behaves like tendon. A stress-strain curve for passive muscle may look something like that shown in figure 12.29.

Fiber arrangement
in tendon

Fiber arrangement
in ligament

Figure 12.27   Parallel arrangement of collagen fibers in tendon, and nearly parallel arrangement of collagen fibers in ligament.

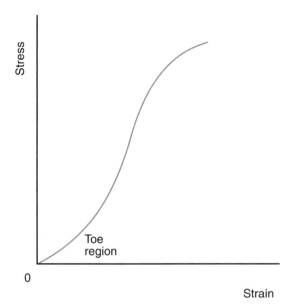

Figure 12.28 General shape of stress-strain curve for a tendon.

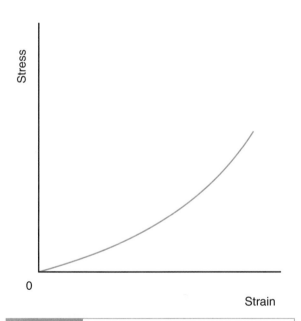

Figure 12.29 General shape of stress-strain curve for passive muscle.

Muscle stiffness varies as a function of the number of active contractile elements.

The toe region of this curve is much longer due to the very low stiffness of the passive contractile elements. Stiffness increases when the contractile filaments no longer overlap and the connective tissue begins to strain. The failure strain for muscle is

much larger due to the ability of the contractile filaments to slide past each other as the muscle increases its length by 50%. Only after the muscle has stretched this much will the connective tissue begin to be strained. The total strain this tissue can withstand is similar to that of tendon or ligament: 8% to 15%. The failure strain for muscle is thus 50% plus 8% to 15%, or a total strain of 58% to 65%. Ultimate strength is much weaker than that of tendon because a cross section includes mostly the contractile elements of the muscle (which are composed of contractile filaments not bound to each other in the passive state).

Summary

Knowledge of stress and strain and the mechanical properties of musculoskeletal tissues will assist in preventing injury, analyzing injury mechanisms, and evaluating rehabilitation or training exercises. The external forces acting on the body are ultimately carried by the bones, cartilage, ligaments, tendons, and muscles of the musculoskeletal system. These external loads cause stresses and strains. Stress is internal force per cross-sectional area. Strain is the ratio of change in length (or deformation) to the undeformed length. Compressive stress is caused by loads that push molecules together, causing compressive strain as the object shortens. Tensile stress is caused by loads that tend to pull molecules apart, causing tensile strain as the object lengthens. Shear stress is caused by loads that tend to slide molecules past each other, causing shear strain as the object's shape distorts.

Uniaxial compression and shear are examples of loads that result in uniform stress. Bending results in tensile stress on one side of a beam and compressive stress on the other. Torsion results in shear stresses that vary in magnitude across a section. Combined loads may result in compressive, tensile, and shear stresses all in the same cross section.

The strength of a material may be characterized by its stiffness (the ratio of stress to strain during elastic deformation); yield stress (the stress it can withstand without plastic deformation); ultimate stress (the largest stress it can withstand before it breaks); failure strain (the strain it can withstand before it breaks); or toughness (the energy it can absorb before it breaks). These properties may all be estimated from a stress-strain curve of a material.

Bone is the strongest of the connective tissues as determined by its stiffness, ultimate strength, yield strength, and toughness. Its strength is dependent on the rate of loading, however. Ligaments, tendons, and cartilage all have similarly shaped stress-strain curves due to their collagenous composition. Under low stresses, these materials are pliant, but as the stresses increase past a certain threshold, they become much stiffer. The mechanical properties of muscle are not as easily examined due to its contractile ability. The ultimate stress of muscle is less than that of tendon, ligament, or bone, whereas its failure strain is much greater.

Key Terms

anisotropic *(p. 230)*

bending load *(p. 220)*

cancellous bone *(p. 231)*

combined load *(p. 224)*

compact bone *(p. 231)*

compression *(p. 217)*

cortical bone *(p. 231)*

creep *(p. 232)*

elastic *(p. 227)*

elastic cartilage *(p. 232)*

elastic limit *(p. 229)*

elastic modulus *(p. 228)*

failure strain *(p. 229)*

failure strength *(p. 229)*

fibrous cartilage *(p. 232)*

hyaline cartilage *(p. 232)*

isotropic *(p. 230)*

modulus of elasticity *(p. 228)*

plastic *(p. 228)*

Poisson's ratio *(p. 227)*

proportional limit *(p. 229)*

shear *(p. 217)*

spongy bone *(p. 231)*

stress *(p. 214)*

tensile stress *(p. 215)*

torsion load *(p. 223)*

toughness *(p. 230)*

trabecular bone *(p. 231)*

ultimate strength *(p. 229)*

yield strength *(p. 229)*

Young's modulus *(p. 228)*

REVIEW QUESTIONS

1. A 1 cm-long section of the patellar ligament stretches to 1.001 cm when it is subjected to a tensile force of 10,000 N. What is the strain in this segment of ligament?

2. The section of the patellar ligament in the previous question is 50 mm² in cross section. What is the stress in this section of ligament as a result of the 10,000-N tensile force?

3. When an athletic trainer pulls laterally on your ankle while pushing medially on your knee, a bending load is placed on your lower extremity. Which side of your knee undergoes compressive stress and which side undergoes tensile stress as a result of this manipulation?

4. The modulus of elasticity (for compression) for a section of compact bone in the femur is 12 GPa (12×10^9 Pa). If this bone is subjected to a compression stress of 60 MPa (60×10^6 Pa), what strain results from this compression?

5. The Achilles tendon is subjected to a large tensile stress that results in a strain of 6%. If the unloaded tendon is 10 cm long, how much does it elongate as a result of this strain?

6. Compact bone from the tibia has an ultimate tensile stress of 150 MPa and a tensile strain at failure of 1.5%. The anterior cruciate ligament has an ultimate tensile stress of 35 MPa and a strain at failure of 30%. Which is probably mechanically tougher, the bone or the ligament?

7. A snowboarder catches an edge and falls. Her board twists in one direction as her body twists in the opposite direction. The torsional load places large shear stresses on her tibia and knee. If the load on the tibia is axial torsion only, where on a cross section of the tibia will the shear stress be greatest?

8. Which can withstand greater tensile stress—bone or ligament?

9. Which can withstand greater tensile strain—bone or ligament?

10. What type of stress is bone
    a. strongest in resisting?
    b. weakest in resisting?

11. Describe the various measures of strength for biological materials.

The Skeletal System: The Rigid Framework of the Body

Corbis-Bettmann

Imagine what life would be like without your skeleton. What would you look like? How would you move? Would you be able to move? If you were in water, you might be functional, but few terrestrial creatures exist without some sort of skeleton. Earthworms and slugs don't have skeletons. They can't stand up, and they move very slowly and inefficiently. Your skeleton allows you to stand up, and its bones and the joints between the bones allow you to move rapidly. These and other functions of the skeletal system and joints will be presented in this chapter.

OBJECTIVES

When you finish this chapter, you should be able to do the following:

▌ Identify the parts of the skeletal system

▌ Describe the functions of the skeletal system

▌ Describe the anatomical features of bones

▌ Classify bones as long, short, flat, irregular, or sesamoid

▌ Describe the growth process of long bones

▌ Describe the structural and functional classification systems for joints

▌ Classify synovial joints as gliding, hinge, pivot, ellipsoidal, saddle, or ball and socket

▌ Describe the anatomical features of synovial joints

▌ Describe the functions of articular cartilage

▌ Describe the functions of synovial fluid

▌ Identify the factors that contribute to the stability of synovial joints

▌ Identify the factors that contribute to the flexibility of synovial joints

The skeletal system includes the bones of the skeleton, the joints where these bones meet, and the cartilage and ligaments associated with the joint structures. The skeleton is sometimes partitioned into the axial skeleton and the appendicular skeleton, as shown in figure 13.1. The axial skeleton includes the bones of the trunk, head, and neck: the skull, vertebral column, and rib cage. The appendicular skeleton includes the bones of the extremities along with the shoulder girdle (clavicle and scapula) and pelvic bones (ilium, ischium, and pubis). An adult has 206 bones: 126 in the appendicular skeleton, 74 in the axial skeleton, and 6 auditory ossicles (bones of the inner ear).

Mechanically, the skeletal system may be thought of as an arrangement of rigid links connected to each other at joints to allow specific movements. Muscles that attach to the bones provide the forces that may cause changes in the positions of the bones relative to each other. This is the primary mechanical function of the skeletal system. The importance of the skeletal system in sports is that it provides the long levers that enable movement. The rigid framework of the skeleton also supports the soft tissues and organs of the body. Various bony structures of the skeleton provide protection to vital organs as well. The skull and vertebrae encase and protect the central nervous system (the brain and spinal cord). The ribs and sternum protect the heart, lungs, major blood vessels, liver, and spleen. The pelvis protects the uterus in women and the bladder. The skeletal system also has metabolic functions. The red marrow in some bones produces red blood cells, white blood cells, and platelets. In addition, bones store calcium and phosphorous. The functions of the skeletal system are summarized in table 13.1.

> **Mechanically, the skeletal system may be thought of as an arrangement of rigid links connected to each other at joints to allow specific movements.**

Bones and joints are the basic components of the skeletal system. The structure and development of these components are presented in this chapter.

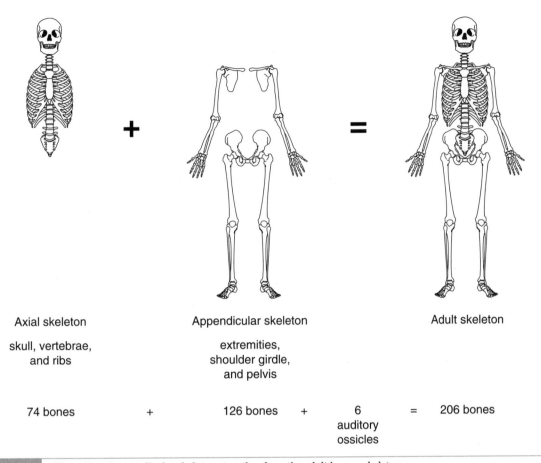

Axial skeleton

skull, vertebrae, and ribs

Appendicular skeleton

extremities, shoulder girdle, and pelvis

Adult skeleton

74 bones   +   126 bones   +   6 auditory ossicles   =   206 bones

Figure 13.1   The axial and appendicular skeletons together form the adult human skeleton.

TABLE 13.1	Functions of the Skeletal System
Mechanical functions	Physiological functions
Support Protection Movement	Hemopoesis (blood cell production) Mineral storage

Bones

Bones, which account for about 16% of an adult's total body weight, are composed of two different types of bony tissue: cortical or compact bone and cancellous or spongy bone. As their names imply, cortical bone is more dense and compact and cancellous bone is more porous and spongy in appearance. As discussed in the previous chapter, the composition of bone makes it the strongest and stiffest of all body tissues. It has great compressive, tensile, and shear strength. These qualities of bone make the skeleton very well suited to its function as a supporting structure.

Bones are living tissues and much different from the dried bones or plastic models of the human skeleton used in anatomy classes and laboratories. Bones are innervated and supplied with blood. As living tissue, bone is adaptable. It responds to the stresses imposed on it by growing thicker in the areas of local stress or becoming more dense. Bone density decreases if it is not stressed regularly. Astronauts returning from extended flights in space thus suffer from loss of bone density, as does anyone who does not maintain a minimal level of activity.

Bone Anatomy and Classifications

Bones come in a variety of shapes and sizes. The mechanical stresses imposed on a bone and its function determine its form. Bones are classified according to their shape. Long bones generally have long, hollow shafts with knobby ends. They are designed for large movements. Long bones occur in the extremities and include the humerus, radius, ulna, femur, tibia, fibula, metacarpals, metatarsals, phalanges and clavicle. Short bones are small, solid, and blocklike. These bones are well suited to transferring forces and shock absorption, but they are not very mobile. The wrist and ankle bones (the carpals and tarsals) are all short bones. As their name implies, flat bones have flat surfaces and are thinner in one dimension. These bones are designed for protection. The ribs, skull, scapula, sternum, and pelvic bones are flat bones. Irregular bones are those that fit in none of the other categories. These bones are designed for support, protection, and leverage. The vertebrae (including the sacrum and coccyx) and the facial bones are examples of irregular bones. Some bones are also described as sesamoid bones. These bones develop within tendons, often in an effort to decrease stress or increase leverage. The patella is an example of a sesamoid bone.

> **The mechanical stresses imposed on a bone and its function determine its form.**

Long bones are the bones most involved in motion. Knowledge of the structure of long bones is thus appropriate for sport biomechanics. Figure 13.2 shows a typical long bone and its structure. A thin, fibrous membrane called the periosteum covers the external surface of the bone except at articular surfaces. The shaft or body of a long bone is a hollow tube with walls of cortical bone. The hollow core of long bones is the medullary or marrow cavity, which contains the red marrow responsible for blood cell production. In adults, most of the red blood cell production takes place in the flat bones, and yellow marrow occupies the medullary cavity of long bones. Yellow marrow is composed primarily of fat cells. As noted in the previous chapter, the long, hollow, cylindrical shape of long bones makes them lightweight, yet still quite strong in resisting bending loads.

The ends of long bones are typically the location of articulations with other bones. Articular cartilage covers the joint surfaces at the ends of long bones. Again, the walls of the ends of long bones are composed of cortical bone, but the interior spaces are filled with cancellous bone rather than marrow.

Specific terms are used to describe the surface irregularities of bones. Some of these terms are listed and defined below and shown graphically in figure 13.3:

- Condyle—a rounded projection that articulates with another bone

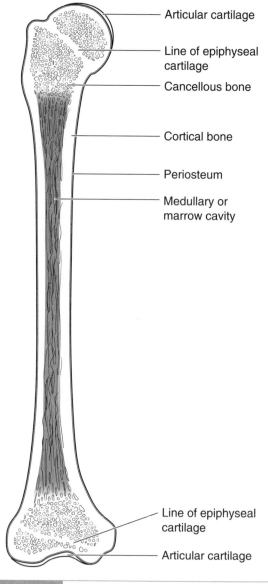

Figure 13.2   Longitudinal section of a typical long bone.

- Epicondyle—a rounded projection near the ends of a long bone but lateral to the axis and not necessarily a part of the articulation
- Facet—a small, smooth, and usually flat articular surface
- Foramen—a hole, usually for nerves or vessels to pass through
- Fossa—a hollow depression or pit
- Fovea—a smaller hollow depression or pit
- Head—the spherical articular end of a long bone
- Line—a raised line or small ridge
- Neck—the part of the bone that joins the head to the shaft
- Process—a projecting part of bone

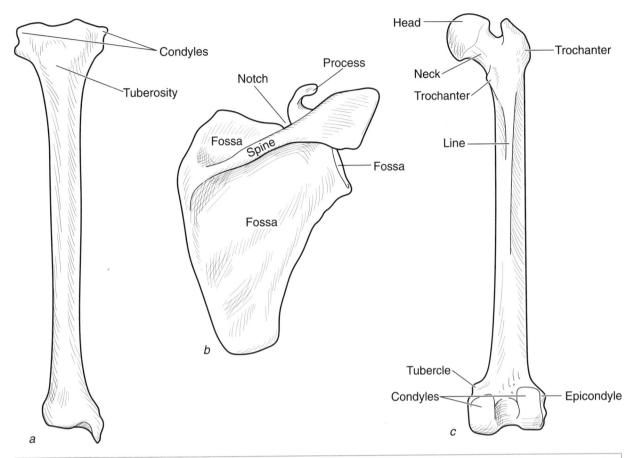

Figure 13.3   Surface irregularities of the femur (*a*), the scapula (*b*), and the humerus (*c*).

- Spine—a sharp projecting part of bone
- Trochanter—a large, knobby projection
- Tubercle—a small, knobby projection
- Tuberosity—a knobby projection

Growth and Development of Long Bones

Long bones develop via endochondral ossification: cartilage is replaced by bone. In a human fetus, a cartilage model of the skeleton forms. Endochondral ossification of this cartilage skeleton begins prior to birth. At birth, the shaft of a long bone has ossified, but the ends are still composed of cartilage. In the metacarpals, metatarsals, and phalanges of the hands and feet, only one end of these bones remains cartilaginous at birth. The cartilage ends of long bones ossify soon after birth, except for the cartilage that separates the ends from the rest of the bone. This cartilage is called the epiphyseal cartilage (epiphyseal plate, epiphyseal disk, growth plate), and the separate end of the bone is called the epiphysis. The remaining part of the bone on the other side of the epiphyseal cartilage is called the diaphysis. As a single long bone in children may actually consist of two or three separate bones, children have more bones than adults. Illustrations of the long bones of an adult and child are shown in figure 13.4.

> **Long bones develop via endochondral ossification: cartilage is replaced by bone.**

The epiphyseal cartilage is responsible for the growth in length of long bones, as shown in figure 13.5. As this cartilage grows, the cartilage nearest the diaphysis ossifies. If the rate of these processes is equal, longitudinal bone growth occurs. If the rate of ossification exceeds the rate of cartilage growth, however, the entire epiphyseal cartilage ossifies, joining the diaphysis with the epiphysis, and longitudinal growth stops. Epiphyseal plate closure, as this is called, occurs naturally at specific ages for each bone. Most epiphyses ossify during puberty, but some may not close until after age 25.

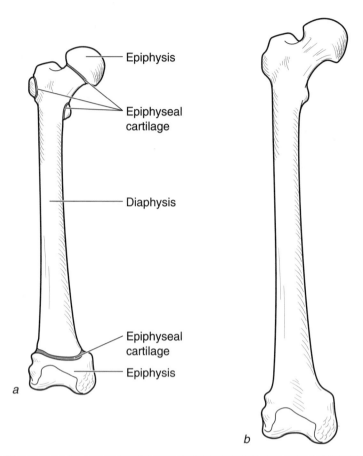

Epiphysis

Epiphyseal
cartilage

Diaphysis

Epiphyseal
cartilage

Epiphysis

a

b

| Figure 13.4 | Femurs of (*a*) a child and (*b*) an adult. The child's femur consists of several bones connected via epiphyseal cartilage, whereas the adult's femur consists of one bone. |

Growth at the epiphyseal plates is affected by the stresses occurring there. A certain amount of mechanical stress is necessary to stimulate growth, but too much may cause premature closure. Certain hormones also affect growth at the epiphyseal plate. Too little growth hormone will retard growth, whereas too much may extend growth past the normal age of closure. Sex hormones increase the rate of cartilage replacement and thus lead to closure of the epiphyseal plates. This is the primary reason for the closure of most growth plates during puberty.

Growth in the diameter of long bones occurs where the periosteum (the membrane covering the surface of the bone) interfaces with bone. New bone is deposited there to increase the thickness of the walls and the diameter of the bone. At the same time, however, bone is absorbed at the inner surface of the wall, and the central cavity of the bone is enlarged. Growth in the diameter of bones ceases with closure of the epiphyseal plates; however, throughout life, bone will continue to adapt to changes in mechanical stress with increases or decreases in wall thickness and density.

Joints

A joint or articulation is any place where two bones meet or join. Joints have a variety of functions. Their primary function is to join bones together while controlling the motion allowed between them. Joints can provide rigid or highly mobile connections between bones, depending on their individual functions. In addition to joining bones together, another joint function is to transfer forces between bones. These two competing functions, force transferral and motion control, lead to interesting structural designs of joints.

> A joint or articulation is any place where two bones meet or join. The joint's primary function is to join bones together while controlling the motion allowed between them.

Joint Classifications

Joints have been categorized in a variety of ways, but most classification schemes are based on joint

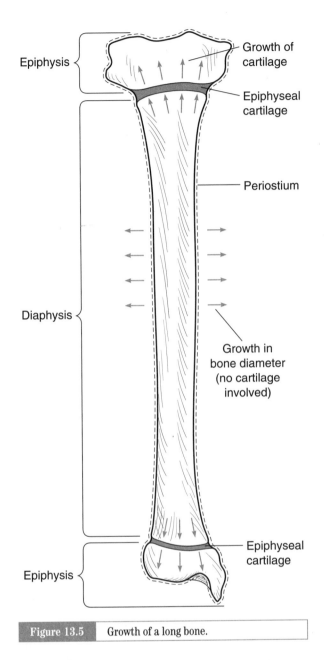

Epiphysis

Growth of cartilage

Epiphyseal cartilage

Periostium

Diaphysis

Growth in bone diameter (no cartilage involved)

Epiphyseal cartilage

Epiphysis

Figure 13.5 Growth of a long bone.

structure or function (mobility). Structurally, joints can be classified into three general groups or types of joints. These groups may be further subdivided into subgroups or types of joints. The structural classifications of joints are fibrous (sutures and syndesmoses), cartilaginous (synchrondoses), and synovial. Bones connected by fibrous connective tissue form a fibrous joint. These joints are typically rigid (although not necessarily so). The sutures of the skull are examples of fibrous joints. Bones connected by cartilaginous tissue form a cartilaginous joint. This type of joint may be rigid or may allow slight movement. The pubic symphysis between the left and right pubic bones of the pelvis is an example of a

cartilaginous joint. The joint between the diaphysis and epiphysis in an immature skeleton is another example of a cartilaginous joint. Bones connected by ligaments and separated by a joint cavity form synovial joints. Synovial joints are highly mobile. Their distinguishing characteristic is a joint cavity that encloses the space between the joints. Most of the joints of the appendicular skeleton are synovial joints.

Functionally, joints can be classified by how much movement they allow. The functional classifications for joints are synarthrodial (immovable), amphiarthrodial (slightly movable), and diarthrodial (freely movable). Some functional classification systems group immovable and slightly movable joints together as synarthroses. Fibrous and cartilaginous joints are classified as synarthrodial and amphiarthrodial, respectively, in the structural classification scheme, whereas synovial joints are classified as diarthrodial joints. The synovial joints are of most interest to us because these are the joints where movement occurs. Examples of joints classified by function and by structure are shown in figure 13.6.

Synovial (or diarthrodial) joints are subclassified into six different types according to the movements allowed and the structure of the joint: gliding, hinge, pivot, ellipsoidal, saddle, and ball and socket. The shapes of the articulating surfaces often determine the movements allowed and thus the joint type. As few joints are exactly like the idealized representations in figure 13.7, joints are classified based on which joint type they most resemble.

Gliding joints are also called irregular, plane, or arthrodial joints. The articulations are flat and small, and planar sliding movements are allowed at these joints. The intercarpal (wrist), intertarsal (ankle), and acromioclavicular (shoulder girdle) joints are examples of gliding joints.

Hinge joints are uniaxial and allow only one degree of freedom of movement (only one number, say the angle between the two bones of this joint, would be needed to fully describe the orientation of the bones with respect to each other). Hinge joints are also called ginglymus joints. The pair of articulating surfaces in a hinge joint approximate a round cylinder (oriented perpendicular to the long axis of the bone) that fits into a matching shallow trough. The movements allowed at a hinge joint are flexion and the return movement of extension (or plantar flexion and dorsiflexion at the ankle). The humero-ulnar (elbow), tibiofemoral (knee), talotibial and talofibular (ankle), and interphalangeal (finger and toe) joints are all examples of hinge joints.

Pivot joints are also uniaxial, allowing only one degree of freedom of movement. These joints may

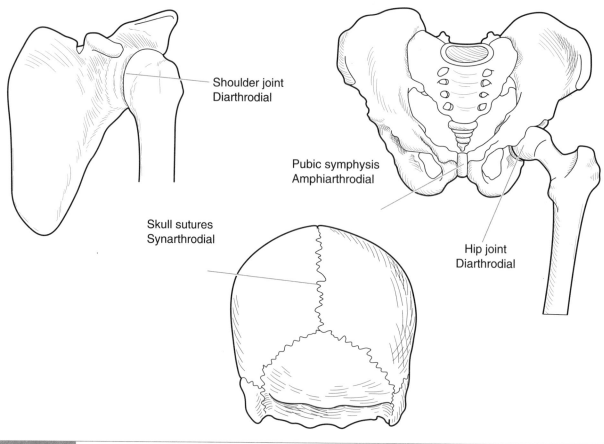

Shoulder joint
Diarthrodial

Pubic symphysis
Amphiarthrodial

Hip joint
Diarthrodial

Skull sutures
Synarthrodial

Figure 13.6   Joints as classified by structure and function.

also be called trochoid or screw joints. The articulating surfaces of a pivot joint may approximate a pin inserting into a hole or a cylinder (aligned with the long axis of the bone) that fits into a shallow trough. Rotation about a longitudinal axis is allowed at a pivot joint. The proximal radioulnar joint (between the bones of the forearm) and the atlantoaxial joint (between the first and second cervical vertebrae) are examples of pivot joints. The rotary movements of the proximal radioulnar joint are called supination and pronation. The rotary movements of the atlantoaxial joint are called rotation to the right or left.

Ellipsoidal joints are biaxial and allow two degrees of freedom of movement. These joints may also be called condyloid or ovoid joints. The articulating surfaces of an ellipsoidal joint approximate the shape of an ellipse (or egg) that fits into a matching oval depression. These joints have also been described as oval ball-and-socket joints. The movements allowed at an ellipsoidal joint are flexion and extension, abduction and adduction, and circumduction, a combination of these movements. The radiocarpal (wrist), metacarpophalangeal (fingers),

metatarsophalangeal (toes), and occipitoatlantal (head and neck) joints are all examples of ellipsoidal joints.

Saddle joints are also biaxial and allow two degrees of freedom of movement. These joints are also called sellar joints. The articulating surfaces of a saddle joint look like a pair of saddles turned 90° to each other. Cup the fingers of each hand, then put your hands together, rotating one of them 90°. This approximates a saddle joint. The U-joints in the driveshaft of a car are like saddle joints. A saddle joint also allows flexion and extension, abduction and adduction, and circumduction. The first carpometacarpal joint (at the base of the thumb) is an example of a saddle joint.

Ball-and-socket joints are triaxial and allow three degrees of freedom of movement. These joints are also called enarthrodial, spheroidal, or cotyloidal. The articulating surfaces of these joints look like a ball and a socket. Ball-and-socket joints are the most freely movable of the synovial joints. They allow flexion and extension, abduction and adduction, and internal and external rotation. The glenohumeral (shoulder) and hip joints are examples of

Figure 13.7 Idealized representation of the six diarthrodial (synovial) joint types.

ball-and-socket joints. The sternoclavicular joint (between the shoulder girdle and the axial skeleton) is classified as a ball-and-socket or gliding joint.

Structure of Synovial Joints

The distinguishing characteristic of a synovial joint is the joint cavity formed by the articular capsule. The articular capsule is a sleeve of ligamentous tissue that surrounds the joint. It attaches to the bones on either side of the joint, connecting them together. The articulating ends of the bones are covered with a thin layer of hyaline cartilage called the articular cartilage. The other exposed bony surfaces within the articular capsule and the inner surface of the articular capsule are lined with a synovial membrane. The synovial membrane and the articular cartilages thus seal the joint cavity. A synovial joint can be compared to sealed bearings, which aren't exposed to the elements and thus rarely

need to be regreased. But unlike sealed bearings, a synovial joint is self-lubricating. The synovial membrane secretes synovial fluid, which fills the joint cavity. In some synovial joints, such as the knee or sternoclavicular joint, a disk or partial disk of fibrocartilage separates the articulating surfaces of the bones. Figure 13.8 shows the various structures of a synovial joint.

> **The distinguishing characteristic of a synovial joint is the joint cavity formed by the articular capsule. The articular capsule is a sleeve of ligamentous tissue that surrounds the joint.**

The synovial fluid excreted by the synovial membrane is a viscous fluid similar in appearance to egg whites (*syn* 'like' and *ovium* 'egg'). One function of the synovial fluid is to lubricate the joint and reduce friction. Under pressure, the articular cartilage

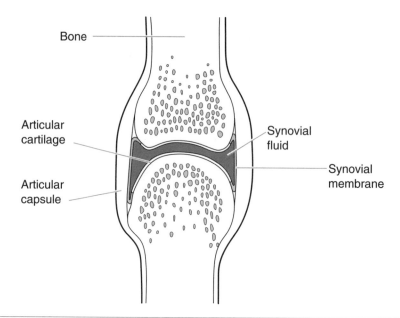

also secretes synovial fluid that lubricates the joint. Another function of the synovial fluid is to nourish the articular cartilage, as articular cartilage has no nerve or blood supply. The synovial fluid also cleanses the joint cavity and imparts some hydrostatic shock-absorbing properties to the joint.

The articular cartilage is the bearing surface between moving bones. It improves the bone-to-bone fit at the joint, increasing the joint's stability and reducing pressure when the joint is loaded. It also reduces friction and prevents wear. Because hyaline cartilage is more resilient than bone, the articular cartilage also provides some shock absorption at the joint. In joints that have disks or partial disks of fibrocartilage, the disks perform similar functions: improving bone-to-bone fit and absorbing shock.

The articular cartilage is the bearing surface between moving bones.

Although the term *cavity* implies that a space of some sort exists between the ends of the articulating bones in a synovial joint, this space is quite small, and in normal situations the articular cartilages actually contact each other. If a joint is injured, however, the swelling caused by increased fluid volume may greatly enlarge the cavity.

Stability of Synovial Joints

Stability was previously defined in chapter 8 as "not easily moved." Joint stability refers to movements of the bones in planes other than those defined by the degrees of freedom of movement for the joint or to movement of the articulating surfaces away from each other through shear dislocation (sliding laterally) or traction dislocation (pulling apart). Stability of a hinge joint thus refers to its ability to resist abduction and adduction, internal and external rotation, or dislocation. Joint flexibility refers to the range of motion possible in the planes of motion defined by the degrees of freedom of movement for the joint and how easily these motions can occur. Flexibility of a hinge joint thus refers to its range of motion in flexion and extension and the ease of these movements.

> **Joint stability refers to movements of the bones in planes other than those defined by the degrees of freedom of movement for the joint or to movement of the articulating surfaces away from each other through shear dislocation (sliding laterally) or traction dislocation (pulling apart).**

The reciprocal convex and concave shapes of the articulating ends of bones in a synovial joint are primarily responsible for determining the planes of motion allowed at the joint. Compressive and shear forces develop between matching parts of articulating bones to resist any shear dislocations or rotations in planes other than those for which the joint was designed, as shown in figure 13.9. The tighter the bone-to-bone fit and the deeper the matching

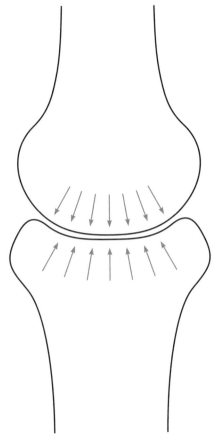

Figure 13.9 Tight bone-to-bone fit allows for compressive and shear force development to prevent dislocations.

convex and concave surfaces of the joint, the more stable the joint. The hip is more stable than the shoulder because the hip socket (the acetabulum) is much deeper than the shoulder socket (the glenoid fossa). Articular cartilage and fibrocartilage disks increase bone-to-bone fit and also assist with joint stability. The menisci of the knee give the condyles of the femur a deeper depression to sit in than the relatively flat tibial plateau.

Whereas the articulating bones resist compressive and shear forces at joints, ligaments provide the tensile forces to resist traction dislocations when something tends to pull the bones apart at a joint. Their tensile strength is also required when a bending load acts on the joint in a plane different from the plane of motion of the joint. In this case, the joint must act like a beam, with compression stress on one side and tensile stress on the other. The articular surfaces of the bones on one side of the joint will resist the compression component of the bending load, whereas the ligaments on the opposing side will resist the tensile component of the bending

load. The articular capsule itself provides some of this tensile strength, but most joints have ligaments as well to strengthen the joint so it can resist dislocating torques or traction forces. The ligaments may occur as thickened bands in the articular capsule itself or as external ligaments separate from the articular capsule. The locations of ligaments relative to the joint they protect thus determine the stability of the joint to bending loads that may cause dislocating abduction, adduction, or rotation. Figure 13.10 illustrates how the medial collateral ligament of the knee carries a tensile load to prevent the knee from abducting when a medial force acts on the knee.

> **Whereas the articulating bones resist compressive and shear forces at joints, ligaments provide the tensile forces to resist traction dislocations when something tends to pull the bones apart at a joint.**

Tendons and the muscles attached to them also resist tensile forces and thus contribute to joint stability in a manner similar to ligaments. The line of pull of most muscles is such that a component of the force generated during a contraction tends to pull the bones of a joint more tightly together, thus affording resistance to any traction-dislocating forces that tend to pull the bones apart. This action of muscles and ligaments is shown in figure 13.11a. In certain joint positions, muscles may actually create forces having components that tend to dislocate rather than stabilize a joint. This action is shown in figure 13.11b. The muscles, tendons, and tendinous sheets (aponeuroses) that cross joints also provide some lateral support to the joints they cross.

Another factor that contributes slightly to the stability of synovial joints is the pressure within the joint cavity. This internal pressure is less than the pressure external to the joint cavity, so a suction force is created. The articulating bones are thus sucked together as long as the joint cavity is sealed and intact.

Flexibility of Synovial Joints

Joint flexibility refers to the range of motion in the planes in which the joint is designed to move. Range of motion of a limb is limited by some of the same factors that contribute to joint stability: the bones, ligaments, and muscles. Range of motion is limited by other factors as well.

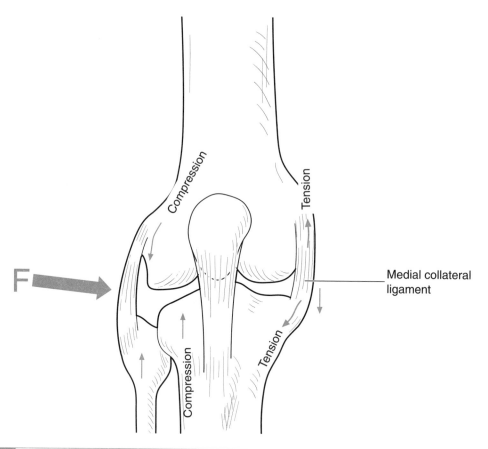

Figure 13.10   Tensile loading of the medial collateral ligament prevents medial dislocation of the knee joint when it is struck from the lateral side.

> **Joint flexibility refers to the range of motion in the planes in which the joint is designed to move.**

Range of motion is limited by the extensibility of the muscles that cross a joint. Muscles that cross more than one joint (multiple-joint muscles) may not be able to stretch far enough to allow full range of motion at a joint if the positions at each joint the muscle crosses stretch the muscle.

Most muscles that cross only one joint (single-joint muscles) are extensible enough that they do not limit range of motion at that joint. With the exception of the rectus femoris, the quadriceps muscles are all single-joint muscles. Flex your hip and knee at the same time. The quadriceps muscles are not stretched to their limit at the knee's range-of-flexion limit. Other factors limit the range of knee flexion.

Ligaments limit range of motion also. Ligaments that do not align radially with respect to an axis of the joint will become looser during one joint action

and tighter as the reverse joint action progresses, until the ligament stops the movement. At ball-and-socket and pivot joints, twisting of the ligaments limits the range of rotation. The articular capsule itself also limits range of motion.

The shapes of the articulating bones may also limit range of motion. Elbow extension stops when the olecranon process of the ulna becomes constrained by the olecranon fossa of the humerus, or vice versa. The lateral malleolus of the fibula limits eversion of the ankle when it contacts the calcaneus.

The bulkiness of soft tissues or clothing around a joint may also limit range of motion. Someone with large elbow flexor muscles (biceps brachii and brachialis) may not have as much range of elbow flexion as someone with less-developed muscles. A child bundled up for outdoor winter weather in a bulky overcoat is an extreme example of how clothing may limit range of motion.

Mobility, or the ease of movement through a range of motion, is affected by friction within the joint and by the inertia and tension in the muscles

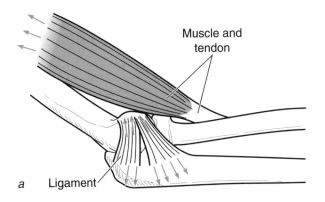

Muscle and tendon

a Ligament

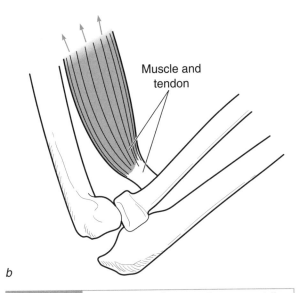

Muscle and tendon

b

Figure 13.11 Tensile forces from muscles may contribute to joint stability (*a*) or instability (*b*).

around the joint, especially those that oppose the joint motion in question (antagonist muscles). Any wearing or damage to the articular cartilage increases joint friction and thus reduces mobility. Wear and tear of the articular fibrocartilage disks reduces mobility as well in joints with these structures. Damage to the articular cartilage or bone within the joint cavity may produce loose particles that decrease mobility. Damage to the synovial membrane and its capacity for producing synovial fluid will also result in loss of joint mobility. People suffering from arthritic diseases will often have one or more of these problems.

Summary

The skeletal system provides the body with a framework of rigid links that enables movement. The 200-plus bones of the skeleton provide support and protection for internal organs. The bones also function as red blood cell producers and as a storage site for minerals.

The bones of the skeleton are joined together at joints that may allow a great range of movement, slight movement, or none at all. Joints may be classified according to their function or structure. Diarthrodial or synovial joints are the joints of interest to us because they allow free movement. The six types of synovial joints are gliding, hinge, pivot, ellipsoidal, saddle, and ball and socket. All synovial joints have a joint cavity lined with a synovial membrane that secretes synovial fluid. The bearing surfaces of the bones in a synovial joint are coated with a thin layer of articular cartilage that is nourished

SELF-EXPERIMENT

Lie on your back. Flex your right hip as far as you can while keeping your right knee extended as shown in figure 13.12a. Now allow your knee to flex and see how much farther you can flex your hip (see figure 13.12b). In the first case, your range of hip flexion was limited by your hamstring muscles. These muscles are two-joint muscles that cross posterior to the hip joint and the knee joint. They are stretched if the knee extends and stretched farther if the hip then flexes (as in the first case). The tension in the hamstring muscles limited your hip flexion when the knee was extended. When the knee flexed, the tension in these muscles was relieved and you were able to flex your hip farther.

Figure 13.12 Muscle extensibility affects joint range of motion, as illustrated in this test of hip joint flexibility. In (*a*), the hamstring muscles limit the range of motion of hip flexion. In (*b*), tension in the hamstring muscles is released by flexing the knee, resulting in greater range of hip flexion.

and lubricated by the synovial fluid. A ligamentous sleeve surrounds synovial joints and connects the bones to form the articular capsule.

Joint stability is affected by bone-to-bone fit, articular cartilage and fibrocartilage disks, ligaments, muscles and tendons, and negative pressure (suction). Joint flexibility or range of motion is affected by muscle extensibility, ligaments, bone-to-bone contact, and the bulk of soft tissue and clothing. Joint mobility or ease of range of motion is affected by antagonist muscle tension and friction within the joint cavity.

Key Terms

appendicular skeleton *(p. 240)*

articular capsule *(p. 247)*

articular cartilage *(p. 247)*

axial skeleton *(p. 240)*

cartilaginous joint *(p. 245)*

diaphysis *(p. 243)*

epiphyseal cartilage *(p. 243)*

epiphysis *(p. 243)*

fibrous joint *(p. 245)*

joint *(p. 244)*

sesamoid bone *(p. 242)*

synovial joint *(p. 245)*

synovial membrane *(p. 247)*

REVIEW QUESTIONS

1. What are the mechanical functions of the skeletal system?

2. What are the metabolic functions of the skeletal system?

3. What type of bone is
   a. the femur?
   b. the third thoracic vertebra?
   c. the calcaneus?
   d. the first metacarpal?
   e. the patella?

4. What component of bone is primarily responsible for resisting compression?

5. At what age do most epiphyseal plates ossify?

6. What are the six types of synovial joints?

7. What type of synovial joint is the glenohumeral joint?

8. What are the functions of synovial fluid?

9. What keeps the synovial fluid in the joint cavity?

10. What are the functions of articular cartilage?

11. What factors influence the stability of synovial joints?

The Muscular System:
The Motors of the Body

© Ray Malace

You are watching a bodybuilding show as the athletes pose to music. You are struck by the number and sizes of the muscles displayed and by how well defined and large they are. You have always thought of muscles as the active force producers within the body, but now you realize what artists have known for centuries: muscles give our bodies the shape and form that make it aesthetically pleasing. What is the anatomy of muscles that allows them to take on such a variety of shapes? How do muscles produce force, and what factors affect this force production? This chapter will attempt to answer these and other questions concerning the structure and function of muscles.

OBJECTIVES

When you finish this chapter, you should be able to do the following:

■ Describe the three types of muscle tissue: smooth, skeletal, and cardiac

■ Discuss the functions of skeletal muscle

■ Describe the microstructure of skeletal muscle

■ Describe the macrostructure of skeletal muscle

■ Discuss the three different types of muscular actions: concentric, eccentric, and isometric

■ Describe the roles that muscles can assume

■ Discuss the factors that influence the force developed during muscular activity

■ Understand the relationship between power output and velocity of contraction of a muscle

Muscles are the active components of the musculoskeletal system. Although the bones and joints of the skeleton form the framework of the body, this framework would collapse without the active force generation of muscles providing stiffness to the joints. Muscles are the motors of the musculoskeletal system that allow the levers of the skeleton to move or change position.

The distinguishing characteristic of muscle, which gives it the capacity to move or stiffen joints, is its ability to actively shorten and produce tension. This characteristic is not unique to the muscles of the musculoskeletal system alone, however. All muscle tissue has the ability to contract. Three different types of muscle tissue are present in the human body: smooth (visceral) muscle, cardiac muscle, and striated (skeletal) muscle. The walls of vessels (veins and arteries) and hollow organs (stomach, intestines, uterus, bladder, etc.) are smooth muscle. This type of muscle is smooth in appearance, as its name implies, and is greatly extensible. Cardiac muscle is irregularly striped (striated) in appearance. The walls of the heart are cardiac muscle. Cardiac and smooth muscle are innervated by the involuntary (autonomic) nervous system. Skeletal muscle has regularly-spaced parallel stripes that give it its striated appearance. Unlike cardiac or smooth muscle,

skeletal muscle cells are multinucleated due to their length. Skeletal muscle contractions are largely voluntary, and thus skeletal muscle is controlled by the somatic (voluntary) nervous system.

> **The distinguishing characteristic of muscle is its ability to actively shorten and produce tension.**

The Structure of Skeletal Muscle

Skeletal muscles serve a variety of functions, including movement and posture maintenance, heat production, protection, and pressure alteration to aid circulation. More than 75% of the energy used during a muscular contraction is released as heat. Muscles are the body's furnace. Muscles may serve as shock absorbers to protect the body. The walls of the abdomen and chest are covered with muscles that protect the underlying organs. A final function of skeletal muscle is pressure alteration. This is primarily a function of cardiac or smooth muscle, but the contraction of skeletal muscles may also alter pressure in veins, thus assisting with the venous return of blood. Because skeletal muscle attaches to bone and is responsible for controlling joint movements, this is the function in body movement that is most interesting. The more than 400 skeletal muscles in the human body are of various shapes and sizes, but their common functions lead to some general similarities in their structure.

Microstructure of Skeletal Muscles

A single muscle cell is a muscle fiber. A muscle fiber is a long, threadlike structure 10 to 100 μm (10 to 100 millionths of a meter) in diameter and up to 30 cm long. Covering the muscle fiber is a thin cell membrane called the sarcolemma. External to the sarcolemma is the endomysium, the connective tissue sheath that encases each muscle fiber and anchors it to other muscle fibers and connective tissue and eventually to the tendon.

Within each muscle fiber are hundreds of smaller (1 μm in diameter) threadlike structures lying parallel to each other and running the full length of the fiber. These are the myofibrils, whose number may vary from less than 100 to more than 1000, depending on the size of the muscle fiber. Transverse light and dark bands appear across each myofibril and align with the same bands on adjacent myofibrils. These bands of light and dark repeat every 2.5 μm and give skeletal muscle its striated appearance. Within a single myofibril, the repeating unit of the myofibril between stripes is called a sarcomere.

The sarcomere is the basic contractile unit of muscle. Thick (myosin) and thin (actin) protein filaments, or myofilaments, overlap within the sarcomere. The thin actin filaments are free at one end, where they overlap with the thick myosin filaments, and at the other end, they anchor to adjacent sarcomeres in series with each other at the transverse Z line or Z band (for *Zwischenscheibe*, or "between disk"). A sarcomere is thus the part of a myofibril from Z line to Z line. The region that includes only actin filaments and the Z band (or the region where the actin filaments do not overlap with the myosin filaments) is called the I band (for isotropic—named for its refraction of one wavelength of light) and appears as a light band. The darker band or region that includes the full length of the myosin filaments along with the region of overlap with the actin filaments is called the A band (for anisotropic—named for its refraction of more than one wavelength of light). The region within the A band where the actin and myosin do not overlap is called the H band (for

Hellerscheibe, or "clear disk"). At the middle of the H band is the M band (for *Mittelscheibe*, or "middle disk"), the transverse band that connects adjacent myosin filaments with each other. The ends of the myosin filaments have projections that give it a brushlike appearance. These projections are the cross bridges that link to the actin filaments and create the active contractile force during a muscular contraction. Figure 14.1 illustrates conceptually the structure of a muscle fiber, its myofibrils, and a sarcomere. A transverse or cross section through a myofibril shows that each myosin filament is surrounded by six actin filaments in the overlap zone, whereas each actin filament is surrounded by only three myosin filaments.

The sarcomere is the basic contractile unit of muscle.

Macrostructure of Skeletal Muscles

Muscle fibers are bundled together in groups of 100 or more to form a fascicle (or fasciculus). Each fascicle is encased by a connective tissue sheath called the perimysium. The endomysium of each muscle fiber is bound to this perimysium. Several fascicles are then bound together within a connective tissue sheath called the epimysium to form a whole muscle, as shown in figure 14.2.

Myofilaments run the entire length of a muscle fiber and attach to the endomysium at their ends. In long muscles, individual muscle fibers are not long enough to run the entire length of the muscle and may not be as long as the fascicle in which they are bound. The endomysium of a muscle fiber thus attaches to the endomysium of the next muscle fiber in series (aligned with it end to end) or to the perimysium of the fascicle. The perimysium of a fascicle then attaches to the perimysium of the next fascicle in series or to the epimysium of the entire muscle. Thus the connective tissue sheaths of the muscle become bound together at the ends of the muscle. This connective tissue extends beyond the region of the muscle containing the contractile elements. It weaves itself into cords or sheets of connective tissue that connect the muscle to bone. These cords are called tendons and the sheets are called aponeuroses. The force of a muscular contraction is thus transmitted from the endomysium of the muscle fiber to the perimysium and epimysium and then to the tendon, which is a continuation of these connective tissues.

Figure 14.1   Skeletal muscle microstructure.

The insertion or attachment of a tendon to bone is similar to the attachment of a ligament to bone. As tendons (and ligaments) are less stiff than bone, there must be a gradual increase in the stiffness of the tendon (or ligament) from the muscle end to the bone end. At the site of insertion, the tendon gradually contains more ground substance so that it becomes fibrocartilage. This fibrocartilage part of the tendon then becomes more mineralized, and thus bony, as the tendon attaches to the periosteum

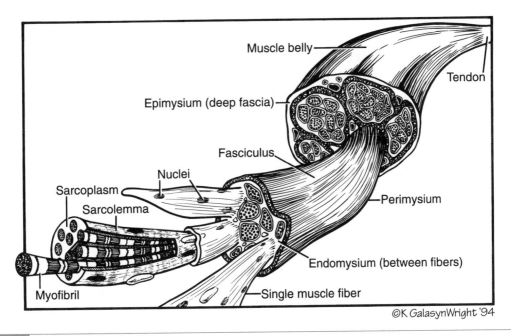

Figure 14.2   Skeletal muscle macrostructure.

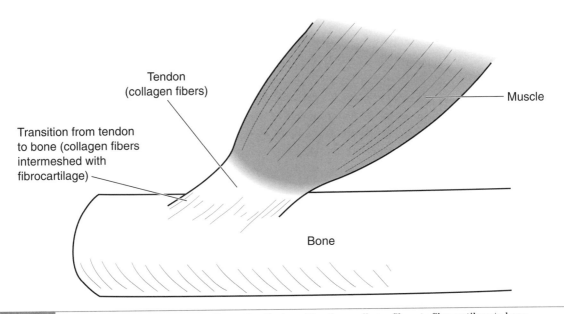

Figure 14.3   The insertion of tendon to bone, illustrating the transition from collagen fibers to fibrocartilage to bone.

of the bone and blends into the bone itself. This attachment and transition at the tendon-bone interface is shown in figure 14.3.

Because muscles attach to movable limbs, the muscles and their attached tendons move when the muscles shorten or lengthen. Movements of the belly of the muscle may create friction between it and adjacent muscles. Loose connective tissue between adjacent muscles functions to reduce this friction. Where tendons or muscles would rub against

bones or ligaments, bursae may be present to reduce friction or prevent damage to the muscle or tendon. A bursa is a sac filled with synovial fluid that pads and lubricates soft tissues on bones. Tendon sheaths are similar to bursae in structure and function, but they wrap completely around a tendon.

Typically, muscles have two ends that are attached by tendons to bones on either side of a joint. When a muscle contracts, it pulls with equal force on each attachment, and each bone will tend to move. Anatomically, the origin of a muscle is its more proximal attachment and the insertion of a muscle is its more distal attachment. Mechanically, the origin of a muscle attaches to the bone that moves less and the insertion attaches to the bone that moves more. The anatomical insertion of the triceps brachii is always the olecranon process of the ulna, the more distal attachment of this muscle. Mechanically, the origin and insertion of the triceps brachii depend on the movement. The olecranon process is the mechanical insertion of the triceps brachii during a bench press, but during a push-up, the olecranon process is the mechanical origin of the triceps. The anatomical terminology is most common, as it is understood even if no movement is referred to.

> **When a muscle contracts, it pulls with equal force on each attachment.**

A muscle's line of pull refers to the direction of the resultant force produced at an attachment. The direction of the resultant force is along a line from the origin to the insertion of the muscle. This will be true for muscles that have single points of origin and single points of insertion. Some muscles, such as the triceps brachii, have more than one insertion or origin; other muscles, such as the trapezius, have broad origins or insertions. In these situations, no single line of pull is apparent, and the line of pull of these muscles depends on which fibers are active.

> **A muscle's line of pull refers to the direction of the resultant force produced at an attachment.**

As alluded to in the previous paragraph, the arrangement of fibers in a muscle may affect its function. Muscles may have more than one head or anatomical origin (e.g., biceps, triceps, quadriceps). Long muscles whose fibers all align parallel to the line of pull of the muscle are called longitudinal, strap, or fusiform muscles. Muscles that have shorter fibers not aligned with the line of pull are called pennate (for feather) muscles. The fibers of these muscles insert at an angle to a tendon and thus resemble the barbs of a feather. Pennate, or penniform, muscles may be unipennate, bipennate, or multipennate. In general, pennate muscles are able to exert greater forces than similar-sized longitudinal muscles, but the longitudinal muscles are able to shorten over a greater distance. Figure 14.4 shows examples of the various shapes and forms of muscles.

Muscle Action

The distinguishing characteristic of muscle is its ability to contract. The development of tension within a muscle causes it to pull on its attachments. This action of a muscle is usually referred to as a muscular contraction; however, this use of the word *contraction* is confusing because it implies that the muscle shortens in length during this activity. But a muscle may be contracting and not changing length at all or may even be lengthening. A more accurate word to describe muscle "contraction" is muscle action. When a muscle is active, it develops tension and pulls on its attachments. The active muscle may shorten, stay the same length, or lengthen.

Types of Muscular Actions

When a muscle is active and its attachments draw closer together, the muscle is active concentrically (or contracting concentrically). The muscle action is a concentric action (or a concentric contraction). This occurs when the torque created by the active muscle on either limb at a joint is in the same direction as the rotation of the limb. Concentric muscle action occurs when

- the flexor muscles are active and flexion occurs,
- the extensor muscles are active and extension occurs,
- the abductor muscles are active and abduction occurs,
- the adductor muscles are active and adduction occurs,
- the internal rotator muscles are active and internal rotation occurs, or
- the external rotator muscles are active and external rotation occurs.

Because the force and torque created by the muscle are in the same direction as the movement of the muscle's points of attachment or the rotation of the limbs, positive mechanical work is done by the muscle. Recall from chapter 7 that work done causes a change in energy. Positive work results in an increase in kinetic and/or potential energies. If the muscle force is the only force acting on the limb (other than gravitational forces), then a concentric muscle action results in an increase in the limb's kinetic or potential energy. Generally, if a limb's velocity is speeding up or its position is being elevated, concentric muscular activity is likely to be

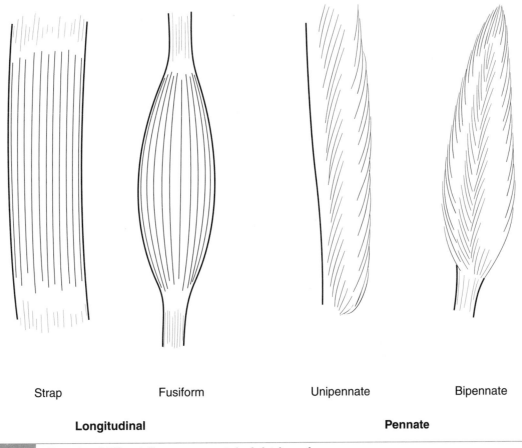

Strap   Fusiform   Unipennate   Bipennate

Longitudinal   **Pennate**

| Figure 14.4 | Examples of different fiber arrangements in skeletal muscles. |

occurring. During the up phase of a push-up, potential energy is increased because the body is being elevated. Elbow extension occurs, and the elbow extensor muscle (the triceps brachii) is active concentrically. Figure 14.5 illustrates examples of concentric muscle activity.

> **When a muscle is active and its attachments draw closer together, the muscle is active concentrically.**

When a muscle is active and its attachments are drawn farther apart, the muscle is active eccentrically (or contracting eccentrically). The muscle action is an eccentric action (or an eccentric contraction). This occurs when the torque created by the active muscle on either limb at a joint is in the direction opposite the rotation of the limb. Eccentric muscle action occurs when

- the flexor muscles are active and extension occurs,
- the extensor muscles are active and flexion occurs,

- the abductor muscles are active and adduction occurs,
- the adductor muscles are active and abduction occurs,
- the internal rotator muscles are active and external rotation occurs, or
- the external rotator muscles are active and internal rotation occurs.

Because the force and torque created by the muscle are in the direction opposite the movement of the muscle's points of attachment or the rotation of the limbs, negative mechanical work is done by the muscle. Negative work results in a decrease in kinetic and/or potential energies. If the muscle force is the only force acting on the limb (other than gravitational forces), an eccentric muscle action results in a decrease in the limb's kinetic or potential energy. Generally, if a limb's velocity is slowing down or its position is being lowered, eccentric muscular activity is likely to be occurring. During the down phase of a push-up, potential energy is decreased because the body is being lowered. Elbow flexion occurs, and the elbow extensor muscle (the triceps

Figure 14.5   Examples of phases of movements that involve concentric muscular activity.

brachii) is active eccentrically. Figure 14.6 illustrates examples of eccentric muscle activity.

> **When a muscle is active and**
> **its attachments are drawn farther apart,**
> **the muscle is active eccentrically.**

When a muscle is active and its attachments do not move relative to each other, the muscle is active isometrically (or contracting isometrically). The muscle action is an isometric action (or an isometric contraction). As no rotation of the limb occurs, no mechanical work is done and no change in kinetic or potential energy occurs. Because the isometrically active muscle exerts a torque about the joint, some other opposing torque must be exerted about the joint to keep it in static equilibrium and not moving. This opposing torque may be provided by the opposing muscles on the other side of the joint or by some external force such as gravity. Generally, if no movement occurs at a joint, but other forces act on the limbs at that joint, an isometric muscular activity is likely to be occurring. If you push yourself halfway up during a push-up and hold that position, no movement is occurring, but the elbow extensor muscle is active to prevent gravity from flexing your elbows and pulling you back down to the ground. The elbow extensor muscle (the triceps brachii) is active isometrically.

When a muscle is active and its attachments do not move relative to each other, the muscle is active isometrically.

Roles of Muscles

A variety of terms are used to describe the roles of muscles either relative to a joint action or relative to another muscle. An **agonist** muscle is capable of creating a torque in the same direction as the joint action referred to. Thus muscles that are active concentrically are agonists to the action occurring at the joint they cross. The biceps brachii is an agonist for elbow flexion; the triceps brachii is an agonist for elbow extension; the quadriceps femoris is an agonist for knee extension; the gastrocnemius is an

Figure 14.6   Examples of phases of movements that involve eccentric muscular activity.

agonist for ankle plantar flexion; and so on. These muscles may also be referred to as prime movers or protagonists. The terms *agonist, prime mover,* and *protagonist* are undefined unless a joint action is referred to in relation to them. The statement, "The biceps brachii is an agonist," is incomplete.

An **antagonist** is a muscle capable of creating a torque opposite the joint action referred to or opposite the other muscle referred to. The biceps brachii is an antagonist for elbow extension; the triceps brachii is an antagonist for elbow flexion; the quadriceps femoris is an antagonist for knee flexion; the gastrocnemius is an antagonist for ankle dorsiflexion; and so on. In the preceding examples, the term *antagonist* is used in reference to a joint action, and thus muscles active eccentrically are antagonists to the action occurring at the joint they cross. *Antagonist* may also be used in reference to another muscle. The biceps brachii is an antagonist to the triceps brachii; the quadriceps femoris are an antagonist to the hamstrings; the gastrocnemius is an antagonist to the tibialis anterior; and so on. In

these cases, the torque the antagonist creates opposes the torque created by the muscle referred to.

Other terms have been used to describe other roles that muscles play, including *synergist, neutralizer, stabilizer, fixator,* and *supporter.* These terms are all used in reference to another muscle. The terms **stabilizer**, *fixator,* and *supporter* refer to muscles that are active isometrically to keep a limb from moving when the reference muscle contracts. When a muscle is active, it will tend to move both bones to which it is attached, when movement of only one limb may be desired. Other external forces may cause unwanted movement of limbs as well. The isometric action of the stabilizing (fixating, supporting) muscles keeps the limb from moving. During the down phase of a push-up, the shoulder girdle will tend to adduct due to the weight of the trunk and the upward reaction force acting on the shoulder joint. The serratus anterior acts as a stabilizer to prevent this adduction by acting isometrically.

A **neutralizer** is a muscle that creates a torque to oppose an undesired action of another muscle. The torques created by many muscles have components in several planes; thus, when these muscles are active, they create torques about more than one axis of a joint. For example, when the biceps brachii is active, it creates a flexion torque at the elbow as well as a supinating torque. If supination is the desired action, the triceps would act isometrically to neutralize the undesired flexion torque of the biceps.

The word *synergy* is used to describe mutually beneficial actions of two or more things. In terms of muscle function, a synergist muscle has been described as a muscle that assists in producing the desired action of an agonist muscle. Neutralizers and stabilizers could thus be described as synergists. A synergist could also be another muscle whose torque adds to the torque of an agonist. The term *synergist* is not very exclusive and its use should be avoided.

Contraction Mechanics

The action responsible for the contraction of a muscle occurs within the sarcomere. In response to a stimulus from the motor neuron that innervates a muscle, the muscle becomes active and the cross bridges of the thick myosin filament attach, pull, release, and reattach to specific sites on the thin actin filament. The cross bridges of the myosin filament tug on the adjacent actin filament and pull themselves past it, similar to the way you might pull on (or climb up) a rope hand over hand. The myosin filaments thus slide past the actin filaments and the muscle shortens (in a concentric contraction). An eccentric contraction is like lowering a bucket on a rope by letting it drop a short distance, then regripping it, then letting it drop, and so on. The cross bridges of the myosin filament attach and tug, but another force pulls them off that attachment site, and as the actin filament slides away, the myosin cross bridges quickly reattach to another site. The force developed during a contraction is the sum of the pulling forces each myosin cross bridge exerts on the actin filament. The more cross bridges attached to actin filaments, the larger the contraction force.

A single stimulus from the motor neuron (the nerve cell that innervates the fiber) results in a twitch response of the fiber. The cross bridges attach briefly and then release with muscle tension, rising and then falling as shown in figure 14.7a. The duration of tension in the muscle is short. A repeated series of stimuli received from the motor neuron results in a repeated series of twitch responses of the muscle

fiber if the time between each successive stimulus is long enough. With increased frequency of stimulus (and less time between stimuli), there will still be tension in the fiber when the next stimulus occurs. The subsequent tension in the fiber will be greater. If the frequency of stimuli is rapid enough, a tetanic response of the fiber results, as shown in figure 14.7b. The cross bridges attach, release, and reattach, with increasing tension developed until a maximum value is reached. The maximum tension achieved in a tetanic response is much greater than that achieved in a twitch response. Continued stimuli will keep the tension in the muscle high until fatigue occurs.

A single stimulus from the motor neuron results in a twitch response of the fiber.

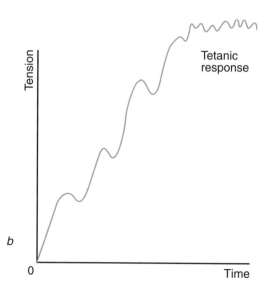

| Figure 14.7 | Tension developed in a muscle fiber as a result of a single stimulus (*a*) is smaller than that developed as a result of repeated stimulation (*b*) if the stimulation frequency is great enough. |

Muscle Contraction Force

The maximum force a muscle is capable of developing depends on several factors concerning the state of the muscle. If all the fibers of a muscle are stimulated to contract, these factors include the physiological cross-sectional area of the muscle, the length of the muscle, and the velocity of muscle shortening, as well as other factors.

Physiological Cross-Sectional Area

The tensile force produced by an active muscle would appear to be determined by the number of myosin cross bridges attached to actin myofilaments throughout the muscle. But this is not exactly the case. Let's consider a single muscle fiber. As described earlier, a muscle fiber consists of a bundle of myofibrils arranged parallel to each other (see figure 14.1). A number of sarcomeres lined end to end in series with each other make up each myofibril, and within each sarcomere are the force-generating myofilaments. If we increased the number of sarcomeres lined end to end in series with each other, we would increase the number of cross bridges and thus increase the number of possible attachments of the cross bridges to the actin myofilaments. The length of the myofibril would increase, and if we did this to each myofibril, the fiber length would increase as well as the length of the resting muscle. But would the muscle be capable of producing any more force?

Series Arrangement of Fibers and Myofibrils

The arrangement of muscle fibers in a muscle or myofibrils in a muscle fiber affects the behavior of the muscle or muscle fiber. Fibers and myofibrils arranged and attached end to end are described as in a series with each other. Fibers and myofibrils arranged alongside each other are described as parallel to each other. The self-experiment described below illustrates the behavior of fibers in series.

Parallel Arrangement of Fibers and Myofibrils

Now suppose that, rather than adding sarcomeres in series to a myofibril, we added them in parallel by increasing the number of myofibrils, or instead of adding sarcomeres, we increased the number of myofilaments within each sarcomere of a myofibril. In these cases, we have increased the number of possible attachments of the cross bridges to the actin myofilaments (as was the case before), but rather than increasing the length of the myofibril or

SELF-EXPERIMENT

Try the following experiment using three rubber bands and two paper clips. Loop one rubber band over the index fingers of your right and left hands, as shown in figure 14.8. Now pull until the rubber band stretches to twice its length. Unhook the rubber band from your left hand and link a second rubber band in series to this one using a paper clip. Now loop the second rubber band over your left index finger and pull until the two rubber bands double in length. Unhook the second rubber band from your left hand and hook the third rubber band to it with the other paper clip, then hook the third rubber band over your left index finger. Pull again until the three rubber bands double in length. By adding one or two rubber bands in series to the first rubber band, did you have to pull with more force to double the length of the rubber bands? No! The force you had to pull with to double the length was about the same. Muscles behave similarly. Longer muscles can stretch and shorten over greater lengths than short muscles, but they are not any stronger, even though they may have more myofilaments. Strength gains in weight training aren't due to athletes growing longer muscles; instead, their muscles grow thicker.

Figure 14.8   Stretching rubber bands in series. The force required to double the length of one, two, or three rubber bands in series is the same in all three cases.

muscle, we have increased its diameter and cross-sectional area. Would the muscle be capable of producing any more force in these cases? The self-experiment described on page 267 illustrates the behavior of fibers in parallel.

Longitudinal Versus Pennate Muscles

Human muscle can produce a tensile force of approximately 30 N/cm² of cross-sectional area of active muscle during an isometric contraction, or the tensile stress produced by muscle during isometric contractions is 30 N/cm² if all the muscle fibers in that cross section are active.

$$F_m = A_m \sigma_m \qquad (14.1)$$

where

F_m = maximal isometric contraction force

A_m = cross-sectional area of muscle

σ_m = maximum stress developed by isometric muscle contraction

Thus a longitudinal muscle with a diameter of 3 cm (and a cross-sectional area of 7 cm²) can produce a maximal isometric contraction of:

$$F_m = (7 \text{ cm}^2) \times (30 \text{ N/cm}^2)$$
$$F_m = 210 \text{ N}$$

SELF-EXPERIMENT

Again, let's use rubber bands for our model of muscle. Loop one rubber band over the index fingers of your right and left hands, as before. Pull until the rubber band doubles in length. Now loop a second rubber band over the index fingers of your right hand and left hands so that both rubber bands are looped over your fingers as shown in figure 14.9. Pull again until the rubber bands double in length. Now loop the third rubber band over the index fingers of your right and left hands so that all three rubber bands are looped over your fingers. Pull again until the rubber bands double in length. By adding one or two rubber bands in parallel with the first rubber band, did you have to pull with more force to double the length of the rubber bands? Yes! Muscles behave similarly. The cross-sectional area perpendicular to the muscle fibers and line of pull of a muscle indicates the maximum tensile force a longitudinal muscle can produce.

| Figure 14.9 | Stretching rubber bands in parallel. The force required to double the length of two rubber bands in parallel is twice the force required to stretch one rubber band the same distance. For three rubber bands, the force required is three times as large. |

Growth in the size of muscles will produce increases in the strength of the muscle as well. Gains in size and strength are thus commonly observed by novice weight lifters.

The maximum tensile force a pennate muscle can produce cannot be estimated by the cross-sectional area perpendicular to the muscle fibers and line of pull of a muscle because the muscle fibers and the line of pull of a pennate muscle are not in the same direction. A cross section taken perpendicular to the line of pull of a pennate muscle would not include all the fibers of the muscle (see figure 14.10). A cross section taken perpendicular to the muscle fibers may not include all the fibers either.

The solution is to take several cross sections perpendicular to the fibers so that all the parallel fibers are included. This area gives an indication of how much total tension the fibers can produce, but this tensile force is not in the same direction as the muscle's line of pull. The component of this total tensile force parallel to the line of pull must be determined. This component is computed by multiplying the total area by the cosine of the angle of pinnation (this product gives the physiological

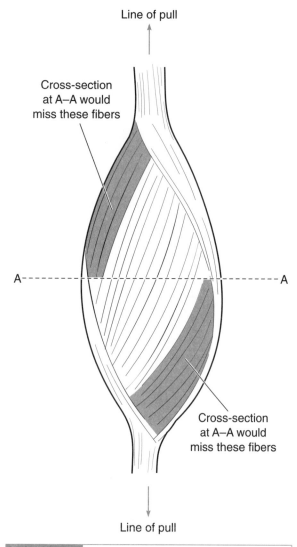

Line of pull

Cross-section
at A–A would
miss these fibers

A - A

Cross-section
at A–A would
miss these fibers

Line of pull

Figure 14.10　A cross section through a pennate muscle and perpendicular to its line of pull may not include all its fibers.

cross-sectional area of the muscle) and then by the maximum tensile stress of the muscle, or:

$$F_m = (A_m \cos \theta)\sigma_m \qquad (14.2)$$

where

F_m = maximal isometric contraction force

$A_m \cos \theta$ = physiological cross-sectional area of muscle

θ = angle of pinnation (the angle of the muscle fibers with the line of pull)

σ_m = maximum stress developed by isometric muscle contraction

For example, let's compare the tension-producing capabilities of two equal volumes of muscle tissue,

one from a longitudinal muscle and one from a unipennate muscle, as shown in figure 14.11.

The maximum isometric tensile force developed in the longitudinal muscle is:

$$F_m = A_m \sigma_m \qquad (14.1)$$
$$F_m = (5 \text{ cm} \times 2 \text{ cm}) \times (30 \text{ N/cm}^2)$$
$$F_m = 300 \text{ N}$$

The maximum isometric tensile force developed in the unipennate muscle is:

$$F_m = (A_m \cos \theta)\sigma_m \qquad (14.2)$$
$$F_m = (11.54 \text{ cm}^2 + 11.54 \text{ cm}^2) \times (\cos 30°)$$
$$\times (30 \text{ N/cm}^2)$$
$$F_m = (23.08 \text{ cm}^2) \times (0.866) \times (30 \text{ N/cm}^2)$$
$$F_m = 603 \text{ N}$$

In this example, the same volume (and mass) of muscle tissue from a unipennate muscle is able to produce 2.0 times as much tensile force during an isometric contraction as tissue from a longitudinal muscle. The shorter fibers of the pennate muscle and their orientation relative to the angle of pull, however, limit the distance over which a pennate muscle can shorten. A comparison of minimum contraction lengths for the longitudinal and unipennate muscles used in the previous example is shown in figure 14.12.

Muscle Length

The physiological cross-sectional area of active muscle will give an indication of the maximum tensile force a muscle is capable of producing, but this maximum is dependent on the length of the muscle during the contraction. The mechanics of contraction within the sarcomere determine the relationship between the length of a muscle and the maximum force of contraction.

> **The physiological cross-sectional area of active muscle will give an indication of the maximum tensile force a muscle is capable of producing.**

Active Tension

The attachment of the myosin cross bridges to the actin filament within a sarcomere is the basis for the active tensile force developed during a muscular contraction. The range of lengths through which the actin and myosin myofilaments can overlap

Longitudinal muscle

Pennate muscle

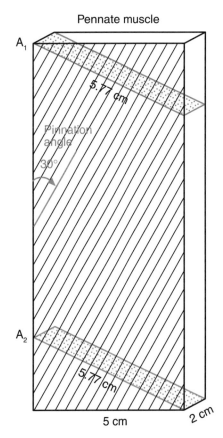

<table>
<tr><td>**Figure 14.11**</td><td>Comparison of total cross-sectional areas of longitudinal and pennate muscles of the same volume. The cross-sectional area, A, of the longitudinal muscle is 5 cm × 2 cm = 10 cm². The cross-sectional area of the pennate muscle is $\Sigma A = A_1 + A_2 = (5.77\ cm \times 2\ cm) + (5.77\ cm \times 2\ cm) = 11.54\ cm^2 + 11.54\ cm^2 = 23.08\ cm^2$.</td></tr>
</table>

determines the range of lengths over which a whole muscle can actively develop tension, as shown in figure 14.13. At maximum tension, the overlap between the actin and myosin myofilaments is maximal. Let's call this the sarcomere's resting length, or ℓ_o (see figure 14.13a). The sarcomere can shorten beyond this point, but then opposing actin filaments begin to overlap, and because the myosin cross bridges cannot attach to these filaments, the tension decreases. Shortening of the sarcomere can still continue until the actin and myosin myofilaments become jammed against the opposite Z disk. At this point, no tension is produced in the muscle and the sarcomere is a little more than half its resting length, or about 60% of ℓ_o (see figure 14.13b). A whole muscle can thus produce tension and shorten to approximately half its resting length (not including the length of its tendons). If the sarcomere is stretched longer than its resting length, fewer myosin cross bridges can attach to the actin myofilaments and the tension decreases. Active tension can be developed at longer lengths until the actin and myosin no longer overlap. At this point, no active

tension is produced and the sarcomere is a little more than 1.5 times its resting length, or 160% of ℓ_o (see figure 14.13c). A whole muscle can thus actively produce tension at lengths 60% to 160% of its resting length.

Passive Tension

Passive tension can be developed in the sarcomere and within the whole muscle by stretching the connective tissue structures: the sarcolemma, endomysium, perimysium, epimysium, and tendon. The passive stretching of these tissues allows a muscle to be stretched and develop tension beyond 160% of its resting length. The tension developed in a whole muscle thus depends on the tension produced by active contraction of the contractile elements (the myofilaments) plus the passive tension developed when the muscle is stretched beyond its resting length. The relationship between muscle length and tension is shown in figure 14.14. Maximum tension can be developed in a whole muscle when it is just a little longer than (about 120% of)

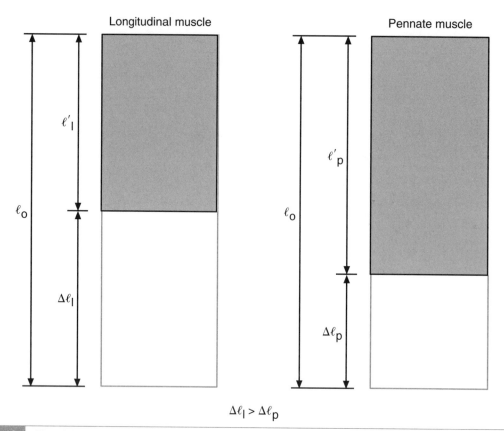

$$\Delta \ell_l > \Delta \ell_p$$

Figure 14.12 The minimum contraction length of a longitudinal muscle (ℓ_l') is shorter than that of a pennate muscle (ℓ_p') if their resting lengths (ℓ_o) are the same. The longitudinal muscle can create tension through a greater range.

its resting length. If the muscle is able to stretch beyond 160% of its resting length, the maximum tension is reached when the muscle is stretched to its maximum length.

> **Maximum tension can be developed in a whole muscle when it is just a little longer than (about 120% of) its resting length.**

Single- Versus Multiple-Joint Muscles

Most single-joint muscles are limited by the range of motion of the joint they cross to operate well within 60% to 160% of their resting lengths. Their maximum tension will be developed at about 120% of their resting length because they cannot be stretched beyond 160% of their length and into the right side of the curve shown in figure 14.14. You will be stronger in some positions than in others, partly because single-joint muscles are able to create the greatest tension at a specific length.

Multiple-joint muscles are not so constrained to operate within 60% to 160% of their resting length. The arrangements of multiple-joint muscles usually

do not allow shortening below 60% of their resting length, but they may be stretched beyond 160% of their resting length. They are able to operate in the right side of the curve shown in figure 14.14. Their maximum tension will be developed at lengths beyond 160% of their resting length due to passive stretching of the connective tissue structures.

The hamstring muscles are another good example of the practical implications of the length-tension relationship. These muscles on the posterior of your thigh create a flexor torque about the knee and an extensor torque about the hip. If you have ever done hamstring curl exercises using a flat bench with weights and a pulley system, you may have noticed that your hamstring muscles are very weak in the fully flexed position shown in figure 14.15a. This is because your hamstrings are as short as you can get them. They are close to the minimum length at which they can develop tension—they are on the left end of the curve shown in figure 14.14. The position you are in, with hips extended and knees flexed, may shorten the hamstrings to less than 60% of their resting length. The hamstring curl exercise, performed on a flat bench, only exercises

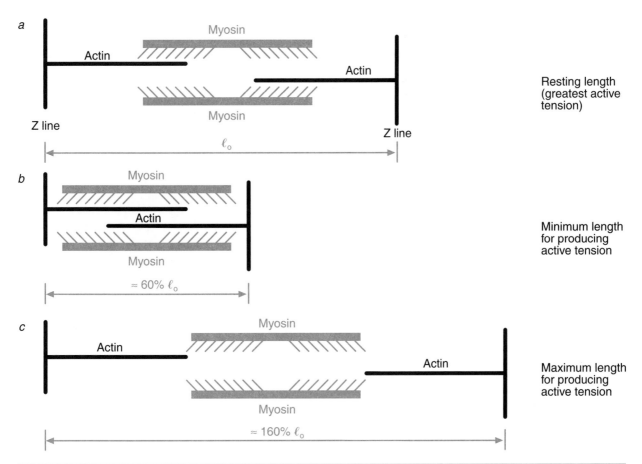

Figure 14.13 Interactions of the actin and myosin filaments with each other and the Z disks determine the magnitude of active tension developed in a sarcomere. The greatest active tension is at resting length (*a*); (*b*) shows the minimum length for producing active tension; and (*c*) shows the maximum length for producing active tension.

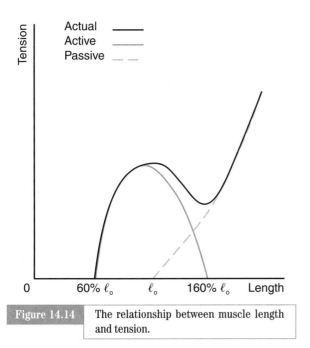

Figure 14.14 The relationship between muscle length and tension.

the muscle in the 60% to 100% of its resting length range. To improve this exercise, you could modify your position so that your hips are slightly flexed rather than extended. This would lengthen the hamstrings a little and thus keep them active in a stronger range of lengths. Rather than using a flat bench, a bench with a hump in it, as shown in figure 14.15b, would allow you to keep your hips flexed while doing the hamstring curl.

Contraction Velocity

The maximum tensile force a muscle is capable of producing is dependent on the velocity of shortening of the muscle as well as its length. Just as the mechanics of contraction within the sarcomere determine the relationship between the length of a muscle and its maximum force of contraction, these mechanics also determine the relationship between the velocity of a muscle's contraction and its maximum force of contraction.

SELF-EXPERIMENT

Consider the muscles that flex or extend your fingers. The finger flexor muscles are located in your anterior forearm, whereas the finger extensor muscles are located in your posterior forearm. Their tendons cross the wrist joint, the carpometacarpal joints, the metacarpophalangeal joints, and the interphalangeal joints. Flex your fingers and grip a pencil as tightly as you can. Now flex your wrist as far as you can. You may notice that you cannot flex your wrist as far with your fingers flexed as you can without your fingers flexed. You may also notice that your grip on the pencil weakened as you flexed your wrist (try to pull the pencil out in both positions). If you push on your hand and cause it to flex further, the pencil may even fall out of your grip. The finger flexor tendons had shortened at each joint they crossed, and thus your finger flexor muscles were unable to produce much tension. The finger extensor tendons were lengthened at each joint they crossed and stretched the extensor muscles beyond 160% of their resting length. Passive tension created by the stretching of the connective tissue resulted in an extension of the fingers when you pushed your hand further into wrist flexion.

a   b

Figure 14.15 Machines for exercising hamstring muscles. In (*a*), muscle length is very short and its ability to create tension is limited. In (*b*), muscle length is longer due to flexion at the hip and its ability to create tension is greater.

> **The maximum tensile force a muscle is capable of producing is dependent on the velocity of shortening of the muscle as well as its length.**

Activity Within the Sarcomere

Remember that the active tension produced during a muscular contraction is a result of the attachment of the cross bridges of the myosin filament to the actin filament within the sarcomere. As the muscle shortens during a contraction, the cross bridges at-

tach to the actin myofilament, pull it toward them, release it, and then reattach to it farther along its length. Thus there are three steps in the cycle for the cross bridges: attach, pull, release. The force developed within a sarcomere is proportional to the number of cross bridges attached and pulling on the actin filaments. When a muscle is shortening, some cross bridges will be in the release step of the process and will not be contributing at all to the tension development. If the velocity of the shortening is slow, only a small percentage of the cross bridges will be in the release step and each cross bridge will spend a small proportion of time in that step, so a large amount of tension will be developed. If the velocity of shortening is faster, more of the cross bridges will be in the release step and each cross bridge will spend a larger proportion of time in that step, so less tension will be developed. Thus, in a sarcomere, greater tension is developed at slower velocities of shortening. The same is true in a whole muscle. It is easy to lift lightweight objects quickly, but as things get heavier, you cannot move them as quickly. Lifting 10 lb with one hand can be done very quickly, but lifting 50 lb with one hand cannot be done as quickly. This is partly due to the inertia of the object, but it is also due to your inability to create large forces if your muscles must shorten quickly. This relationship between contraction velocity and muscle tension is shown in figure 14.16.

Concentric, Eccentric, and Isometric Activity

You may notice that negative velocities of shortening are shown on the graph in figure 14.16. These represent eccentric contractions. A muscle contracting eccentrically or isometrically is capable of producing more force than a muscle contracting concentrically. If you have ever done any weight training, this makes sense. Imagine that you are in the weight room, trying to determine your maximum lift in the bench press. A graphic representation of these lifts on the velocity-tension curve is shown in figure 14.17. You're lying on the bench with your partner spotting you as you try to lift 150 lb. You lower the barbell to your chest and then lift it quickly. The pectoral muscles contract eccentrically during the lowering phase and concentrically during the raising phase. Your partner adds two 25-lb plates to the bar to make the total weight 200 lb. You lower the barbell to your chest and then lift it upward, but this time you can only move the bar slowly. The speed of the muscular contraction is now limited by the size of the force your muscles must develop. You cannot lift 200 lb as quickly as you lifted 150 lb.

> **A muscle contracting eccentrically or isometrically is capable of producing more force than a muscle contracting concentrically.**

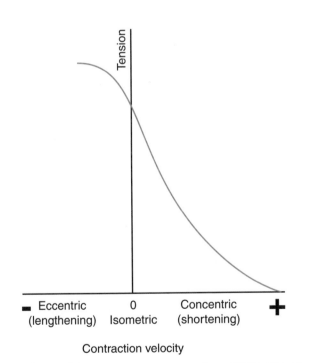

Figure 14.16 The relationship between contraction velocity and tension. Maximum eccentric activity is stronger than maximum isometric activity, which is stronger than maximum concentric activity.

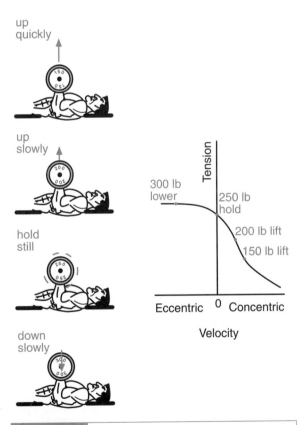

Figure 14.17 Graphic representation of the four bench press lifts on the velocity-tension curve.

Your partner adds two more 25-lb plates to the bar, so the total weight is now 250 lb. You slowly lower the barbell to your chest, but you cannot lift it off your chest (accelerate it upward) without help from your spotter. You and your spotter lift the barbell halfway up, then the spotter lets you handle the weight. You are unable to keep it moving upward (constant velocity with zero acceleration), but you are able to hold the barbell at that position. The pectoral muscles contracting isometrically were able to produce a force large enough to hold the 250 lb statically, but they were unable to produce the same force concentrically to lift the 250 lb. The tension developed in the isometric contraction was greater than the maximum tension that could be developed in the concentric contraction.

Your spotter now adds another 50 lb to the barbell to make a total of 300 lb. You are able to lower the barbell slowly (at constant velocity) to your chest, but you are not able to lift it (accelerate it upward), even with a slight assist from your partner. Your spotting partner helps you lift it halfway, then releases and lets you try again. You cannot even hold it still this time, and the barbell begins to lower slowly (at constant velocity) toward your chest before your spotter helps you complete the lift. The pectoral muscles contracting eccentrically were able to produce a force large enough (300 lb) to lower the 300 lb slowly, but they were unable to produce the same force isometrically to hold the 300 lb still or concentrically to lift the 300 lb. The tension developed in the eccentric contraction was greater than the maximum tension that could be developed in the isometric or concentric contractions.

Other Factors

Several other factors may affect the maximum force produced by an active muscle. Temperature is one of these factors. Increasing the temperature of a muscle slightly increases its ability to produce tension. Muscle temperature can be raised externally through rubbing or heating or internally by engaging in warm-up exercises prior to an activity.

Prestretch

Prestretching a muscle just prior to a concentric contraction may also affect the force of the concentric contraction. The less time between the stretching of the muscle and the subsequent concentric contraction, the greater the force of the contraction. The mechanical bases for this effect are not well understood. The slack in the connective tissue and other noncontractile elements in series with the contractile elements (the myofilaments) may be reduced by prestretching, so that the contraction force is immediately transmitted to the attachments. There is also a neuromuscular basis for this behavior, which will be discussed in the next chapter. In any case, you can demonstrate the effect of prestretch to yourself while performing the self-experiment below.

Stimulus Duration

As discussed in the previous section under "Contraction Mechanics," the frequency of muscle stimulation affects its tension. Maximum tension does not develop within a muscle fiber for a short time (be-

SELF-EXPERIMENT

Try to jump as high as possible using the following two techniques. First, flex your knees and hips slightly, as shown in figure 14.18a, and hold this position for a second before extending your legs and jumping as high as you can. Now try it again, but this time start in an upright position (see figure 14.18b). Quickly flex your knees and hips and then extend them, and jump as soon as you reach the start position of the first jump. You probably jumped higher using the second technique, and the forces exerted by your knee and hip extensors were greater. The muscles were quickly stretched immediately before contracting concentrically in the second technique. The concentric contraction was thus more forceful as a result of the prestretch.

Figure 14.18 Two different standing vertical jump techniques. The second technique utilizes the prestretch of the muscles involved in the jumping action to produce a higher jump.

tween 0.001 and 0.300 s, depending on the muscle type) after it is stimulated. This rate of tension development affects the maximum tension developed in a muscle if a large force is needed in a short time. Thus muscle contractions of very short duration are weaker than contractions lasting longer than 0.001 to 0.300 s.

Fatigue

The maximum tension a muscle can develop is also affected by fatigue. Continuous stimulation of a muscle will result in an eventual decline in the tension it produces. The demand of the contracting muscle for adenosine triphosphate (ATP) eventually exceeds the supply of ATP to the muscle, resulting in diminishing force production by the muscle. You can demonstrate this to yourself, even if the muscle does not produce a maximal contraction. Take the largest weight you can lift (perhaps several books) and hold it at arm's length away from you. How long can you hold the weight in this position? Eventually, the tension your muscles produce weakens and you are unable to hold up the weight.

Fiber Type

Fatigue and the rate of tension development within a muscle are both affected by the type of muscle fiber. Skeletal muscle fibers differ in terms of their fatigue resistance and rate of tension development. Fibers are classified into three types according to these differences: type I, or slow-twitch oxidative (SO); type IIA, or fast-twitch oxidative-glycolytic (FOG); and type IIB, or fast-twitch glycolytic (FG). Type I muscle fibers (SO) have a high density of mitochondria and are thus highly aerobic and fatigue resistant. These fibers have a slow rate of tension development and are smaller in diameter, thus their maximum tension is lower. Type IIB muscle fibers (FG) are glycogen rich and oxygen poor. These fibers have a high anaerobic capacity and a low aerobic capacity, so they fatigue quickly, but their rate of tension development is fast. Type IIB fibers are larger in diameter than type I fibers, so they can produce greater tension, but not for long durations. Type IIA muscle fibers (FOG) have characteristics of type I and type IIB fibers. They have relatively high aerobic and anaerobic capacities, so they develop tension quickly and can maintain it for long durations.

The muscles of the average person contain about 50% to 55% type I fibers, 30% to 35% type IIA fibers, and 15% type IIB fibers, but great variations exist. The more successful athletes engaged in endurance activities will have a higher than normal percentage of type I fibers, whereas those in activities requiring short bursts of power (sprinters, jumpers, throwers, weight lifters, etc.) will have a higher than normal percentage of type II fibers.

Torque Produced

A final consideration in the discussion of maximum force produced by muscle concerns the effectiveness of the force produced. Muscles produce forces that pull on bones and create torques at joints. These torques are used to rotate the limbs and move external loads. The effectiveness of the muscle force thus refers to the effectiveness of the force in producing torque. As torque is force times moment arm ($T = F \times r$), the torque produced by a force depends on the length of the moment arm. The moment arm of a muscle about a joint depends on the point of attachment of the muscle on either limb and the line of action of the muscle force. The maximum length of the moment arm is determined by the distance between the axis of the joint and the muscle's point of attachment closest to the joint. The moment arm of the muscle will equal this distance when the muscle's line of pull is perpendicular to the axis of the bone at this closest attachment. When the limbs are in the position that places the muscle's line of pull at 90° to the bony lever, the torque produced by the muscle is maximum. At all other positions, both the moment arm and the torque will be smaller if the muscle force doesn't change.

Muscle Power

Power is the rate of doing work. In chapter 7, we learned that power could also be expressed as the product of force times velocity. The power output of a muscle is thus the tensile force produced by the muscle times the velocity of shortening of the muscle. Figure 14.16 showed the relationship between the force and velocity of a contraction. From this graph, the relationship between muscle power and velocity of contraction can be obtained by multiplying the velocity values by their corresponding force values to determine muscle power. Figure 14.19 illustrates this relationship.

> **The power output of a muscle is the tensile force produced by the muscle times the velocity of shortening of the muscle.**

Bicyclists utilize this relationship when choosing a gear ratio. In a low gear, they exert small forces on the pedals but spin the crank fast to get the same output as in a high gear, when they exert large forces but spin the crank very slowly. According to the velocity-power curve for muscle, cyclists should choose a gear that allows them to spin the crank at

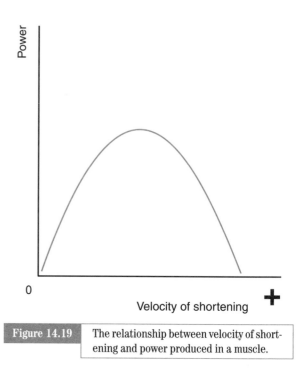

| Figure 14.19 | The relationship between velocity of shortening and power produced in a muscle. |

a moderate speed, thus keeping the muscle's velocity of shortening near the range of values that produces the maximum power output.

Summary

Skeletal muscles are the motors and active force generators of the musculoskeletal system. The forces they produce move the limbs of the body and impart stiffness to the joints. Muscles also produce heat, provide protection, and alter pressure.

Skeletal muscle is composed of numerous muscle cells or fibers arranged in series and in parallel and supported by a complex of connective tissue sheaths. The sarcomere is the fundamental structure of a muscle fiber. Within the sarcomere, the attachments of cross bridges of myosin filaments to actin filaments generate the active tension of a muscular contraction. Fiber arrangements in whole muscles may be longitudinal or pennate. Longitudinal muscles have more muscle fibers in series with each other and may shorten over a greater distance. Pennate muscles have more fibers parallel to each other and may produce greater forces.

A muscular contraction or action may be concentric, isometric, or eccentric. Concentric action occurs when the muscle shortens while it develops tension. Positive mechanical work is done by a concentric muscle action. Isometric action occurs when the muscle does not change length while it develops tension. No mechanical work is done by an isometric muscle action. Eccentric action occurs when the muscle lengthens while it develops tension. Negative mechanical work is done by an eccentric muscle action.

Terms such as *agonist, antagonist, neutralizer,* and *stabilizer* are used to describe muscles when they function in a specific way relative to another muscle or to a joint action.

A number of variables determine the force produced by an active muscle. If all the fibers of a muscle are stimulated, the maximum force developed by a muscular action depends on the physiological cross-sectional area of the muscle, the length of the muscle relative to its resting length, the velocity of contraction, the temperature, and the stimulus rate. The arrangement of a muscle relative to the joint or joints it crosses affects the torque produced and thus determines the effectiveness of the muscular force. Muscle power is determined by the velocity and force of contraction. Maximum power output occurs at a specific contraction velocity.

Key Terms

agonist *(p. 262)*

antagonist *(p. 263)*

aponeurosis *(p. 257)*

concentric contraction *(p. 260)*

eccentric contraction *(p. 261)*

endomysium *(p. 257)*

epimysium *(p. 257)*

fascicle *(p. 257)*

isometric contraction *(p. 262)*

muscle fiber *(p. 257)*

neutralizer *(p. 263)*

perimysium *(p. 257)*

sarcolemma *(p. 257)*

sarcomere *(p. 257)*

stabilizer *(p. 263)*

synergy *(p. 264)*

tendon *(p. 257)*

type I muscle fiber *(p. 275)*

type IIA muscle fiber *(p. 275)*

type IIB muscle fiber *(p. 275)*

REVIEW QUESTIONS

1. **What are the functions of skeletal muscle?**

2. **What is the basic contractile unit of a muscle?**

3. **What is the sarcolemma?**

4. **Describe how the force developed between the actin and myosin filaments of a sarcomere is transferred to the tendon of the muscle.**

5. **During what phase of a baseball throw are the arm muscles' actions concentric? Eccentric?**

6. **During what phase of a vertical jump are the hip muscles' actions concentric? Eccentric?**

7. **What muscles are antagonists to elbow extension?**

8. What muscles are agonists to hip flexion?

9. What muscles are neutralizers for the supination torque produced during concentric activity of the biceps?

10. Discuss the advantages and disadvantages of longitudinal versus pennate fiber arrangements in muscle.

11. In which activity are you able to produce greater forces in your hip muscles: jumping or landing from a jump? Why?

12. In which activity are you able to produce greater forces in your shoulder muscles: throwing a baseball or putting a shot? Why?

The Nervous System: Control of the Musculoskeletal System

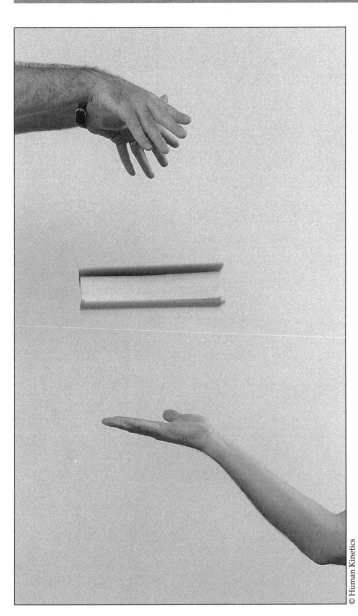

© Human Kinetics

Close your eyes and have someone drop a book into your outstretched hand. Were you able to catch it? How did you know when the book hit your hand? How were you able to detect the weight of the book? How were you able to regulate the amount of force you exerted on the book so that you were able to stop it and catch it? More generally, how does the human body detect external forces acting on it? How does it detect changes in limb positions at joints or changes in orientation of the entire body? How does the body control the contraction force of a muscle? These questions all deal with some aspect of musculoskeletal system control, which is the topic of this chapter.

When you finish this chapter, you should be able to do the following:

▮ List the elements of the nervous system

▮ Describe the different parts of a neuron

▮ List the three types of neurons

▮ Define motor unit

▮ Understand the two strategies (recruitment and summation) used by the central nervous system to control muscle force

▮ Define proprioceptor and list different types of proprioceptors

▮ Describe the function of muscle spindles

▮ Describe the stretch reflex

▮ Describe the function of the Golgi tendon organs

▮ Describe the Golgi tendon organ response

▮ Describe the proprioceptors of the vestibular system

▮ Describe the righting and tonic neck reflexes

▮ Define exteroceptor and list different types of exteroceptors

▮ Describe reflexes initiated by exteroceptors

Control of the musculoskeletal system is administered by the nervous system. The nervous system collects information from external and internal stimuli, processes this information, and then initiates and controls the musculoskeletal system's response to the stimuli. A basic understanding of the nervous system and how it controls movement is thus necessary in our study of the biomechanics of human movement. A thorough treatment of the nervous system and its control of muscular function is the topic of other books, which you are encouraged to read. The purpose of this chapter is to present a simple, brief overview of the nervous system and how it operates to control the musculoskeletal system.

The Nervous System and the Neuron

The nervous system is organized into the central nervous system and the peripheral nervous system. The brain and spinal cord form the central nervous system. Each of these elements is protected by bony structures: the skull and the vertebral column. The brain is the central processor of the nervous system, and the spinal cord transmits signals to or from the peripheral nervous system to the brain. All the nervous tissue that lies outside the skull and the vertebral column forms the peripheral nervous system. The 12 cranial nerves and 31 pairs of spinal nerves are part of the peripheral nervous system. These nerves are actually bundles of nerve fibers, which may include both sensory nerves, which detect information about the external environment and the internal state of the body, and motor nerves, which send stimuli to muscles.

The nervous system is organized into the central nervous system and the peripheral nervous system.

The nervous system is sometimes divided functionally into the somatic nervous system and the autonomic nervous system. The somatic nervous system is involved in conscious sensations and actions and is also referred to as the voluntary nervous system. The autonomic nervous system is involved with unconscious sensations and actions and is also referred to as the involuntary nervous system. It includes sympathetic and parasympathetic nerves. The autonomic nervous system controls and regulates the functions of most internal organs, whereas the somatic nervous system controls and regulates movement.

The fundamental unit of the nervous system is the neuron or nerve cell. A neuron has a cell body, which contains the cell nucleus and other metabolic structures. Numerous branched hairlike projections from the cell body are the dendrites. The axon or what is usually thought of as a nerve fiber is a single, long, threadlike projection that usually extends from the cell body in the opposite direction of the dendrites. The axon often branches into many short fibers at its distal end. Figure 15.1 shows an illustration of two neurons.

The fundamental unit of the nervous system is the neuron or nerve cell.

Basically, there are three types of neurons: (1) sensory, or afferent, neurons, (2) motor, or efferent, neurons, and (3) interneurons, or connector neurons. The sensory and motor neurons are situated primarily in the peripheral nervous system, whereas the interneurons are situated within the central nervous system. The sensory neurons are responsible for sensations. They receive stimuli from our external or internal environment and send this

Dendrites

Cell body

Axon

Terminal branch

a

Axon (central fiber)

Cell body

Dendrites

b

Figure 15.1   Features of a typical (*a*) motor neuron and (*b*) sensory neuron.

information back toward the central nervous system, where they interface with interneurons and occasionally motor neurons. The cell bodies of sensory neurons lie close to but outside the spinal cord. Motor neurons receive stimuli from interneurons or sensory neurons and send signals to muscles, away from the central nervous system. The cell bodies of motor neurons are located within the spinal cord. Because the cell bodies of the sensory and motor neurons are situated close to the spinal cord, whereas the terminal endings of their axons or dendrites may be located in the distal ends of the extremities, nerve fibers (axons and fibers to peripheral dendrites) may be more than 1 m long.

> There are three types of neurons:
> (1) sensory, or afferent, neurons,
> (2) motor, or efferent, neurons, and
> (3) interneurons or connector neurons.

The cell membrane of a neuron is an excitable membrane whose electrical potential (voltage across the membrane) can change as the result of stimulation. Normally, neurons may be stimulated at their cell body and at their numerous dendrite endings. A stimulus can be excitatory (facilitating) or inhibitory, depending on whether the stimulus results in an increase or decrease in the electrical potential of the cell membrane. If the net effect of all the excitatory and inhibitory stimuli results in an increase in the membrane potential above a threshold level at the base of the axon (where it leaves the cell body), an action potential is generated within the neuron

at the base of the axon. If the net effect of all the excitatory and inhibitory stimuli does not increase the membrane potential above the threshold level, no action potential is generated. The action potential or neural impulse is a depolarization of the cell membrane at a specific site. The action potential is propagated along the membrane, away from the cell body, and down the axon to its terminal endings. At the terminal endings, the axon synapses with the dendrites of other neurons if it is a sensory or interneuron, or it synapses with muscle fibers if it is a motor neuron. The neural impulse is then transmitted by chemical means across the synapse and presents a stimulus to the next membrane. Figure 15.2 illustrates how neurons communicate with each other and the direction of neural impulses.

The Motor Unit

The fundamental unit of the neuromuscular system is the motor unit, which is composed of a single motor neuron and all the muscle fibers with which it synapses. The number of muscle fibers innervated by a single motor neuron may be smaller than 20 or larger than 1000. This number represents the number of branches at the end of an axon, and the larger motor neurons usually will have larger threshold potentials. The ratio of muscle fibers to motor neurons in a whole muscle indicates the degree of control one has over a muscle contraction (see figure 15.3). The smaller the number of muscle fibers per motor unit, the more precise the control of the

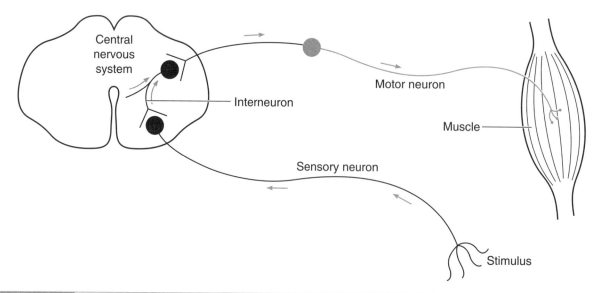

Figure 15.2 Schematic diagram of the pathways of a neural impulse from external stimulus to muscle action.

muscle. The larger the number of muscle fibers per motor neuron, the coarser the control of the muscle.

> **The fundamental unit of the neuromuscular system is the motor unit, which is composed of a single motor neuron and all the muscle fibers with which it synapses.**

The muscle fibers of a single motor unit are not grouped together within the whole muscle but are dispersed throughout a local area of the muscle. Thus adjacent muscle fibers are usually parts of different motor units.

The neural action potential propagated in the axon of a motor neuron will be transmitted to the synapses of each axon motor end plate with each muscle fiber of the motor unit. The threshold potential of the membrane (sarcolemma) of an individual muscle fiber is usually small relative to the chemical stimuli provided by the neural action potential at the synapses, so a neural action potential will almost always generate a muscle action potential. All the muscle fibers of a motor unit thus contract together as a result of an action potential in the motor neuron.

One method of controlling the size of the force produced in a whole-muscle contraction is to control the number of active motor units. If only a small force is desired, only a small number of motor units are recruited, and only a small number of muscle fibers contract to produce the tension. If a larger force is desired, more motor units are recruited, and a larger number of muscle fibers contract to produce more tension.

Another method of controlling the size of the force produced in a whole-muscle contraction is to control the rate of stimulation. A single action potential will generate a single twitch response or contraction in a muscle, as shown in Figure 15.4. If another action potential is received by the muscle fiber before the tension developed in the first twitch response has subsided, the tension developed in the second twitch response will sum with the tension of the first twitch response. If the time between action potentials is short enough, a tetanic contraction occurs, and the maximum tension developed is much larger than that of a single twitch. Both

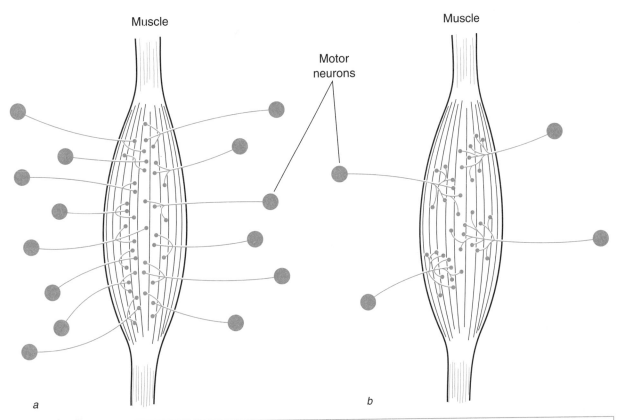

Muscle   Muscle

Motor
neurons

a   *b*

Figure 15.3   Idealized example of a muscle with many motor neurons and fine control (*a*) versus one with fewer motor neurons and coarse control (*b*).

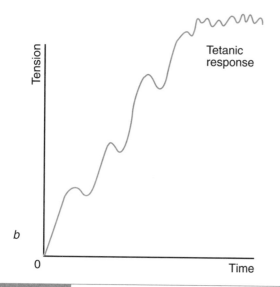

| Figure 15.4 | Tension developed in a muscle fiber as a result of a single stimulus (*a*) is smaller than that developed as a result of repeated stimulation (*b*) if the stimulation frequency is great enough. |

methods (recruitment and summation) of controlling muscle force occur simultaneously.

A certain pattern of motor unit recruitment seems to be followed as the desired tension in a muscle increases. The first motor units recruited are the small motor units, and the stimulation rate is low. These motor units have the least number of muscle fibers and a larger percentage of slow-twitch fibers. As greater tension is desired, larger motor units with more fast-twitch fibers are recruited. At maximal tension, all the motor units have been recruited and the stimulation rate is high.

During a maximal voluntary contraction of a muscle, all of that muscle's motor units may not be recruited. Larger contraction forces may still be elicited from the muscle by electrical stimulation or other means. This result indicates that some of the strength gains associated with a weight-training program (especially the initial strength gains) may

be due to training of the nervous system and its improved ability to recruit more motor units. You may have experienced this yourself if you have ever done any weight training. Some strength gains occur without any weight gain or increase in muscle mass.

Whether or not a motor unit generates an action potential (and thus causes its muscle fibers to contract) depends on the stimuli it receives from neurons with which it synapses. The dendrites of a motor neuron synapse with thousands of other neurons. Some of these synapses may be inhibitory, whereas others may be excitatory. Volitional control of a motor unit occurs only through the synapses between the motor neuron and the neurons whose fibers descend from the higher-order structures in the central nervous system. The other synapses transmit neural impulses that may augment or detract from the desired volitional response. Through feedback from the proprioceptors, the proper intensity of the volitional neural impulse occurs at the motor neuron synapses with the neurons whose fibers descend from the central nervous system. This process of constant feedback and control determines the behavior of a motor unit. The following section describes some of the receptors and the reflex actions they elicit through excitation or inhibition of the motor neurons with which they synapse.

Receptors and Reflexes

Sensory neurons or afferents receive stimuli from specialized receptors. **Exteroceptors** respond to stimuli that come from sources external to the body. These include the receptors for the five senses: sight, hearing, taste, smell, and touch. **Interoceptors** respond to stimuli from sources within the body. These include the receptors associated with the internal organs, or visceroceptors, and the receptors associated with the musculoskeletal system, or proprioceptors.

Some receptors initiate reflexes. Reflexes are involuntary responses that result from sensory input and little if any input from the higher-order structures in the central nervous system. Reflexes may have some protective function during development, and a few continue to have a protective function in adults. A simple spinal reflex may involve only two neurons: a sensory neuron that detects a stimulus, and a motor neuron that synapses with this sensory neuron and is thus activated by it. A sensory stimulus to this spinal reflex arc causes the sensory neuron to excite the motor neuron and elicit a muscular contraction. Most reflexes are much more complex and involve more than two neurons, but these rudimentary spinal reflex arcs may serve to enhance or inhibit certain movements.

Reflexes are involuntary responses that result from sensory input and little if any input from the higher-order structures in the central nervous system.

Proprioceptors and Proprioceptive Reflexes

Proprioceptors may be thought of as the sensory organs that monitor the status of the musculoskeletal system. Receptors are located within joint capsules to give feedback about the joint position (flower-spray or Ruffini endings) or rapid changes in joint position (pacinian corpuscles). Receptors are located within muscles and tendons to give feedback about increases in muscle length (muscle spindles) or muscle tension (Golgi tendon organ). The vestibular system of the inner ear also has specialized receptors that give feedback about the head's position and changes in its position.

The Muscle Spindle and Stretch Reflex

The muscle spindle is a proprioceptor that detects the stretch of a muscle or its relative changes in length. Numerous muscle spindles are present in a muscle. Each muscle spindle includes some short muscle fibers that are arranged in parallel to the other fibers of the muscle. These muscle spindle fibers have several sensory and motor neurons connected to them. When the whole muscle is stretched, the muscle spindle is also stretched. The muscle spindle's sensory neurons are stimulated by this stretch. Slow stretches result in a slow rate of stimulation, whereas faster stretches result in a faster rate of stimulation. Once a muscle has changed its length, the motor neuron to the muscle spindle itself is activated to reset the tension in the muscle fibers of the spindle. The muscle spindle can thus respond to increases in muscle length, no matter what the length of the muscle is.

The muscle spindle is a proprioceptor that detects the stretch of a muscle or its relative changes in length.

The sensory neurons of the muscle spindle synapse with the motor neurons that innervate the whole muscle. If the stimulation of these sensory neurons is great enough, the motor neurons will also be stimulated, and contraction of the stretched muscle is facilitated. This reflex is called the stretch reflex. The response is stronger if the increase in length occurs quickly rather than slowly.

You may have experienced an embarrassing demonstration of the stretch reflex if you have ever fallen asleep during a lecture. As you begin to doze off, your neck extensor muscles relax and your head begins to fall forward. Before your chin reaches your chest, the stretch reflex has been evoked by the quick stretch of the neck extensor muscles. These muscles then contract strongly and pull your head up. This action is usually vigorous enough to startle you out of your dreams and back to the lecture.

Athletes also use the stretch reflex when they incorporate a backswing or a windup or any sort of prestretch prior to a movement. The quicker this prestretch or backward movement, the larger the contraction of muscles creating the torque for the forward movement. The windup of a baseball pitcher during a pitch, the backswing of a tennis player during a forehand stroke, and the backswing of a golfer during a golf drive are all examples of athletes using the stretch reflex to enhance performance.

SELF-EXPERIMENT

Physicians test the stretch reflex when they strike your patellar ligament. The resulting stretch of the quadriceps muscle elicits a stretch reflex, and this muscle contracts, causing extension at the knee. You can try this yourself. Sit on the edge of a table or in a chair that's high enough that your legs hang freely and do not touch the floor. Relax your muscles, especially your quadriceps. Now gently but quickly strike your patellar ligament (on the anterior part of your leg just below your patella but above your tibial tuberosity) with the side of your hand or the spine of a book. Try it several times or have someone else do it if you're unsuccessful. Can you elicit this reflex?

The sensory fibers from the muscle spindle also synapse with an interneuron that synapses with the motor neurons of the antagonist muscle. This synapse is an inhibitory synapse, so the activation of these motor neurons is inhibited. Thus the stretch reflex also inhibits contraction of the antagonist muscle. This effect is called reciprocal inhibition.

The response of the stretch reflex to slow stretching of a muscle is used to control posture and limb position involuntarily. When you sway forward while standing, the posterior posture muscles are slowly stretched. This stretch initiates the stretch reflex, and those posterior muscles contract, stop the forward sway, and pull you back so you remain in balance. The response of the muscles to a slow stretch is not as large as the response to a fast stretch.

The stretch reflex functions to protect the joints crossed by a muscle. A quick lengthening of a muscle occurs when limb positions at a joint change quickly. Such changes may lead to dislocation at the joint unless the limb movements are slowed down. The stretch reflex causes the stretching muscle to contract eccentrically and slow the movement. During the follow-through of many throwing or striking movements, the stretch reflex may be evoked to slow down the throwing or striking limbs.

The stretch reflex causes the stretching muscle to contract eccentrically and slow the movement.

The stretch reflex also influences flexibility exercises. Muscles are stretched most effectively if they are relaxed. A quick stretch of a muscle elicits the stretch reflex and causes it to contract. The slower the stretch, the slower the rate of stimulation of the muscle spindle sensory fibers, and the smaller the stretch reflex response. Slow, static stretching is thus more effective.

The Golgi Tendon Organ and Tendon Reflex

The Golgi tendon organ is another proprioceptor associated with muscle function. A Golgi tendon organ is typically located within the tendon, close to the muscle, and is in series with the muscle. The sensory fibers from this organ are stimulated by tension within the tendon, whether caused by stretching or by contraction of the muscle. The greater the tension, the greater the stimulation. These sensory fibers from the Golgi tendon organ synapse with the motor neurons of the muscle, but rather than transmitting excitatory impulses, the

Golgi tendon sensory fibers transmit inhibitory impulses across the synapses. The contraction of the tensed muscle is thus inhibited by the response of the Golgi tendon organ. If the tension is great enough, the Golgi tendon organ will completely inhibit contraction of the tensed muscle and cause it to relax. This reflex is sometimes referred to as the tendon reflex.

The contraction of the tensed muscle is inhibited by the response of the Golgi tendon organ.

The tendon reflex protects the muscle from rupturing or tearing by turning off the active development of tension if the tensile stress within the muscle is too great. This reflex works in opposition to the stretch reflex. The stretch reflex response is usually greater unless the tension within the muscle is very great. An example of the tendon reflex is occasionally seen when the leg of a high jumper or triple jumper buckles or collapses due to the extreme forces in the knee extensor muscles at takeoff.

Some of the strength gains observed in resistance training may occur due to the increased ability of the central nervous system to provide enough excitatory stimulation to the motor neurons to override the inhibitory stimulation by the Golgi tendon organs.

The Vestibular System and Its Related Reflexes

The vestibular system is composed of the sense organs of balance. Each inner ear contains three proprioceptors: the semicircular canals, the utricle, and the saccule. These proprioceptors are bony tunnels filled with fluid called endolymph. The walls of these tunnels are lined with sensory hair cells that are surrounded by a gelatinous substance. There are three semicircular canals arranged in three mutually perpendicular planes roughly corresponding to the sagittal, frontal, and transverse planes. The endolymph fluid in these canals will move relative to the head when the head is accelerated. This movement bends the sensory hair cells, thus giving feedback about changes in the motion of the head or its acceleration. The utricle and saccule are bulbous lumps in the bony tunnels that contain otoliths, tiny calcium carbonate stones that are embedded in the gelatinous substance along with the hair cells. The otoliths bend the hair cells according to the direction in which they are pulled by gravity, thus giving feedback about the position of the head relative to the gravitational force.

> **The vestibular system, or the labyrinths of the inner ear, is composed of the sense organs of balance.**

The vestibular system and the proprioceptors associated with the joints of the neck give rise to several rudimentary reflexes associated with the head and neck positions. In adults, these reflexes are normally difficult to observe due to overriding volitional control. Righting reflexes occur when the vestibular system detects a nonupright position of the head. Reflex actions of the limbs and the trunk and neck musculature are attempts to correct the head position and maintain upright posture. Figure 15.5 shows examples of righting reflex responses.

Reflexes resulting from the stimulus of the proprioceptors of the neck are called tonic neck reflexes. These proprioceptors communicate with the motor neurons of the upper extremity muscles. Neck flexion initiates a tonic neck reflex response that facilitates contraction of the muscles that cause pulling actions of the arms, whereas neck extension facilitates contraction of the muscles that cause pushing actions of the arms. Rotating the head to the right initiates a tonic neck reflex response that facilitates contraction of the elbow extensors and shoulder abductors on the right side and the elbow flexors and shoulder adductors on the left side. Figure 15.6 illustrates the positions facilitated by the tonic neck reflexes.

Exteroceptors and Exteroceptive Reflexes

The exteroceptors that significantly influence the movement of the body include the receptors for sight, hearing, touch, and pain. The receptors for touch are pacinian corpuscles, which are also the receptors for change in joint position. These receptors are sensitive to pressure, but only when changes in pressure occur. Large changes in pressure result in a large response from the pacinian corpuscles, but only during the change.

The pacinian corpuscles beneath the skin on the soles of your feet or the palms of your hands are responsible for the extensor thrust reflex. The sensory fibers from the pacinian corpuscles synapse with the motor neurons of the extensor muscles of the limb, thus facilitating contraction of these muscles and extension of the limb. The large change in pressure occurring when you land on your feet after a jump initiates the extensor thrust reflex. The extensor muscles of the leg contract eccentrically and stop you from falling. This contraction is facilitated by the extensor thrust reflex.

Cutaneous pain receptors are also involved in reflex actions of the musculoskeletal system that function to protect the body. The flexor reflex, or withdrawal reflex, occurs when pain is sensed at some distal location on a limb, as illustrated in figure 15.7. The response of this reflex is withdrawal or flexion of the affected limb. Placing your hand

Figure 15.5 An example of the righting reflex.

Figure 15.6   Use of tonic neck reflexes to facilitate specific actions.

Figure 15.7   An example of the withdrawal reflex.

on a sharp tack or hot surface will initiate the flexor reflex. The hand will be pulled away from the source of pain.

The crossed extensor reflex is an example of the combined effects of the flexor reflex and the exten-sor thrust reflex. This reflex is initiated by pain in the supporting limb. The response is flexion of that limb and contraction of the extensor muscles of the opposite limb. When you walk barefoot and step on a sharp stone, you quickly shift your weight to the opposite leg by withdrawing the foot and con-tracting the extensor muscles of the opposite leg so that it can support your weight.

Summary

The nervous system is structurally organized into the central nervous system and the peripheral ner-vous system. Nerves are bundles of nerve fibers (axons or dendrites) from individual nerve cells. A neuron is a nerve cell. There are three types of neu-rons: sensory or afferent neurons, motor or efferent neurons, and interneurons. Sensory neurons detect the status or change in status of the external envi-ronment via exteroceptors and of the internal environment via interoceptors and deliver this in-formation encoded as neural impulses to the cen-tral nervous system. Motor neurons receive stimu-lus from the central nervous system and from sensory neurons and send neural impulses to muscle fibers. Interneurons transmit neural im-pulses between neurons.

The motor unit is a motor neuron and all of the muscle fibers it innervates. Motor units may have as few as 20 or as many as 1000 muscle fibers. Muscles with very few fibers per motor unit have precise control. The contraction force of a muscle may be controlled by the recruitment of motor units and the rate of stimulation of the individual motor units.

Receptors are the special sensory organs that detect the status or change in status in the body's environment. Reflexes are involuntary actions that result from sensory input. Proprioceptors detect the status or change in status of the musculoskeletal system. Muscle spindles detect increases in muscle length. The stretch reflex is initiated by a lengthening of the muscle spindle and results in facilitation of the contraction of the muscle being stretched. Golgi tendon organs detect tension in tendons. The tendon reflex is initiated by tension in the Golgi tendon organ and results in inhibition of the contraction of the muscle under tension. The vestibular system and neck joint proprioceptors detect position and changes in position of the head and of the head relative to the neck. Righting reflexes facilitate the maintenance of an upright body and head posture. Tonic neck reflexes facilitate various upper extremity positions when the neck is flexed, extended, or rotated to either side. Pacinian corpuscles and Ruffini endings detect changes in pressure and absolute pressure. The extensor thrust reflex, flexor withdrawal reflex, and crossed extensor reflex are initiated by these receptors.

Volitional control of movement requires proprioceptive feedback to determine intensity of the neural impulses to the motor neurons. Most reflexes can be suppressed through volitional control if the reflex detracts from the desired action. Sometimes reflex responses will enhance the desired action, and in these situations, it may be beneficial to elicit a more intense reflex response.

Key Terms

autonomic nervous system *(p. 281)*

central nervous system *(p. 280)*

endolymph *(p. 286)*

exteroceptor *(p. 284)*

Golgi tendon organ *(p. 286)*

interneuron *(p. 282)*

interoceptor *(p. 284)*

motor neuron *(p. 282)*

motor unit *(p. 282)*

muscle spindle *(p. 285)*

pacinian corpuscle *(p. 287)*

peripheral nervous system *(p. 280)*

proprioceptor *(p. 284)*

sensory neuron *(p. 281)*

somatic nervous system *(p. 281)*

tonic neck reflex *(p. 287)*

REVIEW QUESTIONS

1. **Is a motor neuron part of the central nervous system or the peripheral nervous system?**

2. **Which direction does an action potential travel along a motor neuron axon: toward or away from the cell body?**

3. **Describe the two methods employed by the central nervous system to increase the active tension in a muscle.**

4. **What proprioceptor detects change in muscle length?**

5. **What proprioceptor detects muscle tension?**

6. **What proprioceptors are positioned parallel to muscle fibers?**

7. What proprioceptors are positioned in series with muscle fibers?

8. Give three examples of activities in which a preparatory movement evokes a stretch reflex to enhance the desired motion.

9. What position of your head and neck would facilitate a biceps curl exercise?

10. What stimulus does a pacinian corpuscle detect?

11. A stretching routine that utilizes the actions facilitated by certain proprioceptors is called PNF (for proprioceptive neuromuscular facilitation) stretching. Describe how muscle spindles and Golgi tendon organs may enhance or detract from the effectiveness of a stretching exercise.

Applying
Biomechanical Principles

The previous chapters of this book have presented material about how the body as a whole responds to external forces acting upon it and about how the various components of the musculoskeletal system create internal forces in response to those external forces. Several examples of applications of mechanical principles were also presented in these chapters. The purpose of part IV is to introduce you to several methods that may be used to conduct qualitative biomechanical analyses of activities. Recall from chapter 1 that the goal of biomechanics is performance improvement and a secondary goal is injury prevention and rehabilitation. Further recall that biomechanics can be used to improve performance through improved technique, training, or equipment. Part IV describes qualitative biomechanical analysis methods directed toward these goals. Chapter 16 presents an overview of qualitative biomechanical analysis for improving technique. Chapter 17 presents an overview of qualitative biomechanical analysis for improving training. Chapter 18 concludes the book and presents an overview of qualitative biomechanical analysis for diagnosing and preventing injury.

16

Qualitative Biomechanical Analysis to Improve Technique

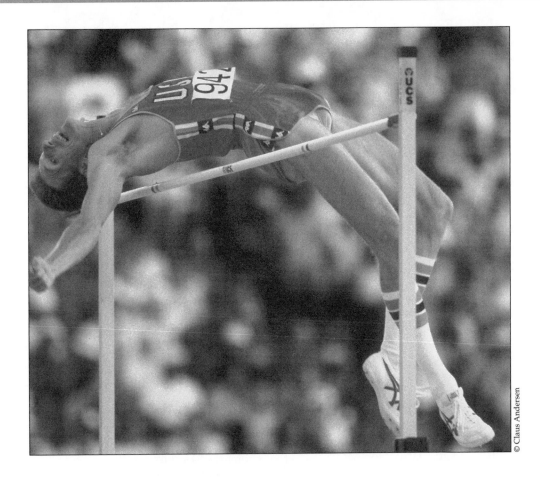

© Claus Andersen

You are watching a high jumper attempt to set a new personal record. Some aspects of the jumper's technique don't seem so odd now that you have some knowledge of biomechanics. His arm action at takeoff increases his vertical velocity at the instant of takeoff. His back layout puts his center of gravity much closer to the bar, so the bar can be higher. But can you explain the basis for the other actions in the jumper's technique? This chapter may help you by outlining a structured approach to a qualitative biomechanical analysis of technique.

OBJECTIVES

When you finish this chapter, you should be able to do the following:

▌ Understand the difference between a qualitative and quantitative bio-mechanical analysis

▌ List the steps involved in a qualitative biomechanical analysis for technique improvement

▌ Understand the process of developing a theoretical, mechanically-based cause-and-effect model of a skill

▌ Describe the important guidelines for observing a performance

▌ Understand how to identify and evaluate errors in technique

▌ Understand how to give instruction to students or athletes that will help them correct errors in technique

Analysis involves breaking something into smaller parts and then examining those parts. A qualitative analysis involves breaking something into smaller parts and then examining those parts without measuring or quantifying their characteristics. A qualitative biomechanical analysis of a movement or sport skill is thus breaking down the movement into its basic elements and then qualitatively examining those elements from a biomechanical perspective. Conducting a qualitative biomechanical analysis requires you to use all the material you have learned from this book, but not in a haphazard way. Applying the mechanical and anatomical knowledge presented in the previous chapters requires an organized approach if the resulting qualitative biomechanical analysis is to be successful. The approach may differ depending on the goal of the analysis. Is the goal to improve technique? To improve training? To prevent injury? To improve equipment? This chapter presents the framework for conducting a qualitative biomechanical analysis with the goal of technique improvement. Subsequent chapters will focus on training improvements and injury prevention.

Types of Biomechanical Analysis

Before getting into the details of a qualitative biomechanical analysis to improve technique, we should first learn more about the difference between qualitative and quantitative biomechanical analyses. The adjectives *qualitative* and *quantitative* describe how the characteristics of the performance are observed and analyzed by the coach, teacher, or clinician. If the performance or any of its aspects is quantified or measured (described with numbers), the resulting analysis based on these measurements will be a quantitative biomechanical analysis. If the performance or any of its aspects is evaluated using only the senses of the observer, the resulting analysis will be a qualitative biomechanical analysis. This and subsequent chapters focus on qualitative biomechanical analysis methods.

Qualitative Biomechanical Analysis

Teachers and coaches often perform qualitative biomechanical analyses, but they rarely perform any quantitative biomechanical analyses. They observe their athletes and students performing and describe the mechanical characteristics of the performance subjectively. Comparative descriptors (faster, slower, higher, lower, shorter, longer, larger, smaller, etc.) may be used to describe these characteristics. The sense of sight, or visual observation, is the basis for most qualitative analyses.

How a coach or teacher observes a performance affects the subsequent qualitative biomechanical analysis. The details of how and from what vantage point to observe performances are discussed later in this chapter.

Quantitative Biomechanical Analysis

Comprehensive quantitative biomechanical analyses are usually limited to performances by elite athletes; however, teachers and coaches may make some performance measurements and thus do limited quantitative biomechanical analyses. A stopwatch and a tape measure may be used to measure and thus quantify many biomechanical parameters. Counting steps and timing how long it takes to take that many steps gives the coach a measure of step rate. Measuring a specific distance and timing how long it takes to move that distance gives a measure of speed. If assistants record where the subsequent footfalls land, step length can be measured. These types of measurement allow the coach or teacher to do limited quantitative biomechanical analyses, but taking such measurements prevents the coach or teacher from observing the whole performance.

A comprehensive quantitative biomechanical analysis requires specialized and expensive equipment for recording and measuring the biomechanical variables of interest. Biomechanists or trained technicians, rather than teachers or coaches, usually conduct comprehensive biomechanical analyses. Most quantitative biomechanical analyses rely on some sort of visual recording of movements, usually on film or videotape. The movements are recorded in such a manner that position and temporal information can be determined from the video or film record. The positional and time data then can be used to determine other kinematic measures such as displacements, velocities, and accelerations—both linear and angular. Radar guns and other devices may also be used to measure velocity. Acceleration may be measured directly using accelerometers. Force platforms (basically sophisticated bathroom scales) or other devices may be used to measure external reaction forces. Electrical activity of muscles may be detected and recorded using EMG. Most of this equipment is interfaced with a computer.

As the cost of completing a comprehensive quantitative biomechanical analysis of a performance is high, these types of analyses are usually done only for elite performers. The other reason comprehensive quantitative biomechanical analyses are usually limited to elite performers relates to the observer's ability to detect errors. As the level of performance increases, the magnitude of the errors in performance decreases. Errors made by novices

are large and easy to detect visually using qualitative biomechanical analysis techniques. With improved performance, the errors decrease in size and become more difficult to detect, and at the elite level, a comprehensive quantitative biomechanical analysis may be necessary to detect them.

Qualitative Biomechanical Analysis to Improve Technique

A variety of procedures for conducting qualitative biomechanical analyses exist (Arend and Higgins, 1976; Brown, 1982; Hay, 1984; Hay and Reid, 1982; Knudson and Morrison, 1997; McPherson, 1988; Norman, 1977). The method presented here is the one recommended by the author. It is not novel, but contains procedures common to existing methods and provides a systematic way of biomechanically analyzing human movements. A qualitative biomechanical analysis to improve technique involves four steps:

1. Description. Develop a theoretical model of the most effective technique and describe what it would look like. Determine what you want to see when you observe your students or athletes.
2. Observation. Observe the performance of your student or athlete to determine what their technique actually looks like.
3. Evaluation. Compare the ideal technique to the observed performance. Identify and evaluate the errors.
4. Instruction. Educate the student or athlete by providing feedback and the instruction necessary to correct those errors.

The following sections describe and illustrate the four steps of a qualitative biomechanical analysis.

Describe the Ideal Technique

Why undertake a biomechanical analysis? As a coach, teacher, therapist, or other human movement professional, when evaluating a student's or athlete's performance, you should be able to distinguish between what is important and what is unimportant, what is correct and what is incorrect, what is possible and what is impossible, what is effective and what is ineffective, what is safe and what is unsafe, and so on. The process of making these distinctions is part of a biomechanical analysis.

As a student or athlete, you probably experienced this when your teacher or coach watched you perform and then provided you with feedback to correct deficiencies in your performance. Biomechanical or movement analysis is often thought of as just that—the process of observing (or measuring) the performance of a skill, identifying faults in the performance, and providing feedback to the performer to help correct those faults. Indeed, these steps are part of a biomechanical analysis. But what you didn't see the teacher or coach do is the most important part of a biomechanical analysis. The reason you didn't see this part of the analysis is that it doesn't occur in the gym or on the playing field. It occurs before any observations and corrections are made.

When coaches or teachers come before their athletes or students, they already have an image of the ideal technique for performing a skill. This ideal becomes the standard against which observed performances are compared. Proposing and conceptually developing the ideally effective technique for a skill is the first and most important step of a biomechanical analysis. It also requires the most use of biomechanical knowledge.

Fundamental Knowledge of the Skill

Prior to any evaluation of a sport skill or other human movement activity, you should have some familiarity with the skill. This is absolutely necessary if you are going to propose the ideally effective technique for that skill. At the very least, if you are analyzing a sport skill, you should know the rules for the activity.

Rules impose constraints. If you don't know what those constraints are, you may suggest improvements in technique or equipment that are considered unfair or cheating and illegal according to the rules. Some rules are obvious. For example, in baseball, your analysis of pitching techniques indicates that the pitcher could increase the velocity of the pitch by taking several running steps toward home plate before releasing the ball. But this technique would violate the rule that the pitcher's back foot must be on the pitching rubber prior to releasing the ball.

Let's look at an example where equipment is changed. Suppose your analysis of goaltending techniques in ice hockey indicates that the goalie's equipment needs improving, and you design a stick with a much longer and wider blade (figure 16.1). Such a stick would obviously make the goalie more effective, but it would put the other team's offensive players at a great disadvantage. The dimensions of a goalie's stick in ice hockey are thus specified by the rules.

Other rules of competition are not always as well known or obvious. The sport of triathlon is a race that includes swimming, bicycling, and running. An analysis of the cycling portion of this sport may indicate that a competitor could ride more effectively by drafting behind the bicycle of another athlete, thus reducing the aerodynamic drag forces. In fact, this is a common tactic in bicycle racing, but it is not allowed in a triathlon according to the rules of that sport.

Knowledge of the techniques and equipment traditionally used in a sport may also be valuable for the biomechanical analysis, but such knowledge is not absolutely essential. The sport skill techniques developed and taught by coaches and teachers are generally effective. Usually, their effectiveness has been proven through the trial-and-error method of coaching. You might use these techniques as the basis for proposing the most effective technique, but to do so, you must know what the traditional techniques are.

Where can you find information about sport techniques? This information can be found in coaching journals or magazines, textbooks, and videotapes. You can also learn by talking with coaches and athletes, watching successful athletes perform, or trying to perform the skill yourself. While gathering this information, be curious but skeptical. Learn to "question authority," as the bumper sticker says. Why is the skill performed that way? Should everyone swing their arms in that manner? What purpose does that type of follow-through serve? Why are the hips turned that way during the preparatory phase? Why, why, why? Realize also that completely different ways may be described (or actually demonstrated by elite athletes) for performing certain aspects of the traditional technique. That is why you must propose and develop the most effective technique yourself, based on your knowledge and understanding of the underlying biomechanics of the movement.

Purpose or Goal of the Skill

Now that you have some fundamental knowledge about the skill, the next step is to identify the purpose or goal of the skill and interpret that purpose or goal in mechanical terms, if possible. The purpose of a skill is its desired outcome, the measure of performance or success, or the performance criterion. What is the athlete striving to achieve dur-

Figure 16.1   A large goalie's stick may be much more effective in ice hockey, but it gives the goalie an unfair advantage and thus the rules don't allow it.

ing performance of the skill? The purpose of some sport or movement skills is easy to define; for others, it's much more difficult. Some sport skills may have only one purpose; others may have more than one. The complexity of this task depends on the skill being analyzed.

Generally, it is easier to identify the purpose of those skills whose outcomes are determined objectively. The outcomes of all track and field skills are determined objectively by answering the questions how fast, how far, or how high? Similarly, the purpose of any racing activity is to complete the course (whatever it may be) in the least time. The purposes of other skills may not be as easy to identify or define mechanically. What is the purpose of a tennis serve, a football block, a triple axel in ice skating (see figure 16.2), a reverse one-and-a-half in a layout position in diving, and so on?

The purpose of the tennis serve may be to score an ace by putting the ball past the receiver or to gain an advantage by putting the receiver in a poor position to return the ball effectively. In either case, the speed and accuracy of the serve are two parameters that may be used to define the purpose mechanically. The purpose of the football block may

be to prevent the defensive end from reaching the quarterback and tackling him. Mechanically, this purpose can be defined as decreasing the stability of the opponent and decreasing his velocity in the direction of the quarterback. The purpose of the triple axel in ice skating may be to impress the judges. Mechanically, the purpose is to jump as high as possible, complete three spins in the air, and land on one skate without falling. The purpose of the dive is to score as many points as possible with the judges. Mechanically, the purpose is to jump as high as possible with just enough angular momentum to complete one and a half revolutions in a layout position, then enter the water in a position perpendicular to the surface with minimal splashing.

Although identifying the purpose of these activities wasn't too difficult, defining that purpose in terms of mechanical parameters may be. Most skills have goals that can be defined mechanically, however. Think of the sports or movement activities you are interested in and try to define the goals of the specific skills used in those sports. Then try to translate those goals into mechanical terms. The more precisely you can identify the purpose of a skill, the better your analysis of that skill will be.

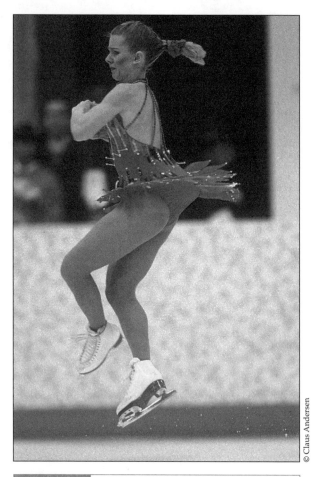

© Claus Andersen

| **Figure 16.2** | What is the goal of a triple axel in figure skating? |

Characteristics of the Most Effective Technique

The next step in the biomechanical analysis is to identify the characteristics of the most effective technique. This step definitely requires your biomechanical knowledge and some common sense. Two different methods may be used to accomplish this task. The first involves an assessment of traditional techniques to determine the characteristics of the most effective technique. The second involves development of a deterministic (cause-and-effect) model that determines the characteristics of the most effective technique.

The first method of determining the characteristics of the most effective technique is based on the assumption that the traditional technique probably exhibits many of these characteristics. The task here is to determine which aspects of the traditional technique are important and which aren't. What matters?

This question can be answered in two ways. First, observe as many performances of the skill by elite performers as possible. Try to identify the actions and positions that are common to all the athletes. Then, from the start of the skill to its end, assess each action or position by asking how it contributes mechanically to achieving the purpose of the skill. Do the actions or positions contribute to the achievement of the goal, or are they detrimental? Those actions or positions that contribute to the achievement of the goal should be included as characteristics of the most effective technique. Those actions or positions that are detrimental to the achievement of the goal should definitely not be included. And those actions or positions that neither contribute to nor detract from the goal are perhaps manifestations of style and also should not be included.

The second way of answering the question "What matters?" is to assess the actions or positions common to the techniques illustrated in most of the coaching and teaching materials you have reviewed. Follow the same procedure as before by asking how each action or position mechanically contributes to the purpose of the skill.

In both of these assessments, it may be easier to break down the purpose of the skill into parts. Then, during the assessment of specific actions or positions, ask how the action or position contributes to achieving a specific part or subgoal of the purpose. For example, the purpose of high jumping is to clear a crossbar set as high as possible. The performance measure is defined mechanically as crossbar height cleared. This can be subdivided into parts. The height of the crossbar cleared (h_b) is determined by the height of the jumper's center of gravity at takeoff (h_{to}), the vertical displacement of the center of gravity from takeoff to maximum height (h_f), and the difference between the maximum height of the jumper's center of gravity and the crossbar height (Δh). Crossbar height cleared can thus be expressed as an equation:

$$h_b = h_{to} + h_f - \Delta h \qquad (16.1)$$

Various aspects of the traditional high jumping technique may then be assessed according to whether or not they contribute to any of these subgoals.

High jumpers swing their arms forward and upward during the takeoff phase of a high jump, but some jumpers "block" their arms at shoulder height before leaving the ground and others continue to move the arms upward above shoulder height until they have left the ground. Are these differences in arm actions just manifestations of style, or is one technique better than the other? The positions of a jumper's arms affect the takeoff height, h_{to} in equation 16.1. The jumper's center of gravity should be as high as possible at the instant of takeoff. The technique of continuing to move the arms above shoulder height before takeoff moves the center of gravity slightly higher, thus contributing to achievement of a high center-of-gravity height at takeoff. The movements of the arms during the takeoff phase contribute to flight height, h_f in equation 16.1. The jumper becomes a projectile once takeoff has occurred. The vertical displacement of a projectile (flight height, h_f, in this case) is determined by its vertical velocity at takeoff. The vertical velocity of the jumper's center of gravity at the instant of takeoff is determined by the vertical velocities of the jumper's body parts. The arms have a vertical velocity at takeoff in the technique of continuing to move the arms above shoulder height before takeoff. They contribute to the vertical velocity of the jumper's center of gravity, thus contributing to achieving a high flight height. In the other technique, the arms stop contributing to the vertical velocity of the center of gravity because they stop their upward swing before takeoff. This aspect of technique is detrimental to high jumping performance.

The second method of determining the characteristics of the most effective technique is to develop the mechanically-based foundation for achievement of the skill's purpose. In doing so, the characteristics of effective techniques will be revealed. This method uses deterministic, cause-and-effect relationships based on mechanical principles. The mechanical purpose of the goal or skill is broken down into subgoals, as in the high jump example. These subgoals are then further reduced by identifying the mechanical bases for each of them. The mechanical bases of the subgoals define the characteristics of the most effective technique.

Let's look at the high jump again. We've defined the mechanical purpose of the skill as maximizing the height of the crossbar cleared, and we've defined this height as h_b. This mechanical goal was then broken down into three subgoals: maximize the height of the jumper's center of gravity at takeoff (h_{to}), maximize the vertical displacement of the center of gravity from takeoff to maximum height (h_f), and minimize the difference between the maximum height of the jumper's center of gravity and the crossbar height (Δh). To define the characteristics of the most effective technique, we must investigate each of these subgoals.

The jumper's center-of-gravity height at the instant of takeoff (h_{to}) is affected by the position of the jumper at the instant of takeoff. The rules of high jumping indicate that the jump must be made from one foot, so the problem is to determine what body position maximizes the center-of-gravity height while standing on one foot. Standing up is obviously better than lying down. An upright standing position with full extension of the takeoff leg is better than one in which the jumper is leaning in any direction. Raising both arms as high as possible raises the center of gravity. Raising the free leg as high as possible also raises the center of gravity. These are thus characteristics of an effective technique.

The center of gravity is higher at takeoff if the jumper is tall, which is why most elite high jumpers are tall. The first man to high jump 8 ft was Javier Sotomayor, who is 6 ft 4.75 in. (1.96 m) tall. The 1996 Olympic Games gold medalist and world recordholder in the women's high jump is Stefka Kostadinova, who is 5 ft 11 in. (1.80 m) tall. Both high jumpers are well above average height for their gender.

The second subgoal in the high jump is to maximize the vertical displacement of the center of gravity from takeoff to maximum height (h_f). The jumper becomes a projectile as soon as the takeoff foot leaves the ground, so the vertical displacement of the jumper's center of gravity after takeoff is determined by the vertical velocity of the jumper's center of gravity at the instant of takeoff. To make h_f as large as possible, the jumper needs to make the vertical velocity of the center of gravity at the instant of takeoff (v_{yto}) as high as possible. What causes a high velocity? Recall from chapter 6 and equation 6.14 that impulse (average net force times the duration of force application) causes a change in momentum (mass times velocity). This equation can be rewritten as equation 16.2, which describes the conditions in the high jump:

$$\Sigma \overline{F} \Delta t = m(v_f - v_i) \qquad (6.14)$$

$$(\overline{R} - W)\Delta t = m(v_{yto} - v_{ytd}) \qquad (16.2)$$

where,

$\Sigma \overline{F} = \overline{R} - W$ = the average net vertical force acting on the jumper

$\overline{R}$ = the average vertical reaction force exerted by the ground on the takeoff foot

W = the weight of the jumper

Δt = the duration of force application or the duration of takeoff foot contact with the ground m = the mass of the jumper

$v_f = v_{yto}$ = the final velocity or vertical velocity of the center of gravity at the instant of takeoff

$v_i = v_{ytd}$ = the initial velocity or vertical velocity of the center of gravity at the instant of touchdown at the start of the takeoff phase

If equation 16.2 is rewritten and solved for v_{yto}, the parameter we wish to maximize, it appears as:

$$v_{yto} = \frac{(\overline{R} - W)\Delta t}{m} + v_{ytd} \qquad (16.3)$$

To maximize vertical takeoff velocity, the jumper should maximize the average vertical reaction force acting on the takeoff foot ($\overline{R}$), the duration of takeoff foot contact (Δt), and the vertical velocity at touchdown (v_{ytd}). The jumper's mass (m) or weight (W) should also be minimized (i.e., the jumper should be fairly lean). Javier Sotomayor weighs 180 lb (his mass is 82 kg). Stefka Kostadinova weighs 132 lb (her mass is 60 kg).

To maximize the average vertical reaction force acting on the jumper's takeoff foot, the jumper's run-up velocity should be fast. An increase in run-up velocity will increase the size of the reaction force acting beneath the takeoff foot at the end of the run-up. The jumper should also strive to accelerate the arms and free leg upward during the takeoff phase. These actions will cause the takeoff leg to push down on the ground with greater force.

To maximize the duration of takeoff foot contact (and force production) during the takeoff phase, the jumper's center of gravity should be low at the start of the phase when the takeoff foot touches down and high at the end of the phase just before the takeoff foot leaves the ground. The takeoff phase is prolonged by swinging body parts through a large range of motion vertically. Thus, when the takeoff foot first touches down, the arms should be extended and in a low position, the free leg should also be in a low position, and the jumper should be leaning backward and away from the bar. These actions will put the jumper's center of gravity in a low position.

Finally, to maximize the vertical velocity of the jumper's center of gravity at touchdown, the last step should be taken quickly. Actually, the jumper's center of gravity is unlikely to have an upward vertical velocity when the takeoff foot hits the ground. It will probably have a downward vertical velocity (or negative velocity, as we have been describing upward as positive). Thus, the task is to minimize

the downward vertical velocity of the center of gravity at the instant the takeoff foot touches down.

The third subgoal in an effective high jump technique is to minimize the difference between the maximum height of the jumper's center of gravity and the crossbar height (Δh), or the bar clearance. To achieve this goal, the jumper's center of gravity needs to be as close to the crossbar as possible, or even below it, when the center of gravity reaches maximum height. But the jumper doesn't want to touch the crossbar or dislodge it. If the jumper goes over the bar feet first in an upright position, the difference (Δh) is large. If the jumper goes over the bar in a lying-down position, Δh is much smaller. If the jumper arches or drapes the body over the bar, Δh is even smaller.

Longer-limbed individuals may be at an advantage here because they can drape their limbs farther below the crossbar while their trunk is above it, making Δh smaller or even negative in value.

From the previous mechanical assessment of the purpose of high jumping, the following characteristics of an effective technique are proposed:

1. Approach the bar with a fast run-up speed.
2. Make the last step of the run-up quick by putting the takeoff foot down early.
3. Lower the center of gravity during the last step, and especially at the start of the takeoff phase, by positioning the arms and free leg low and leaning back and away from the bar.
4. Use the takeoff leg to push down against the ground with as much force as possible by accelerating the arms and free leg upward for as long as possible.
5. Raise the center of gravity as high as possible at the instant of takeoff by achieving an upright trunk position, fully extending the takeoff leg, and reaching high upward with both arms and the free leg.
6. Clear the crossbar with the body in a draped position to minimize the distance between the crossbar and the jumper's center of gravity.

Your model of the most effective technique for high jumping would include these characteristics. You are now ready for the next step of the biomechanical analysis.

Observe the Performance

The next step of a qualitative biomechanical analysis to improve technique is to observe the actual performance. Visual observation is the primary basis for most qualitative biomechanical analyses, but simply watching a performance is not enough. Your observation must be planned in advance considering the following: Who will you observe? What conditions will the subject perform under? Where will the subject perform? Where will you observe the performance? What will you look for?

Who?

Who you observe, especially when considering their skill level, may affect the manner in which you observe and conduct your analysis. As skill level increases, the size of observable errors in technique decreases, as does the variability in performance. Novice performers of a skill will exhibit errors that are easily recognized but that may not be repeated from one performance to the next. Their performances may be so variable that a detailed analysis would be painstaking and futile. Instead, your observations of these students or athletes should be directed at identifying their gross errors in each performance.

More care must be taken in observing and analyzing experienced, highly skilled athletes. Their errors will be minor, so as the observer, you must pay more attention to the details of their technique. In addition, their performances will probably be less variable, so your observation of these athletes should be directed at identifying errors that are repeated from performance to performance.

What Conditions?

Ideally, the environment in which you watch your students perform should be carefully controlled, if possible. Try to duplicate as closely as possible the environment in which the performances normally occur. An actual competition may be the best environment for an athlete because it is one in which the athlete normally performs; however, it may not be best for the observer. The best locations for viewing the performances may not be available to you in a competition, or your view of the performances may be obstructed by officials or other athletes.

A class or team practice session is not ideal either. The activities of other students or athletes may distract you (especially if you are responsible for their supervision) and the performer, thus affecting both your ability to observe and the athlete's ability to perform. It may be difficult to simulate the normal performance environment in a class or practice session as well.

Creating the ideal conditions for observing a performance is difficult, if not impossible, so you'll have to compromise. In any case, do your best to simulate the normal performance environment and minimize distractions. For novices or beginners, a class or practice session may be the normal performance environment.

Where to Observe?

Your vantage point when observing the performance is important because it affects what parts of the performance will be visible to you. First, determine if the activity has a principal plane of movement. Do most of the movements occur in the same plane? If so, your line of sight should be perpendicular to this plane. Running, long jumping, gymnastics vaulting, gymnastics high bar routines, basketball free-throw shooting, and backstroke swimming are all examples of activities whose movements occur primarily in the sagittal plane. To observe these movements, position yourself to the side so that you view the sagittal plane. The sagittal view may be observed from either the left or right side, so choose the position that provides the least obstructed view of the subject or body parts of interest. You must also decide how close you should be to view the action. A more distant view is best for evaluating the quality of the performance as a whole. A closer position is desirable if you are evaluating only certain aspects of the performance. In any case, you should be far enough away that you can see the movements of interest in your field of view without having to move your eyes too quickly to track them.

Even if the activity occurs primarily in one plane, several viewing positions may be necessary if the performance involves multiple skills. But because you are only one person, multiple viewing positions require multiple trials or performances by the student or athlete (unless you have several assistants equipped with video cameras). For example, the three separate parts of a long jump performance—the approach run, takeoff, and landing—may require three separate vantage points, as shown in figure 16.3.

Most activities have multiple planes of motion and usually require multiple viewing positions. Your initial position for observing these activities will be determined by the part of the skill that is most important, which should have been identified by your model. Determine the primary plane of movement for this part of the skill. Your viewing direction should be perpendicular to this plane.

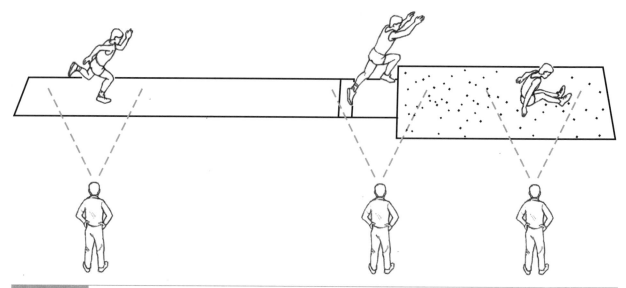

Figure 16.3   Three different viewing positions for observing the long jump approach run, takeoff, and landing.

Subsequent observations may then be from various vantage points that give better views of the other important movements in the skill. Two viewing positions for observing a high jumper are shown in figure 16.4. In this example, the primary viewing position gives a good view of the approach and take-off, whereas the secondary viewing position gives a good view of the bar clearance.

One last note about viewing position. If possible, videotape the performance from each of the viewing positions. Preferably, have an assistant (another teacher or coach, or another student or athlete) videotape the performance. With video recordings of performances, you can view the same trials repeatedly without fatigue affecting your subjects (or your evaluations of them). In addition, the video record will be useful when giving the student or athlete feedback about their performance.

What to Look For?

Now that you've set up the ideal conditions for the performance, you're in the best position to view the performance, and you've got an assistant to video-tape the performance, how do you observe the performance? What do you look for? Generally, you will have to view several performances or several replays of the videotaped performance. During the first trial (or video replay), get a general view of how the student or athlete performs. (If the subject you are observing is a novice or beginner, this may be all it takes to identify technique errors.) During subsequent trials (or video replays), the movements or actions you have previously identified as char-

acteristics of the most effective technique should guide your observation. The following are some of the distinguishing features of these actions:

1. The position of the body or body segments at specific instants (usually at the start and end of some force-producing phase). For high jumping, is the jumper leaning back and in a low position at the start of the takeoff phase (when the takeoff foot first touches the ground), and is the jumper upright with the arms and free leg as high as possible at the end of the takeoff phase (when the takeoff foot leaves the ground)?
2. The duration and range of motion of the body and its segments during specific phases of the skill, especially the force-producing phases. For high jumping, are the duration of the takeoff phase and vertical range of motion of the arms and free leg during this phase as long and as large as possible?
3. The velocity and acceleration (along with their directions) of body segments during specific phases of the skill. For the high jump, are the vertical velocities of the arms and free leg as fast as possible during the takeoff phase, especially at the end of the phase?
4. The timing of body segment motions relative to each other. For the high jump, do the arms and free leg swing upward together during the takeoff phase?

These are the details you should examine and compare to your model of the most effective technique.

Figure 16.4   Two different viewing positions for observing the approach, takeoff, and bar clearance in the high jump.

The position of the body and segments at specific instants indicates the direction of force application as well as the range of motion through the phase. Examples of instants when positions should be examined include the instant of release in throwing activities (football passing, baseball pitching, shot putting, javelin throwing, basketball free-throw shooting, etc.); the instant of takeoff in jumping activities (running, hurdling, high jumping, long jumping, volleyball spiking, diving, figure skating, basketball shot blocking, etc.); and the instant of contact in striking activities (kicking, serving, and hitting in all racket sports, punching in boxing, spiking in volleyball, shooting slap shots in hockey, etc.). These are examples of instants that mark the end of a movement phase. The instants that mark the beginning of these movement phases should also be examined, such as the start of initial forward movement of the ball or implement in throwing and striking activities and the instant of takeoff foot contact in jumping activities.

The duration and range of body motions during the force-producing phases should be examined, including propulsion (speeding up) and follow-through (slowing down) phases. These phases of motion can also be described as phases of positive and negative work. What is the duration of the propulsion phase during a throwing, striking, or jumping activity, and what is the corresponding displacement of the ball or implement or body during this phase? What is the duration of the follow-through phase during a throwing or striking activity, and what is the corresponding range of motion of the throwing or striking limb during this phase?

The velocities and accelerations of specific body segments during specific phases of a movement should also be examined. Often the velocity of a limb's distal end determines the outcome of the performance (especially in throwing and striking activities). Rather than just assessing the velocity of the limb qualitatively, it is more useful to assess the acceleration of the limb. Once the forward motion is initiated, does the throwing or striking limb or implement continuously accelerate throughout the movement, or are there pauses or periods where the limb or implement ceases to speed up?

Finally, the timing or coordination of the body segment movements must be examined. In many throwing and striking activities, the movement is initiated by the larger limbs (the legs) and proceeds in sequence to the smaller throwing or striking limb. This sequence of limb movements from largest to smallest increases the work done by increasing the displacement of the throwing or striking limb as well as the magnitude of the average force applied to the limb.

Use Other Senses

Although visual observation is the basis for most qualitative analyses, other senses may be employed as well. Tasting or smelling a performance won't reveal much about its mechanics (if a performance could be tasted or smelled), but listening might. Many activities have specific rhythms that indicate something about the timing of the movements involved. Touch is another sense that may be used in a qualitative analysis. How do you touch a performance? A coach or teacher spotting an athlete in gymnastics may touch the athlete during the routine. The amount of force the coach or teacher must exert on the athlete to correct the stunt can indicate that errors have occurred in the stunt. Finally, you may rely on an athlete's kinesthesia. Ask the athlete what a performance felt like. Did they pull hard? Did they slip at takeoff? Was the landing hard or soft?

Evaluate the Performance

Once you have described the ideal technique and observed the actual technique, the next step in the biomechanical analysis process is to evaluate the observed technique. This evaluation has two steps. First, errors or deficiencies in the actual performance are identified. These errors are then diagnosed to determine the magnitude of their effect on the performance.

Identify Errors

Errors or deficiencies in the performance should be identified while (or immediately after) observing it. Do this by comparing the actual performance to the characteristics of the most effective technique. For each element of the performance you examine, ask yourself: How does this movement, position, or timing of movements differ from the most effective technique?

Once a difference has been identified, you must then decide whether the difference adversely affects the performance. Differences may indicate style variations in the technique, or they may be actual errors. Discrepancies between the observed performance and the most effective technique should be carefully examined by questioning how the action or position being evaluated contributes to (or detracts from) the desired goal of the performance. Sometimes the differences between the observed and the proposed techniques result from modifications the athlete has made to accommodate morphological constraints.

The morphological constraints of the performer should be identified. In other words, how does the performer's anthropometry and strength constrain, limit, or modify their ability to match the proposed technique? Some constraints cannot be changed and are thus true constraints on the performance of the student or athlete. Other constraints can be modified through training. In the high jump example, height was identified as an advantage. Taller jumpers can have higher centers of gravity at takeoff. The height a jumper can raise her center of gravity to at the instant of takeoff is constrained by her height. The height of the jumper cannot be changed once growth has ceased after puberty. Body weight was identified as another anthropometric characteristic that affects high jumpers. A lean jumper may have an advantage over a jumper with a higher percentage of body fat. This morphological constraint can be changed with training. Strength of the takeoff leg affects the high jump performance as well. A stronger jumper may be able to run faster and get into a lower position at the start of the takeoff phase, whereas a weaker jumper will have to sacrifice some speed or not get into such a low position at takeoff. Strength is a morphological constraint that can be modified with proper training as well. Physical training is discussed further in chapter 17.

Evaluate Errors

Once the errors or deficiencies in the performance have been identified, they should be evaluated to determine the focus of correction efforts. During this evaluation, consider the causes of the errors as well as their effects. There are several concerns here:

1. Does the error expose the performer to the danger of injury? Such errors should be corrected immediately. For sports involving high risk, ensure that risks are minimized, especially with beginners, by using trained spotters and all appropriate safety equipment such as safety harnesses, crash pads, helmets, and other protective equipment.

2. Who are your clients? Are they students learning a new skill, or are they athletes re-

fining a skill they have practiced for years? Are they 5-year-olds or 25-year-olds? For students learning a new skill, focus your attention on errors in the basic elements of the ideal technique.

3. How easy is it to correct the error? Is it caused by a strength deficiency that may require training to correct, or is it due to an incorrect starting position? Will it take months, weeks, days, or only one practice session to correct? How much time do you have to correct the error before the next competition or before the most important competition?

4. Is the error a result of another error that occurred earlier in the performance? If so, the earlier error should be the focus of correction efforts.

5. How great an effect does the error have on the performance? Would correcting it dramatically improve performance, or would the change hardly be noticeable?

6. Is the error or deficiency due to poorly designed or inappropriate equipment? Can it be alleviated with improved or redesigned equipment?

After evaluating each error in light of the above concerns, you must then decide what errors to correct and in what order. In all cases, errors that endanger the athlete should be corrected first. Then the errors should be ranked in three categories: (1) from major to minor in terms of their effect on the performance; (2) from earliest to latest in terms of their chronology during the performance; and (3) from easiest to most difficult in terms of the time and effort it will take to correct them. Errors ranked near the top of each category should be attended to first. Those ranked near the bottom should be ignored or attended to last. Specific circumstances may dictate correction of errors with discrepant rankings from category to category. Errors observed in a baseball pitcher's delivery two days before he starts the championship game should not be corrected until after this game and the completion of the season, except for small errors that are easily corrected. On the other hand, the same errors observed during preseason practices should all be corrected, as the corrections are likely to be completed during the season and prior to the most important games at the end of the season.

Instruct the Performer

The final step in a qualitative biomechanical analysis is to instruct the performer by correcting the errors or deficiencies identified and ranked in the previous step. This instruction is what most people think of as teaching or coaching. As the fundamentals of effective instruction come from the fields of motor learning and sport psychology rather than biomechanics, only a brief overview of this step will be presented here.

To correct errors in performance, you must do three things. First, you must clearly communicate to the athlete or student what they did incorrectly (what their error was). Second, you must clearly communicate to the athlete or student what you want them to do (what the ideal technique looks like). Finally, you must devise means for the performer to correct the errors.

Communicate With the Performer

A large part of effective instruction involves effective communication. You must effectively communicate to your athlete or student what their performance looked like (what error in technique you want to correct) and what the ideal technique looks and feels like. This can be accomplished by describing verbally what the athlete or student did and what you want them to do. Or you (or someone else) can demonstrate what the student or athlete did and what you want them to do. Alternatively, you can show the performer photos or a videotape of the performance along with photos or a videotape of an athlete or student performing the skill in the desired way. The most effective method involves all three modes: verbal description, demonstration, and photographic or videotape records.

Keep your descriptions and instructions simple and work on correcting only one error at a time. Keep the student or athlete focused on that one element of the performance during each exercise or performance.

Your attitude during instruction may affect its effectiveness. Be positive. Praise the athlete or student for the positive aspects of the performance. Realize that during the correction process, some decline in performance may result initially, especially with highly skilled performers. Improvements may not occur immediately. This will be frustrating for the athlete, so you must be positive and patient as the coach or teacher.

Correct the Error

Once the performer understands what error they made in their performance and what the performance should look and feel like without the error, you must devise methods to assist the student or athlete in correcting the error. Certain guidelines may be helpful here.

The correction process involves development of a full teaching progression for the skill along with the drills involved in this progression. First, break the skill into parts. Devise drills that duplicate the movements and forces present in each part of the skill. Then use the drills to correct errors in discrete parts of the skill. Have the athlete practice the drills specific to the error slowly at first, then faster, until the speed is the same as that of the actual performance. Extend the drills to link the next part (or the previous part) of the skill to the movements. When the error is consistently absent from the linked drills, perform the entire skill (at a slower speed at first, if possible). Throughout the correction process, the speed of the movements should be manipulated. Begin with drills that emphasize positions. Then follow with drills in which initial and final positions are emphasized. Slow movement from one position to the next emphasizes the sequence of movements. Speed up the movement as the student or athlete becomes more proficient.

Repeat the Analysis

Once the instruction phase is completed and the errors are corrected, it may be worthwhile to repeat the entire qualitative biomechanical analysis procedure. This would include revising the description of the most effective technique based on any morphological constraints you have learned about your student or athlete. The observation, evaluation, and instruction phases would also be repeated. The process is thus a loop that eventually results in improved performances by your athletes or students.

Example Analyses

The process described in the previous pages for conducting a qualitative biomechanical analysis to improve technique used high jumping as an example to illustrate some of the procedures. In the following pages, models of the most effective techniques of other sport skills are developed for use in qualitative biomechanical analyses. The examples include a throwing activity (a fastball pitch), a striking activity (a tennis forehand drive), and a locomotor activity (sprint running).

Fastball Pitch in Baseball

The pitcher is the most important athlete on a baseball team. An effective pitcher can single-handedly shut down the offense of an opposing team. A fastball pitch is one of the basic pitches in a pitcher's repertoire. What techniques are best used to pitch a fastball? Let's do the first step of a qualitative biomechanical analysis to find out.

Theoretical Model of the Ideal Fastball Pitch

Recall that the first step in a qualitative biomechanical analysis is to describe the ideal technique by developing a theoretical model of the technique based on mechanics and cause-and-effect relationships. The first step in developing a model of a fastball pitch is to acquire the necessary fundamental knowledge of baseball pitching. Let's assume we've done this already and get right to developing the model.

The basis of the theoretical model is the purpose or goal of the skill. What is the purpose of a fastball pitch? What is the pitcher trying to achieve with this pitch? The pitcher is trying to pitch a strike, either by having the batter swing at the ball and miss or by putting the ball in the strike zone without the batter swinging at it.

Now let's identify the factors that determine this goal. The difficulty in hitting a pitch depends on how much time the batter has to react to the pitch and where it is placed in the strike zone. So a strike is more likely to occur if the time is short and the path of the ball places it in the strike zone. These are the two mechanical factors that affect the ultimate goal of pitching a strike with a fastball.

The time it takes for the ball to reach the plate is affected by the horizontal velocity of the ball at the instant it leaves the pitcher's hand, the horizontal distance to the plate from this point of release, and the horizontal component of the forces exerted on the ball by air resistance during its flight (primarily drag forces). A pitch with a faster horizontal velocity, a shorter horizontal distance to travel, and smaller drag forces acting on it will take less time to travel from the pitcher's hand to home plate.

Horizontal velocity of the ball at release is the most important factor in determining the time it takes a fastball pitch to reach the batter. Most of the characteristics of a pitcher's technique are directed at maximizing this factor. At the start of a pitch, the ball is stationary in the pitcher's hand or glove; at the end of the pitching motion, when the pitcher releases it, the ball is moving very fast. During the pitching motion, the ball's kinetic energy increases. Remember from chapter 7 that kinetic energy is energy due to the motion of an object, or quantitatively, kinetic energy is one-half the object's mass times its velocity squared. This increase in kinetic energy is a result of the work done on the ball by the pitcher. Because work is a product of force and displacement, the horizontal velocity of the ball at release is determined by the average horizontal force exerted on the ball by the pitcher during the pitching motion and the horizontal displacement of the ball while this force is exerted on it. A faster pitch results if the average horizontal force acting on the ball and the horizontal displacement of the ball are greater during the pitching motion. The horizontal displacement of the ball may be greater if the pitcher is taller and has longer limbs.

> **Horizontal velocity of the ball at release is the most important factor in determining the time it takes a fastball pitch to reach the batter.**

The other two factors that affect the time it takes for the ball to reach the plate are the horizontal distance to the plate from the point of release and the horizontal component of the forces exerted on the ball by air resistance during its flight. The horizontal distance from the point of release to home plate is limited by the distance from the pitching rubber to home plate. This distance is 60 ft 6 in. (18.4 m). The pitcher's foot must remain in contact with the pitching rubber until the ball is released, but the horizontal distance from the point of release to home plate can be less than 60 ft 6 in. (18.4 m), depending on the pitcher's anthropometry and position at release. The flight time of the ball will be shorter if this distance is shorter.

Besides the time it takes for the ball to reach the batter, the path of the ball also affects whether or not the pitch is a strike. The path of the ball after release is determined by its height at release, its horizontal and vertical velocities at release, and air resistance. The height at release is determined by the pitcher's position and anthropometry. Air resistance is affected by the air density, the relative velocity of the ball with respect to the air, the roughness of the ball's surface, and the spin on the ball (the ball's angular velocity at release). The spin on the ball at release is determined by the torque exerted on the ball by the pitcher's fingers during the pitching action and the time during which this torque acts. The factors affecting horizontal velocity at release were discussed previously. Vertical velocity at release is affected by similar factors—the average vertical force exerted on the ball and the vertical displacement of the ball during the pitching action. The theoretical model of the fac-

tors that mechanically affect the outcome of a fastball pitch are shown in figure 16.5.

Application of the Theoretical Model

In developing the ideal model of the most effective fastball pitching technique, certain mechanical factors were identified as desirable. Now let's examine those factors and determine what actions by the pitcher produce these desirable features. From our mechanical assessment of the purpose of a fastball pitch, the following factors were identified as important:

1. Maximize the horizontal velocity of the ball (and pitching hand) at the instant of ball release. This is achieved by maximizing the horizontal force exerted on the ball while maximizing the horizontal displacement of the ball during the pitching action. The drag force due to air resistance also affects this slightly.
2. Minimize the horizontal distance the ball has to travel to get from the point of release to the plate. This is achieved by releasing the ball as far forward as possible. The pitcher's height and limb lengths also affect this.
3. Release the ball such that its path after release intersects the strike zone. The horizontal and vertical velocities of the ball at release affect this, as does air resistance and the height of release.

Maximum horizontal velocity of the ball is the most important factor in determining the success of a fastball pitch. This velocity is maximized by maximizing the horizontal force exerted on the ball and the horizontal displacement of the ball during the pitching motion. The horizontal displacement of the ball during the pitching motion can be lengthened by turning the body (rotating the pelvis and trunk to the right for a right-handed pitcher) so that the pitching arm is away from the batter and then reaching the pitching hand as far back as possible. These actions constitute the pitcher's windup, which places the ball in a starting position from which it will then be moved forward (see figure 16.6). A taller, longer-limbed pitcher can put the ball in a starting position farther from the plate. The release position of the pitch should be as far forward of this starting position as possible. This is accomplished by stepping forward onto the left foot, rotating the pelvis and trunk to the left, flexing the left hip, and extending the throwing arm forward. Again, a taller, longer-limbed pitcher has the advantage of being able to put the ball in a release

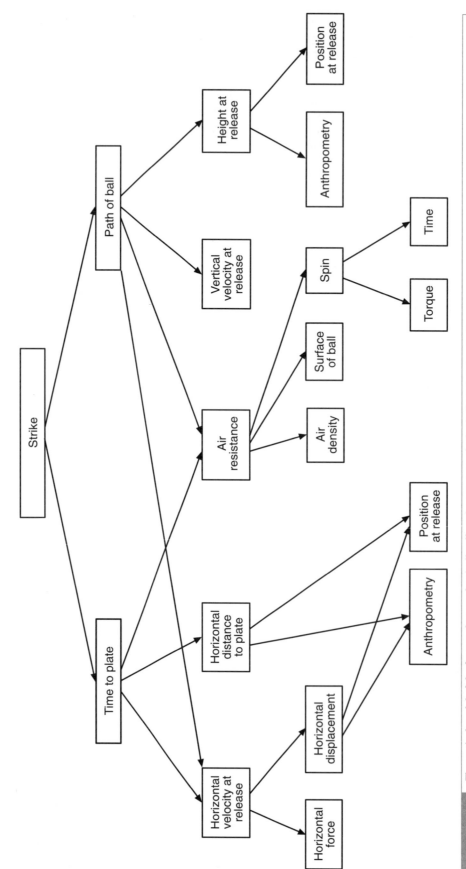

Figure 16.5   Theoretical model of the factors that mechanically affect the outcome of a fastball pitch in baseball.

Figure 16.6 The windup or starting position for the forward movement of the ball during a fastball pitch.

Figure 16.7 Finger positions for pitching a fastball.

position closer to the plate. The release position is constrained by the rules, however, which require that the pitcher keep one foot on the pitching rubber prior to release of the ball. If the pitch is to be a strike, the direction of the ball's velocity at release is restricted, and this restriction also places constraints on the release position.

The finger positions on the ball also affect the horizontal displacement of the ball during the pitching action. Contact can be maintained slightly longer if the ball is held by the fingers rather than in the palm of the hand. Further, the middle finger (the longest finger) should lay across the top of the ball, aligned with the direction of the throw (see figure 16.7). Slower pitches result if the ball is held deeper in the hand because the ball is released sooner, resulting in less horizontal displacement of the ball during the pitching action.

The average horizontal force exerted against the baseball during the pitching motion is maximized by using a kinetic chain with movement proceeding in sequence from proximal segments (those closest to contact with the ground) to distal segments. The forces are generated and transmitted from

proximal to distal and from larger to smaller segments as well. A kinetic chain is similar to a whip. The muscles of the lower extremity and trunk generate large forces initially as the pitcher steps forward. These forces are transmitted to the upper extremity and ultimately to the baseball through the bones, ligaments, and stretched muscles and tendons of the throwing arm. The pelvis rotates forward and the hip flexes as the left foot touches down. This stretches the trunk muscles, which then contract more forcefully due to the stretch reflex and cause the trunk to rotate. The muscles crossing the shoulder joint (especially the internal rotators and extensors) are then stretched maximally and contract forcefully to produce the rapid internal rotation and extension of the throwing arm that occur just prior to release.

Minimizing the horizontal distance from the point of release to the plate will also reduce the time it takes for the fastball pitch to reach the plate. How small this distance is depends on the pitcher's position at the instant of ball release and the pitcher's anthropometry. If the pitcher steps forward onto his opposite foot while moving his pitching hand forward, this distance can be reduced by several feet. A taller, longer-limbed pitcher will be able to stretch farther forward from the pitching rubber while still keeping his foot on it, thus minimizing this distance. The position of the ball at release affects both the horizontal distance from the point of release to the plate and the horizontal displacement of the ball during the pitching action. A release position closer to the plate is beneficial to each of these factors. A good release position is shown in figure 16.8.

The path the ball follows from its release by the pitcher until it crosses the plate is the other major factor affecting the success of a fastball pitch. This

Figure 16.8   The release position for a fastball pitch.

path depends on the direction and speed of the ball at release (its horizontal and vertical velocities), the height of release, air resistance, and the spin of the ball (its angular velocity) at release. Air resistance and horizontal velocity also affect the time it takes for the ball to reach the plate. The effect of air resistance can be minimized by holding the ball so that only two seams lie in the spin plane (see figure 16.7). This will also minimize deviations from the parabolic path the ball would follow if there were no air resistance. The backspin imparted on the ball will cause it to be slightly above this parabolic path due to the Magnus effect. Remember from chapter 11 that the Magnus effect is a ball's deviation from its normal flight path due to the effect of air resistance and the spin of the ball. This deviation can be increased by holding the ball so that four seams lie in the spin plane. A fastball pitch thrown with such a grip is called a rising fastball.

The direction of the ball at release depends on the pitcher's position at release. If the pitcher is slightly ahead of or behind the normal release position, the direction of the ball will be affected. The stride taken during the pitching action is forward and down the front of the pitching mound. This stride flattens the path followed by the ball and

pitching hand prior to release. Due to this flattening of the ball's path, errors caused by early or late releases are minimized.

This analysis has identified some of the fundamental characteristics of the most effective technique for throwing a fastball. Not all aspects of technique were discussed, however. The follow-through (the movements of the pitcher that occur following release of the ball) was not discussed because it does not directly affect the performance criterion. This does not mean that the follow-through is unimportant. The purpose of the follow-through is to slow down the limbs safely and prepare the pitcher for fielding a hit ball. Other aspects of technique may vary and do not necessarily concern the mechanics of the pitch. For instance, it is advantageous for the pitcher to hide the ball for as long as possible to give the batter less time to react to the pitch. Some actions of the pitcher's glove hand are directed at this task. The presence of base runners may also alter the pitcher's technique.

Forehand Drive in Tennis

The forehand drive in tennis is an example of a striking activity in sport. It is also an example of an open skill—a skill performed under conditions that vary and are unknown from one performance to the next. In a forehand drive, some of the conditions that vary are your location on the court, your opponent's location on the court, and the incoming direction and velocity of the ball. Your return may be offensive or defensive, depending on the situation. For this example, let's assume that your position on the court and the incoming velocity and direction of the ball are such that your return is offensive. What are the characteristics of an effective offensive forehand drive? Let's develop a theoretical model of the most effective technique to find out.

Theoretical Model of the Ideal Tennis Forehand Drive

Again, the first step in a qualitative biomechanical analysis is to describe the ideal technique by developing a theoretical model of the technique based on mechanics and cause-and-effect relationships. The first step in developing a model of a forehand drive in tennis is to acquire the necessary fundamental knowledge of tennis. Let's assume we've done this work already and get right to development of the model.

The basis of the theoretical model is the purpose or goal of the skill. What is the purpose of a forehand drive? What is the player trying to achieve with this stroke? Let's assume that the conditions are such that the player is attempting an offensive shot. If this is the case, the player is trying to hit a winner. This can be achieved by hitting the shot so that the ball goes past the opposing player without him having a chance to swing at it. But to be a winner, the ball must also go over the net and bounce in the opponent's court. This is similar to the goal of a fastball pitch in baseball.

Now let's identify the factors that determine this goal. The difficulty in returning a tennis shot depends on how much time the player has to react to the shot and what path the ball follows. So a winner is more likely to result if the time the player has to react and move to the ball is short and if the path of the ball puts it over the net and far from the player in the opposing court. These factors are interrelated because the path the ball takes determines how far the player has to move to intercept it. Most of the time the player has to react and move to the ball is accounted for by the time it takes for the ball to travel from the racket until it passes the baseline after it bounces in the court. Let's call this the racket-to-court time. So the two factors determining a winner in a forehand drive are racket-to-court time and the path of the ball from the racket to the court.

> **The difficulty in returning a tennis shot depends on how much time the player has to react to the shot and what path the ball follows.**

The factors that affect racket-to-court time include the horizontal velocity of the ball after impact, the horizontal distance the ball must travel from racket to baseline, the horizontal component of the forces exerted on the ball by air resistance during its flight, and friction between the ball and the court surface when the ball bounces. The most important of these factors, horizontal velocity of the ball after impact, is determined by the horizontal velocity of the racket head at impact, the velocity and direction of the ball just prior to impact, the angle of the racket face at impact, and the racket characteristics (string tightness, head size, stiffness, etc.). Of these factors, horizontal velocity of the racket head at impact is most important for a forehand drive. If we use work and energy principles in our analysis, the horizontal velocity of the racket head at impact is determined by the work done on the racket. So the factors determining the horizontal velocity of the racket head are the average hori-

zontal force exerted on the racket during the stroke and the horizontal displacement of the racket head while this force is exerted. Greater displacement and greater force produce a faster racket head velocity.

The other factors affecting racket-to-court time include the horizontal distance the ball must travel from the racket to the baseline of the opponent's court, the horizontal component of the forces exerted on the ball by air resistance during its flight, and the friction between the ball and the court when the ball bounces. The horizontal distance the ball must travel is affected by the player's position in the court when hitting the ball and the path followed by the ball as a result of the stroke. Friction between the ball and the court is affected by the court surface, the spin on the ball, and the velocity of the ball when it hits the court.

The second factor determining if a forehand drive will be a winner is the path of the ball from the racket to the court. The factors affecting the path of the ball from racket to court are the horizontal and vertical velocities of the ball after impact, the height of the ball at impact, air resistance, and friction between the court and the ball when the ball bounces. The vertical velocity of the ball is determined by the vertical velocity of the racket head at impact, the velocity and direction of the ball just prior to impact, the angle of the racket face at impact, and the racket characteristics (string tightness, head size, stiffness, etc.). The vertical velocity of the racket head at impact is determined by the average vertical force exerted on the racket during the stroke and the vertical displacement of the racket head while this force is exerted.

Air resistance is the next most important factor affecting the path of the ball after its horizontal and vertical velocities and its height after impact. Air resistance is affected by the density of the air, the relative velocity of the ball with respect to the air, the roughness of the ball's surface, and the spin of the ball. The spin of the ball affects whether or not the ball rises, drops, or curves during its flight and after it bounces. The spin of the ball is affected by the same factors that affect its velocity: the velocity and direction of the ball before impact, the velocity and direction of the racket head at impact, the angle of the racket face at impact, and the racket characteristics. The theoretical model of the factors that mechanically affect the outcome of a tennis forehand drive are shown in figure 16.9.

Application of the Theoretical Model

The model of the most effective tennis forehand drive shown in figure 16.9 identifies the mechani-

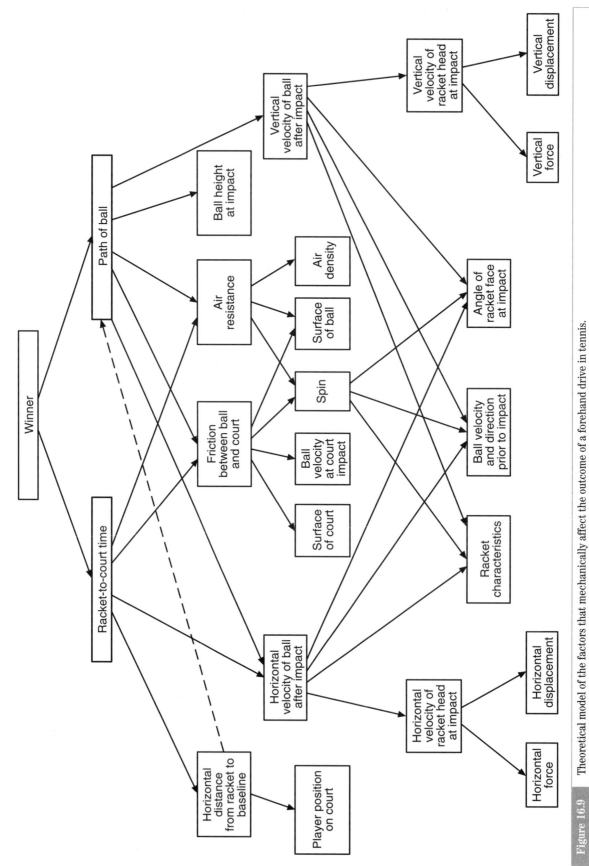

Figure 16.9   Theoretical model of the factors that mechanically affect the outcome of a forehand drive in tennis.

cal factors that produce the desired outcome. Some of these factors (the velocity of the ball prior to impact, the court surface, the height of the ball at impact, and air density) are not under the player's control. Furthermore, since the forehand drive is an open skill (one performed in a relatively unpredictable environment), some of the factors are determined by the game conditions. Certainly where the player decides to hit the forehand drive (the path of the ball and its horizontal displacement from racket to court) is affected by both the player's and the opposing player's positions on the court. Further complicating the application of our model are the interactions among the factors identified. The incoming velocity of the ball, the angle of the racket face, the velocity and direction of the racket, and the racket characteristics all affect the horizontal and vertical velocities of the ball as well as the spin of the ball. These factors, in turn, affect the racket-to-court time and the path of the ball.

The open nature of the forehand drive and the interactions between the determining factors make it difficult to identify the factors that are controlled exclusively by the player and that are obviously important in determining the success of the skill. One such factor is the horizontal velocity of the racket head at impact. To impart a fast horizontal velocity to the ball after impact, the horizontal velocity of the racket head must be fast at impact. Keep in mind, however, that the horizontal velocity of the ball after impact must be such that it will allow the ball to clear the net and land in the court. A fast horizontal velocity is achieved by exerting a large horizontal force on the racket while maximizing the horizontal displacement of the racket head during stroke execution.

The horizontal displacement of the racket head during the stroke can be lengthened by turning the body (rotating the pelvis and trunk to the right for a right-handed player) and extending the racket as far back from the oncoming ball as possible. These actions constitute the player's backswing, which places the racket in a starting position from which it will then be moved forward (see figure 16.10a). A taller, longer-limbed player can start the racket farther back. The racket's impact with the ball should be as far forward of the starting position as possible (see figure 16.10b). Again, taller, longer-limbed players have the advantage of being able to hit the ball in a position farther forward. The total horizontal displacement of the racket during the stroke is constrained by how much time the player has to get in position to execute the stroke. If the time is short, preparation for the stroke is limited and the horizontal displacement of the racket head may not be as long.

The grip position along the racket and the length of the racket also affect the horizontal displacement

a

b

Figure 16.10   The starting (*a*) and ending (*b*) positions of a forehand drive in tennis.

of the racket head during the stroke. With a longer racket, the racket head can be moved through a longer horizontal displacement during the stroke. The rules limit racket length, but the grip position can be used to maximize the horizontal displacement of the racket within the limits imposed by its length. Gripping farther from the racket head and closer to the heel of the handle (see figure 16.11) lengthens the horizontal displacement of the racket head by increasing the effective radius of the stroke. Gripping farther up the handle and closer to the racket head (see figure 16.11) shortens this displacement. There is a tradeoff here, however. The axis of the swing moves farther from the racket head as the grip moves down the handle, increasing the racket's moment of inertia and making it harder to control.

The average horizontal force exerted against the racket during the forehand stroke is maximized by using a kinetic chain, with movement proceeding in sequence from proximal to distal segments, as in the baseball pitch. The forces are generated and transmitted from proximal to distal and from larger to smaller segments as well. For example, a right-handed player initially pivots and shifts the center of gravity toward the rear and over the right (back) foot. The muscles of the lower extremity and trunk then generate large forces as the player shifts the center of gravity forward over the left foot. Depending on the time available and the position of the incoming ball, the player may also step forward rather than just shifting her weight forward. The forces generated by the lower extremities and trunk are transmitted to the upper extremity and ultimately to the racket through the bones, ligaments, and stretched muscles and tendons of the throwing arm. The pelvis and trunk rotate forward as the center of gravity moves forward over the left foot. This movement stretches the trunk muscles, which then contract more forcefully due to the stretch reflex and cause the trunk to rotate. The muscles crossing the shoulder joint (especially the horizontal adductors) are then stretched and contract forcefully to produce the arm and racket movements that occur just prior to ball impact.

The horizontal forces on the racket are generated and transmitted from proximal to distal and from larger to smaller segments.

Another important factor is racket orientation (the angle of the racket face), because it affects the velocity and direction of the ball after impact. The angle of the racket face also affects the spin on the ball, which affects the path taken by the ball. There is no one best racket face angle, because it is influenced by the preimpact velocity and direction of the ball and the racket head. These factors all interact to determine the postimpact velocity and

Figure 16.11   Grip positions on a tennis racket affect the displacement of the racket head during a stroke.

direction of the ball. Let's analyze a simple situation first. Assume the racket is stationary and the ball impacts the racket face at an angle. If the impact between the ball and the racket were perfectly elastic (a coefficient of restitution between the ball and racket of 1.0, meaning the velocity of the ball with respect to the racket face before the collision is equal to the velocity of the ball with respect to the racket face after the collision), the angle of incidence (the preimpact angle of the ball with the racket face) would equal the angle of reflection (the postimpact angle of the ball with the racket face) (see figure 16.12a). In reality, the collision is not perfectly elastic (the coefficient of restitution is less than 1.0), so the postimpact angle is less. This means that if you hold your racket still and let the incoming ball strike it, the racket face should be facing a little more than halfway toward the incoming direction of the ball from the direction you want the ball to go (see figure 16.12b). If the racket is swung and the racket head has some velocity perpendicular to the racket face before impact, the postimpact angle will be equal to or greater than the preimpact angle (see figure 16.12c). Thus, as you swing the racket faster and faster, you should aim the racket face closer and

closer to the direction you want the ball to go. You would never aim the racket face exactly in the direction you want the ball to go, however, unless that is the direction the ball came from.

Spin on the ball influences the path it takes due to differing air pressures on the sides of the ball (remember the Magnus effect from chapter 11). Topspin causes the ball to follow a slightly lower than normal trajectory; backspin causes it to follow a slightly higher than normal trajectory; and clockwise or counterclockwise spin causes it to veer slightly sideways. Spin also affects the bounce of the ball due to friction between the ball and the court.

Spin is imparted to the ball by friction between the racket strings and the ball when it strikes the racket and rolls or slides across the strings. This friction force creates a torque about the center of the ball, which creates the spin. The friction force also acts to change the linear direction of the ball. No spin is imparted to the ball if the velocity of the incoming ball and the velocity of the racket head are exactly opposite in direction and if the racket face is perpendicular to those directions. The angle between the direction of the racket face and the

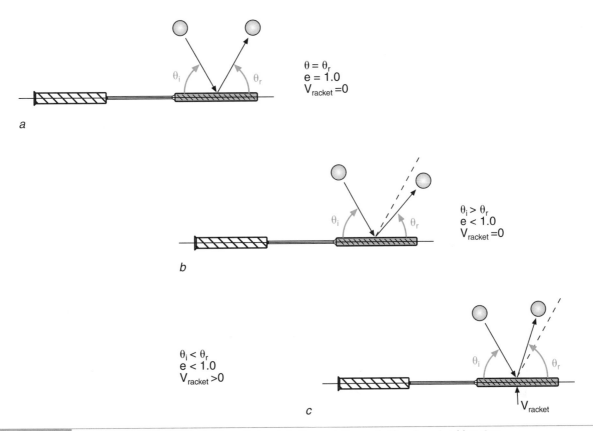

$\theta = \theta_r$
$e = 1.0$
$V_{racket} = 0$

a

$\theta_i > \theta_r$
$e < 1.0$
$V_{racket} = 0$

b

$\theta_i < \theta_r$
$e < 1.0$
$V_{racket} > 0$

V_{racket}

c

Figure 16.12 Angles of incidence and reflection of a tennis ball on a stationary (a-b) and moving (c) racket.

direction of the relative velocity of the ball with respect to the racket head determines the amount of spin imparted to the ball (see figure 16.13). The greater the angle, the more spin imparted to the ball. A swing path from low to high will impart topspin to the ball, and a swing path from high to low will generally impart backspin to the ball. A swing path from left to right will impart counterclockwise spin, whereas a swing path from right to left will impart clockwise spin.

Ideally the forehand drive should have a fast horizontal velocity, so the racket face should be directed close to the desired direction of the ball. This fast horizontal velocity increases the chances that the ball will go past the baseline, so some topspin must be imparted to the ball to keep it in the court. In a good forehand drive, the racket head undergoes some vertical displacement along with its horizontal displacement.

This has been an incomplete analysis of the factors affecting the performance of the forehand drive in tennis. Because the forehand drive is an open skill, it can be performed effectively in many ways based on the numerous combinations of factors the performer is unable to control. This reveals one of the shortcomings of the proposed qualitative biomechanical analysis procedure. Developing a model of the most effective technique is more difficult for open skills. A qualitative biomechanical analysis of a player's forehand drive technique should include observation and consideration of the various conditions under which the stroke is executed.

Sprint Running

Running is a basic human movement. How fast a person can run or sprint may determine the level of success in many sports. The skill of sprinting is the last skill we will analyze. Unlike the previous sport skills, sprinting is a cyclic skill—the movements occur in cycles that are repeated over and over again. For sprinting, one stride represents one cycle. A stride begins and ends with the body parts in the same relative positions. If left foot touchdown is identified as the start of a sprinting stride, the end

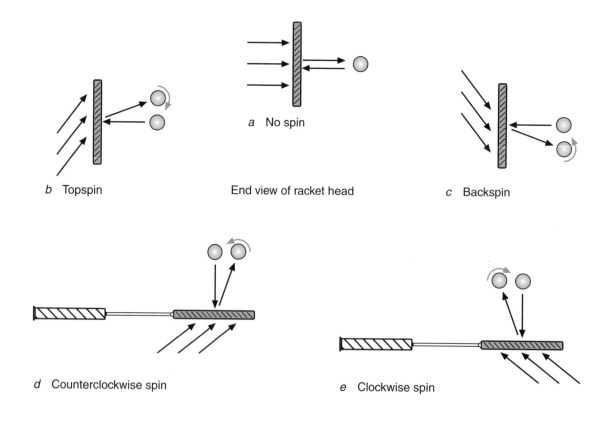

b Topspin   End view of racket head   *c* Backspin

d Counterclockwise spin   *e* Clockwise spin

Top view of racket

Figure 16.13   Directions of the ball velocity, racket head velocity, and racket face determine ball spin.

of the stride or cycle occurs at the next instant the left foot touches down. For our example analysis, let's consider a 100-m dash in track and field. This skill has several parts: the start, the acceleration phase, and the velocity-maintenance phase. Our analysis will focus on the last phase, the velocity-maintenance phase, in which the sprinter attempts to maintain maximum velocity.

Theoretical Model of Sprinting

As stated previously, the first step in a qualitative biomechanical analysis is to describe the ideal technique by developing a theoretical model of the technique based on mechanics and cause-and-effect relationships. The first step in developing a model of sprinting is to acquire the necessary fundamental knowledge of sprinting. Let's assume we've done this work already and get right to developing the model.

The basis of the theoretical model is the purpose or goal of the skill. What is the purpose of sprinting in a 100-m dash? What is the runner trying to achieve? The runner is trying to win the race and get to the finish line first—in the least amount of time—so the performance criterion may be considered the time taken to run 100 m. The time it takes to run 100 m is a function of the average horizontal velocity of the sprinter during the race and the 100-m distance. Thus, the performance criterion for the velocity-maintenance phase of the 100-m dash is the average horizontal velocity of the sprinter during this phase.

A faster sprinter has a faster average horizontal velocity during each step of the velocity-maintenance phase of the 100-m dash. The average velocity of the sprinter is determined by the sprinter's average step length and step rate. Longer steps and more steps per second result in faster sprinting. How are longer steps and a faster step rate accomplished?

> **The average velocity of the sprinter is determined by the sprinter's average step length and step rate.**

Step length is the sum of the sprinter's horizontal displacement during the support phase of a step (when the sprinter is in contact with the ground) and the sprinter's horizontal displacement during the flight phase of a step (when the sprinter is not in contact with the ground). Greater displacements during each of these phases result in a longer step length. The horizontal displacement during the support phase can be further divided into the touchdown distance (how far behind the foot the sprinter's center of gravity is at touchdown) and the takeoff distance (how far ahead of the foot the sprinter's center of gravity is at takeoff). Each of these distances depends on the sprinter's position at these instants and the sprinter's anthropometry. A taller, longer-legged sprinter is able to achieve longer touchdown and takeoff distances.

The horizontal displacement during the flight phase of a step is determined by the horizontal velocity of the sprinter at takeoff and the duration of the phase. The faster the horizontal velocity and the longer the duration of the flight phase, the greater the horizontal displacement during the phase. The horizontal velocity at takeoff is, in turn, determined by the horizontal velocity at touchdown and the work done by the friction force acting on the takeoff foot during the previous support phase. This work could be subdivided into the negative work done during the braking part of the support phase (when the friction force slows down the sprinter) and the positive work done during the propulsive part of the support phase (when the friction force speeds up the sprinter). The negative and positive work done are determined by the average horizontal forces and horizontal displacements occurring during each period. A sprinter who minimizes the negative work done and maximizes the positive work done will accelerate forward. During that part of a 100-m dash where the sprinter maintains the same velocity from step to step, the negative and positive work done during a typical step are equal.

The duration of the flight phase (or flight time) also affects the horizontal displacement during the phase. The flight time depends on the vertical velocity of the sprinter at takeoff, the height of the sprinter's center of gravity (CG) at takeoff, the height of the sprinter's center of gravity at touchdown, and the average drag force due to air resistance. The vertical velocity at takeoff is, in turn, determined by the vertical velocity at touchdown and the work done by the normal reaction force acting on the takeoff foot during the previous support phase. This work could be subdivided into the negative work done when the upward reaction force slows the downward velocity of the sprinter at the beginning of the support phase and the positive work done when the upward reaction force speeds the upward velocity of the sprinter during the second half of the support phase. The negative and positive work done are determined by the average normal reaction forces and vertical displacements occurring during each period. A sprinter must do equal amounts of positive and negative work to maintain the same horizontal velocity from step to step.

The flight time during a step is also affected by the sprinter's center-of-gravity height at the beginning and end of the flight phase (i.e., at takeoff and touchdown). If the takeoff height is greater than the touchdown height, the time of flight will be extended. If the takeoff height is less than the touchdown height, the flight time will be shortened.

Average step length and average step rate determine a sprinter's average horizontal velocity. The step length determinants were discussed earlier. Step rate is determined by the time per step. The shorter the time per step, the faster the step rate. The time per step can be subdivided into support time and flight time. The flight time determinants were also described previously, as flight time is one of the factors affecting horizontal displacement during the flight phase. A greater displacement during the flight phase results if the flight time is longer; however, the time per step is shorter and the step rate is faster if the flight phase is shorter. The theoretical model of the factors that mechanically affect sprinting is shown in figure 16.14.

Application of the Theoretical Model

Sprinting is a closed skill, so it should be easier to interpret and apply the model of the most effective technique to sprinting than it was for an open skill such as the tennis forehand drive. However, because sprinting is cyclic by nature, some of the input factors in the model are outcomes of a previous cycle (a previous step), and thus our model becomes sort of an endless loop. For instance, the horizontal displacement during the flight phase is determined by the flight time and the horizontal velocity at takeoff. Horizontal velocity at takeoff is, in turn, determined by the work done during the previous support phase and the horizontal velocity at touchdown. This horizontal velocity is a performance measure of the previous step. Despite this shortcoming of our model, let's examine it and see how it can be applied.

Prior to applying the model, let's examine how the instantaneous horizontal velocity of a sprinter's center of gravity varies during a single step. Figure 16.15 shows a plot of the horizontal velocity of a sprinter during one step. Notice how the velocity decreases at the start of the support phase and increases during the second half of that phase. In which part of the step is the sprinter's average hori-

zontal velocity faster? The sprinter is obviously faster during the flight phase. The horizontal velocity during the flight phase is always faster than the horizontal velocity during the support phase. When the sprinter's foot strikes the ground, friction between the ground and the sprinter's foot creates a braking force, and the sprinter slows down. In the second half of the support phase, the sprinter speeds up to regain the velocity lost during the braking phase. Thus, the average horizontal velocity during the support phase of a sprinting step is slower than the horizontal velocities at the beginning or end of the phase, which are equal to the average velocities during the preceding or succeeding flight phases. A sprinter should thus minimize the support time and maximize flight time. Let's keep this in mind as we analyze our model.

In our sprinting model, the factors affecting the step length during the support phase are touchdown and takeoff distance. Although a longer touchdown distance would appear to be beneficial, it increases the braking forces as well as lengthening the support phase. Therefore, touchdown distance should be kept short by having the foot strike the ground almost directly beneath the sprinter. No detrimental effect is associated with takeoff distance, so it should be maximized. By fully extending the takeoff leg and rotating the pelvis toward the side of takeoff, a sprinter can add another 3 or 4 cm to this distance (see figure 16.16).

The factors affecting step length during the flight phase are flight time and horizontal velocity at takeoff. Both factors are affected by the work (positive and negative) done during the previous support phase. The negative work done to slow the vertical velocity of the sprinter at touchdown should have a large average vertical reaction force and a short vertical displacement. The hip, knee, and ankle should flex enough to reduce the reaction force and the force required in the muscles to values below injury threshold. Too much flexion and downward vertical displacement will increase the time spent in the support phase. Another benefit of minimizing the length of this negative work phase is that the larger load and quicker stretching of the lower extremity muscles will evoke a stronger stretch reflex. The muscles will contract more forcefully during the propulsive or positive work phase. The larger vertical reaction force will also cause a larger frictional force between the foot and the ground, increasing the negative work done to slow the runner's forward velocity. Driving the free leg and arms upward during the second half of the support phase increases the vertical reaction force as well

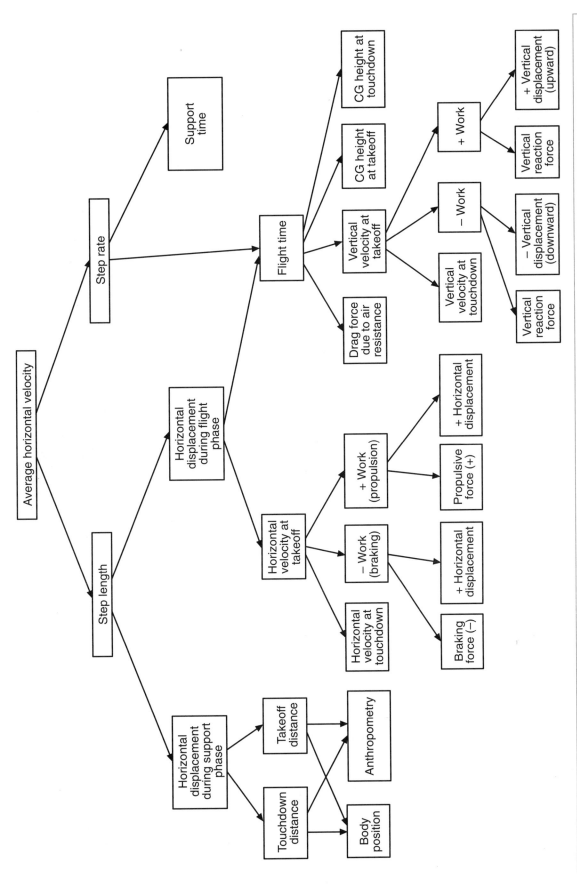

Figure 16.14 Theoretical model of sprinting.

Figure 16.16   A sprinter's takeoff distance is the horizontal distance between the takeoff foot toe and the sprinter's center of gravity.

V₁ = Average velocity during flight phase

V₂ = Average velocity during one full step

Vₐ = Average velocity during support phase

Figure 16.15   (a) Theoretical plot of the anteroposterior ground reaction force acting on a sprinter, and (b) the horizontal velocity of the sprinter.

as the vertical displacement of the sprinter. These actions, along with full extension of the support leg, increase the positive work done to speed up the vertical velocity of the sprinter during the support phase.

The negative work done to slow the horizontal velocity of the sprinter at the beginning of the support phase should be minimized by having the sprinter's foot touch down almost directly beneath the center of gravity. The positive work done to speed up the runner's horizontal velocity should be maximized by thrusting the swing leg forward vigorously, but the knee should be flexed to minimize the leg's moment of inertia about the hip joint so that the movement is quicker. The arms should

swing in opposition to counter the angular momentum generated by the swinging leg. Some elbow flexion is desirable to reduce the arm's moment of inertia about the shoulder joint and thus increase its movement speed. Again, full extension of the support leg as well as rotation and tilt of the pelvis are desirable to increase the horizontal displacement of the runner during the positive work phase.

During the support phase, especially the propulsive portion, the sprinter's trunk should be fairly rigid to keep the propulsive forces from being dissipated and absorbed when the trunk and vertebral column flex. This is the phase in which positive work is done, and the more powerful the sprinter, the faster the sprinter will run. If the upward- and forward-thrusting forces from the legs do not cause upward and forward displacement of the trunk but rather a deflection or collapsing of the trunk and vertebral column, little or no positive work is done and power is not transmitted from the lower extremity to the trunk.

A maximum sprint represents a fine balance between step rate and step length. Both need to be as large as possible, but overemphasizing one will adversely affect the other. Too long a step will decrease step rate by increasing the landing distance and the braking forces at the start of the support phase. These conditions increase the length of the support phase, thus slowing the step rate and reducing the velocity. On the other hand, an increased step rate will require a shorter flight time and support time. If stride length remains the same but the support time is shorter, the sprinter has to be more powerful to absorb the energy of the downward motion when the foot first strikes the ground and then produce the same energy during the propulsive part of the phase. If the support time is too short, less energy will be absorbed or produced and the subsequent takeoff velocity will be slower, resulting in a shorter step length. This analysis reveals the primary physical requirement of sprinting—a sprinter must be powerful. The sprinter's muscles must be able to develop large forces (both eccentrically and concentrically) at high contraction velocities.

ing a qualitative biomechanical analysis to improve technique includes the following steps:

1. Develop a theoretical model of the most effective technique.
2. Observe the actual performance.
3. Evaluate the performance by comparing it to the most effective technique.
4. Instruct the performer by providing feedback about discrepancies between the actual performance and the most effective technique.

This biomechanical analysis may be qualitative or quantitative. A qualitative biomechanical analysis relies on subjective observations of the performance, whereas a quantitative biomechanical analysis uses actual measurements to quantify certain mechanical parameters of the performance. Practitioners (coaches, teachers, clinicians) use this or some similar procedure to evaluate the movement or performance of their students, athletes, or clients.

Summary

The biomechanics presented in the previous chapters of this book can be applied systematically to analyze human movement. One system for conduct-

Key Terms

qualitative biomechanical analysis *(p. 294)*
quantitative biomechanical analysis *(p. 294)*

REVIEW QUESTIONS

1. **What steps are involved in a qualitative biomechanical analysis for technique improvement?**

2. **What is the difference between qualitative and quantitative biomechanical analyses?**

3. **Identify the performance criteria for the following activities:**

   a. **A 110-m hurdle race**

   b. **A basketball pass**

   c. **A football punt**

   d. **A discus throw**

   e. **Batting in baseball**

4. Describe where you should position yourself to observe the following:

   a. The takeoff phase of a long jump

   b. The clearance technique of the trailing leg in a 110-m hurdle race

   c. The start of a 50-m freestyle swimming race

   d. A field goal kick in football

   e. The arrow release in archery

5. How are errors detected in a technique prioritized for correction?

Qualitative Biomechanical Analysis to Improve Training

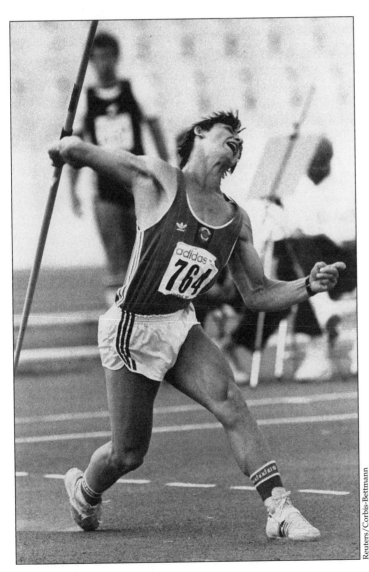

Reuters/Corbis-Bettmann

You are watching the men's javelin competition in the Summer Olympic Games on television. The winning throw flies well beyond 300 ft, and you wonder how the thrower strengthens his arm to generate the forces necessary to throw the javelin so far. Obviously, the thrower uses more than just his arm muscles. What other muscle groups must be well developed to throw the javelin so far? What kind of exercises and drills are effective for training a javelin thrower? This chapter may help you answer these questions by providing you with the basis for completing a qualitative biomechanical analysis to improve training.

When you finish this chapter, you should be able to do the following:

▌ Distinguish between technical training and physical training

▌ Understand how the results of a qualitative analysis to improve technique may also be used to improve technical training

▌ Evaluate a technical exercise or drill and determine whether it is appropriate

▌ Describe the steps involved in a qualitative anatomical analysis

▌ Identify the active muscle groups during any phase of a human movement and whether their contractions are concentric, eccentric, or isometric

In chapter 1, we stated that the primary goal of biomechanics is performance improvement and a secondary goal is injury prevention and rehabilitation. In the previous chapter, we discussed how biomechanics can be used to improve technique and thus improve performance. But what if the performer is not strong or powerful enough to execute the most effective technique? Just as an athlete's technical errors may limit performance, deficiencies in the athlete's physical capabilities may also limit performance or prevent execution of the most effective technique. In this chapter, we will learn how biomechanics can be used to improve training and thus improve performance. Specifically, we will learn how to conduct a qualitative anatomical analysis of a performance or exercise to identify the active muscle groups.

Biomechanics and Training

A fundamental principle of training is the principle of specificity. Exercises and drills must be specific to the sport or activity. If the exercises and drills you use in training closely match the aspects of the skill you are training for, you will have greater gains in performance. For some sports or activities, the type of training that is specific to the sport or activity is obvious. If you want to run a marathon, your training should consist primarily of running long distances. For other sports, the training exercises that are specific to the sport or activity may not be

as obvious. For example, the specific strengths needed to be successful in pole vaulting are not obvious to the casual observer. Biomechanics can contribute to the improvement of training by identifying the specific technical or physical requirements necessary to perform a skill well. Before learning more about how to identify the specific technical or physical requirements of a skill, let's consider the various types of training.

> If the exercises and drills you use
> in training closely match the aspects
> of the skill you are training for,
> you will have greater gains in performance.

Technical Training

Practice and training time for specific sport or movement activities is usually devoted to improving technique (technical training) or physical condition (physical training). The proportion of time allocated to technical training depends in part on the technical difficulty of the activity. Technical training may involve performing the actual skill or performing drills that mimic specific aspects of the skill.

Biomechanics can contribute to improvements in technical training in several ways. First, a qualitative biomechanical analysis of an actual performance should be done to identify deficiencies in technique. We described the procedure for completing this type of analysis in chapter 16. If the athlete is unable to correct the deficiency after being made aware of it, you prescribe drills and exercises that are specific to the aspect of the skill that is deficient.

Another way biomechanics can contribute to improving technical training is by identifying exercises and drills that closely simulate the specific technical aspects of the skill. Examine the drills and exercises used in the technical training of an activity you are familiar with. What is the purpose of the drill or exercise? What aspect of the skill is it specific to? Are the joint positions, velocities, and ranges of motion of the exercise similar to those of the skill? Are the muscle forces and contraction velocities similar? Are the external forces similar? These are questions you should ask yourself when evaluating drills or exercises for your students, clients, or athletes. The greater the fidelity between the exercise and the actual skill, the more specific the exercise is to that skill and the greater its potential for improving the performance of the skill. If few similarities exist between the exercise and the skill, the exercise is inappropriate for developing that skill. The muscle forces, contraction velocities, and magnitudes of the external forces are difficult to assess qualitatively, but you can view the joint positions and ranges of motion and from these assess joint angular velocities. Quantitative biomechanical analyses of exercises and drills for specific sport skills would be more valuable, but few have been reported in the biomechanical literature.

Physical Training

Physical training is the other part of training. Whereas technical training is directed primarily at correcting or improving aspects of technique, physical training is directed at altering performance limitations due to the physical condition of the performer. Be aware that there is usually some overlap between the two types of training. Technical training may have some effect on the performer's physical condition, and physical training may have some effect on the performer's technical proficiency. In any case, physical training is directed at improving the components of physical fitness, including

1. muscular strength,
2. muscular power,
3. muscular endurance,
4. flexibility,
5. cardiovascular fitness, and
6. body composition.

Again, biomechanics is directed at improving the specificity of physical training. A biomechanical analysis of an activity can identify the specific muscle groups whose strength, power, endurance, or flexibility limit performance. Specific exercises can then be chosen that strengthen or stretch these specific muscle groups. Likewise, a biomechanical analysis of an exercise can identify if the muscles used in the exercise are those used in the sport or activity.

In this text, the type of qualitative biomechanical analysis used to identify the active muscle groups during each phase of a movement is called a qualitative anatomical analysis. The next section of this chapter describes the steps involved in a qualitative anatomical analysis of a movement. The remainder of the chapter presents example qualitative anatomical analyses.

Qualitative Anatomical Analysis Method

The purpose of a qualitative anatomical analysis is to determine the predominant muscular activity during specific phases of a performance and to identify instants when large stresses may occur due to large muscle forces or extremes in joint ranges of motion. The coach or teacher may complete such an analysis on a student or athlete or on an elite performer who demonstrates effective technique. The analysis of the elite performance will identify which muscles are involved in performance of the most effective technique, whereas the analysis of the student or athlete will identify the muscles they use in performing their specific technique. In either case, the methods used to identify the muscles involved will be the same.

Let's think for a minute. Do you already know of any methods for identifying what muscles are active during a sport or movement? One way is to touch and feel the performer's superficial muscles during the movement. If the muscle is firm and rigid, it is actively developing force. If the muscle is soft and flabby, it is inactive. This method only works if the activity involves static positions or slow movements (and if the performer is comfortable with being touched). Certain weight lifting exercises and gymnastics routines are examples of movements that may be appropriate for this type of analysis. Obviously, this method is impractical for analyzing dynamic activities such as throwing or running. The method is also intrusive and affects how the subject performs.

Another qualitative method for determining which muscles are active during an activity is to perform the activity vigorously and then wait a day or two to see which muscles become sore. This only works when you haven't been practicing the activity regularly. It may also identify only those muscles that experience large eccentric activity during the performance, as eccentric muscular action is associated with muscle soreness.

Quantitatively, a researcher can wire the performer's muscles with electrodes and use electromyography to monitor the electrical activity of the muscles. The EMG recordings of a muscle will indicate whether or not the muscle was active. Further analysis of the EMG signal will provide a general idea of the magnitude of the muscular action. This type of quantitative analysis is expensive, time consuming, and unavailable to most teachers and coaches. Besides, the electrodes and associated wiring are likely to cause the athlete's performance to differ from a typical performance.

Another quantitative method involves recording the movement on film or videotape and then digitizing the film or video to get a complete kinematic description of the movement. Force platforms or transducers can also be used to measure any external contact forces acting on the subject. Anthropometric measurements are then taken from the subject to estimate body segment masses and moments of inertia. Free-body diagrams of each segment are then used to identify the resultant joint forces and torques that cause acceleration of the segments. Equations of motion based on Newton's second law are solved to determine these forces and torques. The computed resultant joint torques are related to the muscles crossing those joints and indicate which muscles are active. This type of mechanical analysis, called an inverse dynamic analysis, is also one that few coaches or teachers will have the equipment, time, or expertise necessary to complete.

None of these methods are practical or reasonable for coaches or teachers to use. Muscle activity cannot be directly observed; however, we can get a general idea of which muscles are active based on the principles used in an inverse dynamic analysis. This type of qualitative anatomical analysis may be practical for coaches and teachers and provides reasonable results for most activities. The following are step-by-step procedures for completing a qualitative anatomical analysis:

1. Divide the activity into temporal phases.
2. Identify the joints involved and the movements occurring at those joints.
3. Determine the type of muscular contraction (concentric, eccentric, or isometric) and identify the predominant active muscle group at each joint.
4. Identify instances when rapid joint angular accelerations (rapid speeding up or slowing down of joint motions) occur and where impacts occur.
5. Identify any extremes in joint ranges of motion.

The results of a qualitative anatomical analysis can then be used to determine appropriate strength or flexibility exercises for the muscle groups identified. The five steps of a qualitative biomechanical analysis are described and illustrated below.

Temporal Phases

The first step of a qualitative anatomical analysis is to break down the performance into specific phases or motions. For the simplest movements, such as slow movements or those that involve raising and lowering, the end of one phase and the beginning of the next may simply be designated by the change in direction of movement. For example, let's look at a bench press. The bench press involves two phases: the down phase, in which the lifter lowers the barbell to the chest, and the up phase, in which the lifter raises the barbell.

In contrast, faster, more complex movements may require breakdown into more phases than are obvious. The phases of a baseball pitch include the windup, delivery, and follow-through, but breaking the pitch into only these three phases may not provide enough detail about the joint motions to determine the muscle activity that occurs. For such high-speed movements, it is better to make a video recording of the performance. The motion should then be examined frame by frame rather than phase by phase. Comparisons with elite performers can be made by examining videotapes or sequence photos of such performers. Sequence photos of elite performances are often found in textbooks or coaching journals on the sport in question. These publications, as well as those of the National Strength and Conditioning Association, may also include results of qualitative anatomical analyses or indications of what muscle groups are important in the execution of the skill being examined.

Joint Motions

Once the performance has been broken down into specific phases or recorded on video, the next step is to identify which body segments and joints to examine. Which segments and joints are involved in the performance of the skill? Which segments move, and which joints are involved in the movements? For skills involving gross movement of the whole body, most of the major joints will be involved. For example, running involves the ankle, knee, hip, shoulder, and elbow joints on either side of the body. Other skills, especially strength-training exercises, may involve movements of only a few segments and joints. Let's go back to the example of the bench press exercise. Which segments move, and which joints are involved in this exercise? The upper arm and forearm are the moving segments, and the elbow and shoulder are the joints primarily involved in the movement.

Once the segments and joints involved in the activity have been identified, the next step is to identify the motions occurring at each joint during each phase of the activity. Is flexion or extension occurring at the joint? Abduction or adduction? Internal or external rotation? Is more than one motion occurring at the joint? Here you are identifying the motion or change in position of the segments that occur at the joint during the phase or between phases. How did the adjoining segments move relative to each other? Joint motion differs from joint position, and the two must be distinguished from each other. For example, during the down phase in the bench press, flexion (the joint motion) occurs at the elbow joint, and the elbow is in positions of flexion. During the up phase, extension (the joint motion) occurs at the elbow joint, but the elbow is in positions of flexion until the very end of the up phase. At the shoulder joint, horizontal extension (or horizontal abduction) occurs during the down phase and horizontal flexion (or horizontal adduction) occurs during the up phase. In a qualitative anatomical analysis, we are interested in joint motion, not joint position.

Muscle Contractions and Active Muscle Groups

The next step in a qualitative anatomical analysis is to determine what types of muscle contractions occur within the active muscle groups crossing each joint and to identify these active muscle groups. Recall from chapter 14 that concentric muscular contractions produce positive work that results in an increase in mechanical energy; eccentric muscular contractions produce negative work that results in a decrease in mechanical energy; and isometric muscular contractions produce zero work that results in no change in mechanical energy. If no external contact forces act on the body, the easiest way to identify the type of muscular contraction is to identify only the change in mechanical energy that occurs during the phase of motion under examination. Was something lifted (increasing potential energy) or lowered (decreasing potential energy)? Was something sped up (increasing kinetic energy) or slowed down (decreasing kinetic energy)? Was something stretched or deformed (increasing strain energy) or slowly unstretched and re-formed (decreasing strain energy)? If the total mechanical energy (the sum of potential, kinetic, and strain energies) increases, the contraction of the active muscle group is concentric. If the total mechanical energy decreases, the contraction of the active muscle group is eccentric. If the total mechanical energy does not change, the contraction of the active muscle group is isometric or the muscles are inactive.

This general method will usually give correct results, even when external contact forces act on the body. In some cases, however, the direction of the force the body exerts in reaction to the external force and the direction of displacement of the object the body exerts a force against must be examined. If the directions are the same, the force does positive work; the muscles contract concentrically. If the directions are opposite, the force does negative work; the muscles contract eccentrically. If no displacement occurs, no work is done; the muscles contract isometrically or are inactive.

Identifying the active muscle groups is simple once the type of muscular contraction has been determined. If the contraction has been identified as concentric, the active muscles are those that create torque in the same direction as the observed joint motion. For instance, if the observed joint motion is flexion and the contraction is concentric, the flexor muscles are active at that joint (the opposing extensor muscles might be active as well, but the resultant torque is a flexor torque, so the flexor muscles must produce greater torque). If the muscle contraction is eccentric, the active muscles are those that create torque opposing the observed joint motion. For example, if the observed joint motion is flexion and the contraction is eccentric, the extensor muscles are active at that joint (the opposing flexor muscles might be active as well, but the resultant torque is an extensor torque, so the extensor muscles must produce greater torque).

> **Identifying the active muscle groups is simple once the type of muscular contraction has been determined.**

Let's look at the example of the bench press again. During the down phase, what happens to the mechanical energy of the barbell? Its potential energy decreases, as do the potential energies of the arm and forearm segments of the lifter, because they are all lowered. The kinetic energies of these segments increase, however, because they change from no movement to movement downward. The downward acceleration of these segments is not as fast as the acceleration due to gravity that would occur if the decrease in potential energy exactly equaled the increase in kinetic energy—that is, if you just let the barbell drop to your chest. Potential energy decreases more than kinetic energy increases, so there is a net decrease in the total mechanical energy. This indicates that the elbow and shoulder muscles act eccentrically during this phase. Because elbow flexion occurs and the contraction is eccentric, the active elbow muscle group is the elbow extensors. At the shoulder, horizontal extension occurs and the contraction is eccentric, so the active muscle group must be the shoulder horizontal flexors.

During the up phase of the bench press, the potential and kinetic energies of the barbell and arm segments increase, indicating that the elbow and shoulder muscles contract concentrically. Because elbow extension occurs during this phase, the elbow extensors are again the active muscle group. Likewise, because shoulder horizontal flexion occurs during this phase, the shoulder horizontal flexors are again the active muscle group. The pattern of muscular activity observed in the bench press is typical for most movements and exercises that primarily involve changes in potential energy (lifting and lowering). The same muscle groups are active throughout the movement or exercise. The muscle groups contract eccentrically during the lowering phase and then concentrically during the lifting phase.

Generally, if something is raised slowly, concentric muscular contractions are responsible. If something is lowered slowly, eccentric muscular contractions are responsible. If something is held still, isometric muscular contractions are responsible.

Analysis of the forces and displacements during the bench press yields the same result. During the down phase, the lifter exerts an upward force on the barbell, but the barbell's displacement is downward. The force and displacement are opposite in direction, so the work done by the force is negative. The elbow and shoulder muscles contract eccentrically during the down phase. During the up phase, the lifter continues to exert an upward force on the barbell, but in this phase, the barbell's displacement is also upward. The force and displacement are in the same direction, so the work done by the force is positive. The elbow and shoulder muscles contract concentrically during the up phase.

How is the active muscle group determined when the muscle contraction at a joint has been identified as isometric? In this case, the active muscle group can be determined by imagining what joint motion would occur if all the muscles around the joint relaxed. If movement would occur, the active muscle group (the muscle group that is active isometrically) would be the muscle group that produces a torque opposing this movement. In the bench press, for example, suppose you pause during the up phase and hold the barbell halfway up. No movement occurs at the elbow joint, so the muscle contraction is isometric or the muscles are inactive. If the muscles were inactive, the barbell would crash to your chest and your elbow would flex. The active muscle group must be the extensor muscles that prevent elbow flexion.

Rapid Joint Angular Accelerations and Impacts

After identifying the active muscle groups, the next task in a qualitative anatomical analysis is to identify any instances when rapid joint angular accelerations and impacts occur. This task will identify the active muscle groups that must produce the largest forces as well as the positions of the limbs when these large forces must be produced. These are the muscle groups whose strength will be tested and the muscle groups that must be trained. This step is more important when analyzing fast movements, but we can still use the bench press as our example.

At the beginning of the down phase, the barbell accelerates in the downward direction and elbow flexion and shoulder horizontal extension occur. As the force exerted by the lifter is upward at this instant, it is reduced to create this downward acceleration. Less torque is created by the active muscle groups at this instant. At the end of the down phase, the downward movement of the barbell is slowed, so its acceleration is upward. Elbow flexion and shoulder horizontal extension also slow down. The force exerted by the lifter is upward at this instant, so it must be increased to cause the upward acceleration of the barbell. More torque must be created by the active muscle groups at this instant to cause this acceleration. The same is true at the start of the up phase, when the barbell is accelerated upward again. At the end of the up phase, the barbell is slowed and its acceleration is downward. The force exerted by the lifter is upward at this instant, but it is reduced to create this downward acceleration. Thus, in the bench press, the active muscle groups are most stressed at the end of the down phase and the beginning of the up phase.

Extreme Joint Ranges of Motion

The last step in a qualitative anatomical analysis is to identify any extremes in joint ranges of motion. The purpose of this step is to identify those muscles and soft tissues that may be stretched and possibly injured. Flexibility exercises may be appropriate for these muscle groups. As with the previous step, this step is more important when analyzing fast movements.

In the bench press, the elbow is at full extension at the start of the down phase and at the end of the up phase. This is not an unusual position for the elbow joint, however. At the end of the down phase and the beginning of the up phase, the elbow is fully flexed and the shoulder may be close to the limit of its horizontal extension. The muscular torques are greatest at this position as well, so specific flexibility exercises to stretch the elbow extensors and shoulder horizontal flexors may be appropriate.

Charting the Analysis

Keeping track of what has been identified in each step of a qualitative anatomical analysis is difficult without a record of what you have identified in the previous steps. This job can be made easier if you make a chart or table of the results of each step of the analysis. The chart might include columns for the phases of the movement (or the frames of video), the joints analyzed, the joint motions, the muscle contractions, the active muscle groups, the rapid joint angular accelerations or impacts, and the extreme joint ranges of motion. Table 17.1 is an example of such a chart representing our analysis of a bench press.

Example Analyses

The steps in conducting a qualitative anatomical analysis to improve training were described in the previous pages using a simple exercise, the bench press, as an example to illustrate the procedures. A qualitative anatomical analysis of a skill may be simple or complex, depending on the activity being analyzed. Activities that primarily involve changes in potential energy (such as the bench press and most weight lifting exercises) are generally easier to analyze than activities that include faster movements and involve changes in kinetic energy. The following section provides example qualitative anatomical analyses of some of these more difficult-to-analyze activities. The examples include a jumping activity (a vertical jump), a striking activity (a football punt), a locomotor activity (sprint running), and a throwing activity (a javelin throw). The analyses progress in difficulty from simplest (the bench press and vertical jump) to most complex (the javelin throw).

TABLE 17.1   Example Qualitative Anatomical Analysis of a Wide-Grip Bench Press

Joint	Phase of motion	Joint motion	Muscle contraction	Active muscle group	Rapid acceleration or impact	Extreme range of motion
Elbow	Down	Flexion	Eccentric	Extensors	At end of phase	Full flexion at end of phase
	Up	Extension	Concentric	Extensors	At beginning of phase	
Shoulder	Down	Horizontal extension	Eccentric	Horizontal flexors	At end of phase	Full horizontal extension at end of phase
	Up	Horizontal flexion	Concentric	Horizontal flexors	At beginning of phase	

Vertical Jump

A standing vertical jump is only slightly more difficult to analyze than the bench press, as it still involves primarily changes in potential energy. The vertical jump can be broken down into three phases, as shown in figure 17.1: a preparatory (or down) phase, a propulsive (or up) phase, and a flight phase. The first two phases occur while the jumper is on the ground. Because the height reached by the jumper's center of gravity is determined by what the jumper does on the ground, let's analyze only the two ground phases.

The movements that occur during the vertical jump primarily involve movements about the ankle, knee, hip, and shoulder joints. Some movement occurs around the elbow joints, but it doesn't appear to be as important as the other joints, so let's examine just these four joints. Let's assume the jumping action is symmetrical, so the left and right sides move together.

During the preparatory phase, the ankles dorsiflex, the knees flex, the hips flex, and the shoulders hyperextend. During the propulsive phase, the opposite motions occur at each joint. The ankles plantar flex, the knees extend, the hips extend, and the shoulders flex.

During the preparatory phase, the body is lowered, so its potential energy decreases. The segment immediately above the ankle joint is lowered (its

motion is downward relative to the ankle joint), so its potential energy relative to the ankle joint decreases. The contraction of the active ankle joint muscle group is eccentric. The ankle joint motion is dorsiflexion but the muscle contraction is eccentric, so the ankle plantar flexors are the active muscle group. The segment immediately above the knee joint (the thigh) is also lowered relative to the knee joint, so its potential energy relative to the knee joint decreases. The contraction of the active knee joint muscle group is eccentric. The knee extensors are the active muscle group. The segment immediately above the hip joint (the trunk) is also lowered relative to the hip joint, so its potential energy relative to the hip joint decreases. The contraction of the active hip joint muscle group is eccentric and the hip extensors are the active muscle group. The segment distal to the shoulder joint (the arm) is raised (the arm moves upward relative to the shoulder joint), so its potential energy relative to the shoulder joint increases. The action of the active shoulder joint muscle group is concentric. Hyperextension occurs at the shoulder joint, so the shoulder extensors are the active muscle group.

During the propulsive phase, the potential and kinetic energies of all the body segments increase. The contraction of the active muscles at each of the joints is concentric. The ankle plantar flexes, so the plantar flexors are active; the knee extends, so the knee extensors are active; the hip extends,

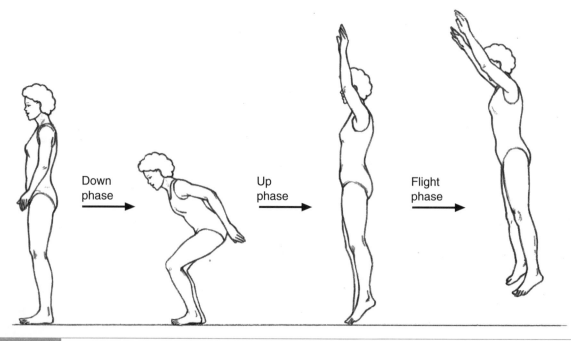

Down phase → Up phase → Flight phase

so the hip extensors are active. Closer examination of the shoulder joint reveals that the arm initially moves downward relative to the shoulder joint and then moves upward. This slight decrease in potential energy is much smaller than the segment's large increase in kinetic energy, so positive work is done and the shoulder flexors contract concentrically.

At the end of the preparatory phase and the beginning of the propulsive phase, the body is accelerated upward rapidly. The joints all experience accelerations at these instances as well. Strength is required in the ankle plantar flexors, knee and hip extensors, and shoulder flexors to vertical jump well. (Further examination of the flight and landing phases would reveal an impact at landing that would also stress all these muscles except the shoulder flexors.) Strength (and power) training exercises specific to these muscle groups may be appropriate.

The only extremes in joint ranges of motion observed occur at the shoulder joint. The hyperextension of the shoulder joint that occurs during the preparatory phase may be limited by the shoulder flexors. Flexibility exercises for the shoulder flexor muscles may be appropriate.

Table 17.2 is a completed chart of the qualitative anatomical analysis of the vertical jump.

Football Punt

The football punt is an example of an activity where the work done by the muscles primarily causes changes in kinetic energy. Thus, it is more difficult to analyze than the previous examples. Because it involves quick movements, a frame-by-frame analysis is required. Sequence drawings of an elite football punter are shown in figure 17.2.

Joint Motions

The most important segments in football punting are the legs, so we should analyze the hip, knee, and ankle joints of both legs. The activity occurs primarily in the sagittal plane, so for simplicity, let's confine the analysis to movements occurring in this plane. We'll begin with the right hip joint, which hyperextends from frames 1 to 2 as the punter steps forward with his left leg and flexes from frames 2 to 5 during the kicking action and follow-through.

The left hip actually extends slightly from frames 1 to 2, even though the punter is stepping forward onto this leg. This extension occurs primarily because the trunk was leaning forward in frame 1 and is upright in frame 2. This extension continues from frames 2 to 3. From frames 3 to 4, slight flexion

Joint	Phase of motion	Joint motion	Muscle contraction	Active muscle group	Rapid acceleration or impact	Extreme range of motion
Ankle	Down	Dorsiflexion	Eccentric	Plantar flexors	At end of phase	
	Up	Plantar flexion	Concentric	Plantar flexors	At beginning of phase	
Knee	Down	Flexion	Eccentric	Extensors	At end of phase	
	Up	Extension	Concentric	Extensors	At beginning of phase	
Hip	Down	Flexion	Eccentric	Extensors	At end of phase	
	Up	Extension	Concentric	Extensors	At beginning of phase	
Shoulder	Down	Hyperextension	Concentric	Extensors	At end of phase	Full hyperextension
	Up	Flexion	Concentric	Flexors	At beginning of phase	

TABLE 17.2   Example Qualitative Anatomical Analysis of the Preflight Phases of a Standing Vertical Jump

Figure 17.2 Sequence drawings of a football punt executed by a skilled performer.

appears to occur at the left hip joint. The position of the right leg in frame 4 prevents a clear view of the left thigh, so the action at the left hip cannot be determined with certainty. From frames 4 to 5, the hip joint appears to maintain the same position, so no joint motion occurs.

From frames 1 to 2, the right knee extends slightly as it pushes off. It then flexes from frames 2 to 3. It appears to maintain position from frames 3 to 4, so there is no joint motion. From frames 4 to 5, the right knee extends rapidly as the ball is kicked.

From frames 1 to 3, the left knee extends. It then flexes from frames 3 to 4. It appears to maintain the same flexed position in frame 5, so no joint motion occurs from frames 4 to 5.

From frames 1 to 2, the right ankle plantar flexes as it pushes off. It appears to maintain the same plantar-flexed position in frames 2 to 5, so no joint motion occurs from frames 2 to 5.

The left ankle plantar flexes from frames 1 to 3, then dorsiflexes from frames 3 to 4 and plantar flexes from frames 4 to 5.

Muscle Contractions and Active Muscle Groups

Now that we've identified the joint motions occurring during the football punt, the next step is to determine the muscle contractions and identify the active muscle groups. At the right hip, hyperextension occurs from frames 1 to 2. This appears to be the end of an extension phase of the right hip, so the muscle contraction is concentric at the beginning of this phase (as the extension speeds up and kinetic energy increases) and eccentric at the end (as the extension slows down and kinetic energy decreases). The hip extensors contract concentrically in the beginning, and the hip flexors contract eccentrically to slow the extension at the end. From frames 2 to 5, the right hip flexes. This flexion is initiated and continued by concentric contraction of the hip flexors from frames 2 to 4 (as flexion speeds up and the kinetic energy of the right leg increases). Concentric contraction of the hip flexors continues during the initial period from frames 4 to 5 (as the flexion continues to speed up), but just after foot contact with the ball, this flexion slows down and the right leg's kinetic energy decreases. Eccentric contraction of the hip extensors causes this slowing of hip flexion.

From frames 1 to 3, the left hip extends due to a concentric contraction of the hip extensors. Although difficult to determine, the extension of the hip appears to speed up from frames 1 to 3, indicating an increase in kinetic energy. From frames 3 to 4, the left hip flexes slightly. In these frames, the left foot is in contact with the ground. The ground reaction force pushing upward and slightly backward is what causes the hip to flex. Also, the kinetic and potential energies of the body decrease when the foot strikes the ground. The muscle contraction is thus eccentric and the hip extensors are the active muscle group. From frames 4 to 5, the hip does not change position, so any muscle contraction that occurs is isometric. If no muscles were active at the left hip joint during this phase, the hip would flex, so the hip extensors must contract isometrically to maintain the joint angle.

From frames 1 to 2, the right knee extends. This extension phase is short, with the extension speeding up at the beginning and slowing down at the end, so the muscle contraction is concentric initially and eccentric finally (corresponding to an increase in kinetic energy followed by a decrease in kinetic energy). Because knee extension occurs, the active muscle groups are the knee extensors initially and the knee flexors finally. From frames 2 to 3, the right knee flexes. Again, this flexion phase is short and involves an initial speeding up of flexion followed by a slowing down. Thus, the muscle contraction is concentric at the beginning of this phase and eccentric at the end. The knee flexors contract at the beginning of the phase and the knee extensors contract at the end. From frames 3 to 4, no motion occurs at the right knee, so any muscle contraction that occurs is isometric. If all the knee muscles were relaxed at this point, the knee would flex, so the extensors must be the active muscle group. From frames 4 to 5, the right knee continues its extension, speeding up until ball contact and then slowing down. The increase and then decrease in kinetic energy indicate that the muscle contractions are concentric and then eccentric. The knee extensors contract concentrically, followed by an eccentric contraction of the knee flexors.

From frames 1 to 3, the left knee extends. The muscle contraction is concentric from frames 1 to 2 because the extension speeds up and the leg's kinetic energy increases. The active muscle group is the knee extensors. From frames 2 to 3, the extension slows down and the leg's kinetic energy decreases, but the foot has contacted the ground. The reaction force from the ground slows down the extension. The muscle contraction is still concentric and the knee extensors are still the active muscle group. From frames 3 to 4, the left knee flexes due to the ground reaction force. The potential and kinetic energies of the left leg decrease, thus muscle contraction is eccentric. The knee extensors are still the active muscle group. From frames 4 to 5, no joint motion occurs at the left knee, so any muscle activity occurring is isometric. If no muscles were active at the left knee joint during this phase, the knee would flex, so the knee extensors must be contracting isometrically.

The right ankle joint plantar flexes from frames 1 to 2. The potential and kinetic energies of the segments above the joint increase, so the muscle contraction is concentric and the plantar flexors are the active muscle group. No change in position occurs in frames 2 to 5, so any muscle activity occurring at the right ankle joint is isometric. From frames 2 to 4, the ankle plantar flexors are probably active to hold the plantar-flexed position. From frames 4 to 5, the ankle dorsiflexors must be active to brace the ankle and foot for impact with the ball.

The left ankle joint plantar flexes from frames 1 to 3. To initiate this action, the muscle contraction is concentric from frames 1 to 2 and the plantar flexors are the active muscle group. As the heel strikes the ground, the ground reaction force accelerates this plantar flexion. To slow down this plantar flex-ion, the muscle contraction is eccentric from frames 2 to 3, so the dorsiflexors are the active muscle group. From frames 3 to 4, the left ankle dorsiflexes. This dorsiflexion is also caused by the ground re-action force, now acting through the forefoot. The potential and kinetic energies of the left leg decrease, which means the muscle contraction is eccentric and the plantar flexors are the active muscle group. From frames 4 to 5, the left ankle plantar flexes and the kinetic and potential energies of the body increase. The muscle contraction is concentric and the plan-tar flexors continue to be the active muscle group.

Rapid Joint Angular Accelerations and Impacts

The next step in analyzing the football punt is to identify any instances of rapid joint angular accel-eration and any impacts. The kicking leg is the focus of attention here. From frames 4 to 5, the right hip rapidly accelerates in flexion and then quickly slows during the follow-through. The right knee rapidly accelerates in extension and then quickly slows dur-ing the follow-through. During this period, the foot also impacts the ball. The right hip and knee exten-sors and flexors must be specifically strengthened to produce the torques necessary to cause these ac-celerations. In addition, the right dorsiflexors must be strong to maintain foot position during impact with the ball. Another impact occurs when the left foot strikes the ground between frames 2 and 3. The left knee and hip extensors must be strong to stabi-lize the body while the right leg swings. Strength and power training exercises are indicated for the hip extensors and flexors as well as the knee exten-sors and flexors. In addition, strength training exer-cises are indicated for the right ankle dorsiflexors.

Extreme Joint Ranges of Motion

The obvious extreme joint range of motion occurs at the right hip and right knee joints in frame 5. Here the right hip is almost fully flexed while the right knee is almost fully extended. The hamstring muscle group that is a hip extensor and knee flexor will be stretched maximally during this activity. Flexibil-ity exercises are warranted for this muscle group.

Table 17.3 is a completed chart of the qualitative anatomical analysis of the football punt illustrated in figure 17.2.

Sprint Running

Sprinting is another activity where the work done by the muscles primarily causes changes in kinetic energy. It is more difficult to analyze than the foot-ball punt because it involves more segments. Again, because it involves quick movements, a frame-by-frame analysis is required. Sequence drawings of an elite sprinter are shown in figure 17.3.

Joint Motions

Let's analyze one full sprinting stride, from takeoff of the right foot to the next takeoff of the right foot. The movements of the right and left sides are simi-lar but out of phase with each other, so we'll only analyze the joints on one side—in this case, the left side. Sprinting occurs primarily in the sagittal plane, so we'll simplify our analysis further by consider-ing only the sagittal plane movements. The joints we'll examine are the left hip, knee, ankle, and shoulder.

The left hip flexes from frames 1 to 5, then ex-tends from frames 5 to 7 and hyperextends from frames 7 to 8.

The left knee flexes from frames 1 to 3, then ex-tends from frames 3 to 6. It flexes slightly from frames 6 to 7 and extends again from frames 7 to 8.

No noticeable joint motion occurs at the left ankle from frames 1 to 2. The ankle dorsiflexes from frames 2 to 5 and then plantar flexes slightly from frames 5 to 6. The left ankle dorsiflexes at foot con-tact with the ground from frames 6 to 7 and then plantar flexes during pushoff from frames 7 to 8.

The left shoulder extends from frames 1 to 2 and then hyperextends from frames 2 to 5. It flexes from frames 5 to 8.

Muscle Contractions and Active Muscle Groups

Similar to the football punt, the muscle contractions in sprinting are primarily due to changes in kinetic energy of the sprinter's segments. From frames 1 to 4, left hip flexion speeds up as the kinetic energy of the leg increases, so the muscle contraction is concentric. The hip flexors are the active muscle group. From frames 4 to 5, the hip flexion slows down as the kinetic energy of the leg decreases, so the muscle contraction is eccentric. The hip exten-sors are the active muscle group. From frames 5 to 7, hip extension speeds up as the kinetic energy of the leg increases, so the muscle contraction is con-centric. The hip extensors are the active muscle group. From frames 7 to 8, kinetic energy increases, but at the end of this phase, hip hyperextension slows down. The muscle contraction is initially con-centric and then eccentric. The hip extensors are the active muscle group initially, followed by the hip flexors.

TABLE 17.3   Example Qualitative Anatomical Analysis of a Football Punt

Joint	Frames	Joint motion	Muscle contraction	Active muscle group	Rapid acceleration or impact	Extreme range of motion
Right hip	1-2	Hyperextension	Concentric then eccentric	Extensors then flexors		
	2-3	Flexion	Concentric	Flexors		
	3-4	Flexion	Concentric	Flexors		
	4-5	Flexion	Concentric then eccentric	Flexors then extensors	Yes	Full flexion
Left hip	1-2	Extension	Concentric	Extensors		
	2-3	Extension	Concentric	Extensors		
	3-4	Flexion	Eccentric	Extensors		
	4-5	No motion	Isometric	Extensors		
Right knee	1-2	Extension	Concentric then eccentric	Extensors then flexors		
	2-3	Flexion	Concentric then eccentric	Flexors then extensors		
	3-4	No motion	Isometric	Extensors		
	4-5	Extension	Concentric then eccentric	Extensors then flexors	Yes	Full extension
Left knee	1-2	Extension	Concentric	Extensors		
	2-3	Extension	Concentric	Extensors		
	3-4	Flexion	Eccentric	Extensors		
	4-5	No motion	Isometric	Extensors		
Right ankle	1-2	Plantar flexion	Concentric	Plantar flexors		
	2-3	No motion	Isometric	Plantar flexors		
	3-4	No motion	Isometric	Plantar flexors		
	4-5	No motion	Isometric	Dorsiflexors	Ball impact	
Left ankle	1-2	Plantar flexion	Concentric	Plantar flexors		
	2-3	Plantar flexion	Eccentric	Dorsiflexors	Ground impact	
	3-4	Dorsiflexion	Eccentric	Plantar flexors		
	4-5	Plantar flexion	Concentric	Plantar flexors		

At the left knee, flexion speeds up from frames 1 to 2 and slows down from frames 2 to 3. The muscle contraction is thus concentric from frames 1 to 2 and eccentric from frames 2 to 3. The knee flexors contract concentrically from frames 1 to 2, and the knee extensors contract eccentrically from frames 2 to 3. Knee extension speeds up from frames 3 to 5 but slows down from frames 5 to 6 (and possibly even during the end of the period from frames 4 to 5). The muscle contraction is concentric from frames 3 to 5 and eccentric from frames 5 to 6 (and maybe even eccentric during the end of the period, from frames 4 to 5). The knee extensors contract concentrically from frames 4 to 5, then the knee flexors contract eccentrically to slow down the extension from frames 5 to 6. From frames 6 to 7, the knee

Figure 17.3   Sequence drawings of an elite sprinter over one full stride.

flexes when the left foot strikes the ground due to the large ground reaction force acting on the left foot. Negative work is done at the knee joint, and the muscle contraction is eccentric. The knee extensors are the active muscle group during this phase. From frames 7 to 8, the left knee extends. Initially, knee extension speeds up, but at the end, it slows down. The knee extensors contract concentrically initially, followed by an eccentric contraction of the knee flexors at the end of the phase.

No noticeable joint motion occurs at the left ankle joint from frames 1 to 2, so any muscle activity that occurs is isometric. The previous motion at this joint was plantar flexion as the sprinter pushed off. An eccentric contraction of the dorsiflexors followed the pushoff, so the dorsiflexors are probably still active from frames 1 to 2. From frames 2 to 5, the ankle dorsiflexes. The muscle contraction is concentric from frames 2 to 4 as the dorsiflexion speeds up. The dorsiflexors are the active muscle group. From frames 4 to 5, the muscle contraction is eccentric as the dorsiflexion slows down. The plantar flexors are the active muscle group. From frames 5 to 6, plantar flexion speeds up, so the muscle contraction is concentric and the plantar flexors are still the active muscle group. From frames 6 to 7, dorsiflexion occurs when the left foot strikes the ground due to the large ground reaction force acting on the left foot. Negative work is done at the ankle joint, and the muscle contraction is eccentric. The plantar flexors continue to be the active muscle group. From frames 7 to 8, the ankle plantar flexes. Initially, the muscle contraction is concentric as the muscles do positive work and kinetic and potential energies are increased. The plantar flexors continue to be the active muscle group. At the end of this phase, the dorsiflexors contract eccentrically to slow plantar flexion.

At the left shoulder, extension or hyperextension occurs from frames 1 to 5. This extension speeds up from frames 1 to 4 and slows down from frames 4 to 5. The muscle contraction is thus concentric from frames 1 to 4 and eccentric from frames 4 to 5. The shoulder extensor muscles are the active muscle group from frames 1 to 4, then the shoulder flexor muscles contract eccentrically in frames 4 to 5. From frames 5 to 8, the left shoulder flexes. Shoulder flexion speeds up from frames 5 to 7 and slows down at the end of the period, from frames 7 to 8. The muscle contraction is thus concentric starting at frame 5 until just before frame 8, when it becomes eccentric. The shoulder flexors contract concentrically from frames 5 to 7, then the shoulder extensors contract eccentrically.

Rapid Joint Angular Accelerations and Impacts

Several instances of rapid joint angular acceleration occur during the sprinting stride. The hip joint rapidly accelerates in flexion from frames 1 to 2 due to the action of the hip flexors. This flexion slows rapidly from frames 4 to 5 due to the eccentric action of the hip extensors. From frames 7 to 8, hip hyperextension speeds up rapidly while the leg is bearing weight due to a large and powerful action of the hip extensors. The hip hyperextension then slows down prior to frame 8 due to eccentric action of the hip flexors.

Knee flexion rapidly accelerates from frames 1 to 2 due to the concentric action of the knee flexors. The eccentric action of the knee extensors rapidly slows this knee flexion from frames 2 to 3. The subsequent concentric contraction of the knee extensors causes rapid knee extension from frames 3 to 4. This extension is suddenly slowed by the eccentric action of the knee flexors at the end of the period, from frames 4 to 5 and frames 5 to 6. From frames 4 to 5, the two-joint hamstring muscle group (a knee flexor and hip extensor) must produce a large force as it acts eccentrically to slow both knee extension and hip flexion. Hamstring muscle pulls usually occur at this instant during the sprint stride. Rapid acceleration of the knee in extension also occurs at the start of frame 7 and continues through frame 8 as the leg pushes off due to the strong concentric action of the knee extensors.

The left ankle joint undergoes a rapid dorsiflexion from frames 6 to 7 as the foot impacts the ground. The strong eccentric action of the plantar flexors prevents the joint from injury. From frames 7 to 8, the ankle joint rapidly accelerates in plantar flexion while bearing weight due to the strong concentric action of the plantar flexors.

Rapid extensor acceleration of the left shoulder joint occurs from frames 2 to 3 due to concentric action of the shoulder extensors. Rapid slowing of this extension occurs in frames 4 to 5 due to eccentric action of the shoulder flexors. The shoulder flexors then act concentrically to accelerate flexion of the shoulder joint from frames 5 to 6.

Appropriate strength and power training exercises are indicated for the hip flexors and extensors, the knee flexors and extensors (especially the hamstring muscle group), the plantar flexors, and the shoulder flexors and extensors.

Extreme Joint Ranges of Motion

Few extreme joint ranges of motion are demonstrated by the sprinter in figure 17.3. The hip joint reaches a position of extreme hyperextension during pushoff in frames 1 and 8, so the hip flexors are stretched. The ankle joint reaches positions of extreme plantar flexion in frames 1 and 8, thus possibly stretching the dorsiflexors. The ankle joint also reaches a position of extreme dorsiflexion at touchdown in frame 7, so the plantar flexors may be stretched. The shoulder joint reaches a position of extreme hyperextension in frame 5, so the shoulder flexors are stretched. Flexibility exercises are indicated for each of these muscle groups.

Table 17.4 is a completed chart of the qualitative anatomical analysis of the sprinting stride illustrated in figure 17.3.

Javelin Throw

As with the football punt and sprinting stride, the javelin throw is an activity where the work done by the muscles primarily causes changes in kinetic energy. It is more difficult to analyze than the previous examples because it involves many segments and is a multiplanar activity. Once again, it involves quick movements, so a frame-by-frame analysis is required. Sequence drawings of an elite javelin thrower are shown in figure 17.4.

Joint Motions

The joints that seem to be important in javelin throwing include the elbow and shoulder of the throwing arm, the trunk (the intervertebral joints), and the right and left hip joints. Let's begin our analysis with the right elbow joint. From frames 1 through 5, the positions of the limbs at the elbow joint do not change, so no joint motion occurs. The elbow begins to flex from frames 5 to 6 and continues flexing through frame 8. From frames 8 to 10, the elbow then extends during the final delivery and release of the javelin.

The shoulder joint also maintains a fairly static position early in the throwing action, from frames 1 through 6. Some external rotation occurs from frames 6 to 8. Rapid internal rotation along with abduction occur during the delivery, release, and follow-through from frames 8 to 10. From frames 9 to 10, extension also occurs at the shoulder joint.

The trunk (or intervertebral joints) is inactive from frames 1 to 2. It rotates slightly to the right as the crossover step is made from frames 2 to 3. From frames 3 to 6, the trunk unwinds and rotates to the left. From frames 6 to 7, it again rotates to the right as the left leg is planted. From frames 6 to 7 (and perhaps even earlier), the trunk also extends and

hyperextends. From frames 7 to 10, the trunk rotates rapidly to the left and flexes during the delivery and follow-through.

The right hip flexes and adducts from frames 1 to 4. From frames 4 to 5, it flexes even more during the right foot plant, and some abduction occurs. From frames 5 to 6, hip extension begins and abduction continues. From frames 6 to 7, hyperextension occurs along with rapid internal rotation. From frames 7 to 10, the right hip flexes. Internal rotation continues from frames 7 to 8.

The left hip extends and adducts from frames 1 to 3. It then flexes, abducts, and externally rotates from frames 3 to 6. From frames 6 to 8, the left hip internally rotates rapidly and flexes as the left foot is planted. From frames 8 to 10, the left hip internally rotates farther as it extends.

Muscle Contractions and Active Muscle Groups

Now that we've identified the joint motions occurring during the javelin throw, the next step is to determine the type of muscle contractions and identify the active muscle groups. No movement occurs at the elbow joint from frames 1 to 5, so the muscle contraction is isometric or there is no muscle activity at all. The arm is acting as a rope or cable pulling on the javelin. To protect the elbow from hyperextending, the elbow flexor muscles contract isometrically. Elbow flexion speeds up from frames 5 to 7 (kinetic energy increases) and slows down from frames 7 to 8 (kinetic energy decreases). The elbow flexors contract concentrically from frames 5 to 7, and the elbow extensors contract eccentrically from frames 7 to 8. Elbow extension occurs from frames 8 to 10, speeding up from frames 8 to 9 and slowing down from frames 9 to 10. The elbow extensors continue to contract concentrically from frames 8 to 9, then the elbow flexors contract eccentrically from frames 9 to 10.

The shoulder is in abduction from frames 1 to 6 but does not change position, so the muscle contraction is isometric or there is no muscle activity at all. To hold this position, the shoulder abductors must be active isometrically. Some external rotation occurs from frames 6 to 8 due to the rapid rotation of the trunk and the inertia of the javelin and arm. The muscle contraction is eccentric. (Here our energy analysis scheme does not work very well or perhaps is difficult to understand. The body and legs seem to be speeding up, but the arm and javelin don't speed up as much, so there is a loss of energy—or perhaps a storage of elastic energy in

TABLE 17.4 Example Qualitative Anatomical Analysis of a Sprinting Stride

Joint	Frames	Joint motion	Muscle contraction	Active muscle group	Rapid acceleration or impact	Extreme range of motion
Left hip	1-2	Flexion	Concentric	Flexors	Yes	Hyperextension
	2-3	Flexion	Concentric	Flexors		
	3-4	Flexion	Concentric	Flexors		
	4-5	Flexion	Eccentric	Extensors	Yes	
	5-6	Extension	Concentric	Extensors		
	6-7	Extension	Concentric	Extensors		
	7-8	Hyperextension	Concentric then eccentric	Extensors then flexors	Yes	Hyperextension
Left knee	1-2	Flexion	Concentric	Flexors	Yes	
	2-3	Flexion	Eccentric	Extensors	Yes	
	3-4	Extension	Concentric	Extensors	Yes	
	4-5	Extension	Concentric	Extensors	Yes	
	5-6	Extension	Eccentric	Flexors	Yes	
	6-7	Flexion	Eccentric	Extensors		
	7-8	Extension	Concentric then eccentric	Extensors then flexors	Yes	
Left ankle	1-2	No motion	Isometric	Dorsiflexors		Plantar flexion
	2-3	Dorsiflexion	Concentric	Dorsiflexors		
	3-4	Dorsiflexion	Concentric	Dorsiflexors		
	4-5	Dorsiflexion	Eccentric	Plantar flexors		
	5-6	Plantar flexion	Concentric	Plantar flexors		
	6-7	Dorsiflexion	Eccentric	Plantar flexors	Ground impact	Dorsiflexion
	7-8	Plantar flexion	Concentric then eccentric	Plantar flexors then dorsiflexors	Yes	Plantar flexion
Left shoulder	1-2	Extension	Concentric	Extensors		
	2-3	Hyperextension	Concentric	Extensors	Yes	
	3-4	Hyperextension	Concentric	Extensors		
	4-5	Hyperextension	Eccentric	Flexors	Yes	Hyperextension
	5-6	Flexion	Concentric	Flexors	Yes	
	6-7	Flexion	Concentric	Flexors		
	7-8	Flexion	Concentric then eccentric	Flexors then extensors		

Figure 17.4   Sequence drawings of an elite javelin thrower.

the shoulder muscles—across the shoulder joint.) The shoulder internal rotators are the active muscle group. From frames 8 to 10, shoulder abduction and rapid shoulder internal rotation occur. From frames 9 to 10, shoulder extension also occurs. Kinetic energy increases greatly as the arm and javelin speed up in these frames, so the muscle contractions are concentric. In frame 10 and beyond (if more of the follow-through were shown), these joint motions continue, but they slow down due to eccentric muscle contractions. The shoulder abductors, internal rotators, and extensors are the active muscle groups until frame 10, when the abductors, external rotators, and flexors become active eccentrically.

The trunk is generally static or rigid from frames 1 to 2, so the normal postural muscles of the trunk contract isometrically. From frames 2 to 3, the trunk rotates to the right as the pelvis rotates to the left. This action is caused by the crossover step of the right leg and thrust of the left leg, as well as the action of the trunk muscles. The kinetic energy of the pelvis increases initially, then decreases as the pelvic rotation stops, so the muscle contraction is first concentric and then eccentric. The right trunk rotators are the active muscle group initially, followed by the left trunk rotators. From frames 3 to 6, the trunk rotates to the left (as the pelvis rotates to the right). This action is partly due to the planting of the right foot in frame 5, but the increase in energy is caused by the concentric muscle contraction that initiates the movement. The left trunk rotators are the active muscle group. From frames 6 to 7, extension and hyperextension along with rotation to the right occur rapidly as the left leg is planted. At this point, the situation is similar to the shoulder joint. The large reaction force from the left leg causes the back to hyperextend as the pelvis rotates to the left, but the rest of the trunk lags behind momentarily. There is a loss of energy within the trunk (or perhaps a storage of elastic energy), so the muscle contraction is eccentric. The trunk flexors and left trunk rotators are the active muscle groups. From frames 7 to 10, the trunk rotates rapidly to the left and flexes. This rotation and flexion speed up from frames 7 to 9, then slow down from frames 9 to 10. The muscle contraction is thus concentric from frames 7 to 9 and eccentric from frames 9 to 10. The trunk flexors and left trunk rotators continue to be the active muscle groups in frames 7 to 9. The trunk extensors and right rotators are the active muscle groups in frames 9 to 10.

The right hip adducts and flexes from frames 1 to 4. These actions speed up from frames 1 to 3 and the muscle contraction is concentric. The hip adduc-

tors and flexors are thus the active muscle groups. The adduction and flexion slow down from frames 3 to 4 and the muscle contraction is eccentric. The hip abductors are thus the active muscle group. The right hip continues to flex as it slows down from frames 4 to 5, so the muscle contraction is eccentric. The hip extensors contract eccentrically from frames 3 to 5. From frames 4 to 6, the right hip abducts. This action speeds up throughout these frames, so the muscle contraction is concentric and the hip abductors are the active muscle group. From frames 5 to 7, the right hip extends and then hyperextends. The muscle contraction is concentric from frames 5 to 6, as the extension speeds up, and eccentric from frames 6 to 7, as the hyperextension slows down. The hip extensors contract concentrically from frames 5 to 6, then the hip flexors contract eccentrically from frames 6 to 7. From frames 6 to 8, rapid internal rotation occurs at the right hip. This action speeds up initially and then slows down, so the internal rotators contract concentrically from frames 6 to 7, then the external rotators contract eccentrically from frames 7 to 8. From frames 7 to 10, the right hip slowly flexes. The flexion speeds up slowly, so the hip flexor muscles contract concentrically.

The left hip adducts and extends from frames 1 to 3. These actions increase the body's kinetic energy as the left foot pushes off the ground, so the hip adductors and extensors contract concentrically until the very end of this phase, when these actions slow down and the hip abductors and flexors contract eccentrically. From frames 3 to 6, the left hip flexes, abducts, and externally rotates. These actions speed up from frames 3 to 5, when the hip flexors, abductors, and external rotators contract concentrically. The hip flexion, external rotation, and abduction slow down from frames 5 to 6, when the hip extensors, internal rotators, and abductors contract eccentrically. From frames 6 to 10, the left hip internally rotates as the pelvis rotates to the left. This action speeds up from frames 6 to 9 and slows down from frames 9 to 10. The muscle contraction is thus concentric from frames 6 to 9, and the internal rotators are the active muscle group. The muscle contraction is eccentric from frames 9 to 10, and the external rotators are the active muscle group. From frames 6 to 8, the left hip flexes. The flexion speeds up from frames 6 to 7, when the hip flexor muscles contract concentrically, and slows down from frames 7 to 8, when the hip extensor muscles contract eccentrically. The left hip extends from frames 8 to 10. This extension speeds up throughout this period due to the concentric contraction of the hip extensors.

Rapid Joint Angular Accelerations and Impacts

The most rapid joint angular acceleration in the javelin throw occurs as the right shoulder joint internally rotates, abducts, and extends from frames 8 to 10. The shoulder abductors, internal rotators, and extensors must be strong and powerful to produce these rapidly accelerated joint motions, and the shoulder adductors, external rotators, and flexors must be strong to slow these motions.

The impact of the left foot with the ground in frame 7 places large stresses on the trunk and lower extremity. The trunk musculature, specifically the trunk flexors and left rotators, produces large torques during this phase.

Appropriate strength and power exercises are indicated for all muscles around the shoulder joint as well as the trunk flexors and rotators.

Extreme Joint Ranges of Motion

Extremes in joint ranges of motion are observed in several instances in the javelin throw. The most extreme example is the position of maximal external rotation of the right shoulder joint in frame 8. The shoulder internal rotators are stretched maximally in this position. Other extremes in joint positions include the hyperextension of the trunk and right hip in frame 7. These extremes stretch the trunk and hip flexors. Appropriate flexibility exercises are indicated for all these muscle groups.

Table 17.5 is a completed chart of the qualitative anatomical analysis of the javelin throw illustrated in figure 17.4.

Summary

Biomechanics can be used to improve performance by improving training for that performance. A basic principle of training is specificity. Biomechanics can improve the specificity of training by: identifying specific aspects of technique that need to be perfected (as discussed in the previous chapter); identifying drills and exercises that mimic specific aspects of the technique or that exercise specific muscle groups used during the performance; and identifying the specific muscles whose strength, power, or flexibility limit the performance. Most of this chapter addressed the last item. A qualitative anatomical analysis of a performance will identify the specific muscles that are active during a movement.

Five steps are involved in a qualitative anatomical analysis:

1. Divide the activity into temporal phases.
2. Identify the joints involved and their motions.
3. Determine the type of muscular contraction (concentric, eccentric, or isometric) and the predominant active muscle group at each joint.
4. Identify instances when rapid joint angular accelerations (rapid speeding up or slowing down of joint motions) occur and where impacts occur.
5. Identify any extremes in joint ranges of motion.

Steps 1, 2, 4, and 5 are self-explanatory and straightforward. Step 3, the heart of the analysis, is more difficult. Muscle contractions are determined by using a work and energy analysis, as concentric muscle contractions produce positive work and increases in energy and eccentric muscle contractions produce negative work and decreases in energy. Identifying the energy changes that occur at each joint or within the body during a movement usually indicates the type of muscle contraction involved at that joint. The active muscle groups are then identified by considering the joint motion and muscle contraction together. If the muscle contraction is concentric, the active muscles are those that produce torque in the same direction as the joint motion. Thus, if the joint motion is flexion and the muscle contraction is concentric, the active muscles are the flexors. If the muscle contraction is eccentric, the active muscles are those that produce torque in the opposite direction from the joint motion. Thus, if the joint motion is flexion and the muscle contraction is eccentric, the active muscles are the extensors. The muscles that are active during instances when large joint angular accelerations or impacts occur need to be strong, so these are identified as muscle groups to train for strength and power. Likewise, the muscles that are stretched during instances of extreme joint ranges of motion need to be flexible, so these are identified as muscle groups to train for flexibility.

Key Terms

physical training *(p. 325)*

qualitative biomechanical analysis *(p. 325)*

technical training *(p. 324)*

TABLE 17.5 Example Qualitative Anatomical Analysis of a Javelin Throw

Joint	Frames	Joint motion	Muscle contraction	Active muscle group	Rapid acceleration or impact	Extreme range of motion
Right elbow	1-2	No motion	Isometric	Flexors		
	2-3	No motion	Isometric	Flexors		
	3-4	No motion	Isometric	Flexors		
	4-5	No motion	Isometric	Flexors		
	5-6	Flexion	Concentric	Flexors		
	6-7	Flexion	Concentric	Flexors		
	7-8	Flexion	Eccentric	Extensors		
	8-9	Extension	Concentric	Extensors		
	9-10	Extension	Eccentric	Flexors		
Right shoulder	1-2	No motion	Isometric	Abductors		
	2-3	No motion	Isometric	Abductors		
	3-4	No motion	Isometric	Abductors		
	4-5	No motion	Isometric	Abductors		
	5-6	No motion	Isometric	Abductors		
	6-7	External rotation	Eccentric	Internal rotators		
	7-8	External rotation	Eccentric	Internal rotators		External rotation
	8-9	Internal rotation	Concentric	Internal rotators	Yes	
		Abduction	Concentric	Abductors	Yes	
	9-10	Internal rotation	Concentric then eccentric	Internal rotators then external rotators	Yes	
		Abduction	Concentric then eccentric	Abductors then adductors	Yes	
		Extension	Concentric then eccentric	Extensors then flexors	Yes	

(continued)

TABLE 17.5 *(continued)*

Joint	Frames	Joint motion	Muscle contraction	Active muscle group	Rapid acceleration or impact	Extreme range of motion
Trunk (Intervertebral joints)	1-2	No motion	Isometric	Postural muscles		
	2-3	Rotation right	Concentric then eccentric	Right rotators then left rotators		
	3-4	Rotation left	Concentric	Left rotators		
	4-5	Rotation left	Concentric	Left rotators		
	5-6	Rotation left	Concentric	Left rotators		
	6-7	Rotation right	Eccentric	Left rotators	Ground impact	Hyperextension
		Extension	Eccentric	Flexors	Ground impact	
		Hyperextension	Eccentric	Flexors		
	7-8	Rotation left	Concentric	Left rotators		
		Flexion	Concentric	Flexors		
	8-9	Rotation left	Concentric	Left rotators		
		Flexion	Concentric	Flexors		
	9-10	Rotation left	Eccentric	Right rotators		
		Flexion	Eccentric	Extensors		
Right hip	1-2	Flexion	Concentric	Flexors		
		Adduction	Concentric	Adductors		
	2-3	Flexion	Concentric	Flexors		
		Adduction	Concentric	Adductors		
	3-4	Flexion	Eccentric	Extensors		
		Adduction	Eccentric	Abductors		
	4-5	Flexion	Eccentric	Extensors		
		Abduction	Concentric	Abductors		
	5-6	Extension	Concentric	Extensors		
		Abduction	Concentric	Abductors		
	6-7	Hyperextension	Eccentric	Flexors		Hyperextension
		Internal rotation	Concentric	Internal rotators		

Left hip

	Movement	Contraction	Muscle group	
7-8	Flexion	Concentric	Flexors	
	Internal rotation	Eccentric	External rotators	
8-9	Flexion	Concentric	Flexors	
9-10	Flexion	Concentric	Flexors	
1-2	Extension	Concentric	Extensors	
	Adduction	Concentric	Adductors	
2-3	Extension	Concentric then eccentric	Extensors then flexors	
	Adduction	Concentric then eccentric	Adductors then abductors	
3-4	Flexion	Concentric	Flexors	
	Abduction	Concentric	Abductors	
	External rotation	Concentric	External rotators	
4-5	Flexion	Concentric	Flexors	
	Abduction	Concentric	Abductors	
	External rotation	Concentric	External rotators	
5-6	Flexion	Eccentric	Extensors	
	Abduction	Eccentric	Abductors	
	External rotation	Eccentric	Internal rotators	
6-7	Flexion	Concentric	Flexors	Ground impact
	Internal rotation	Concentric	Internal rotators	Ground impact
7-8	Flexion	Eccentric	Extensors	
	Internal rotation	Concentric	Internal rotators	
8-9	Extension	Concentric	Extensors	
	Internal rotation	Concentric	Internal rotators	
9-10	Extension	Concentric	Extensors	
	Internal rotation	Eccentric	External rotators	

REVIEW QUESTIONS

1. What other qualitative or quantitative methods (besides the qualitative anatomical analysis described in the previous pages) exist for identifying the specific muscle groups that are active during an exercise or movement?

2. What elbow and shoulder joint muscle groups are exercised during a push-up?

3. What muscle groups are exercised during a squat?

4. What hip, knee, and ankle joint muscle groups are exercised during the landing phase following a vertical jump?

5. Complete a qualitative anatomical analysis of an underhand grip pull-up.

6. Complete a qualitative anatomical analysis of a jumping jack exercise.

7. Complete a qualitative anatomical analysis of a standing long jump.

Qualitative Biomechanical Analysis to Understand Injury Development

Steven T. McCaw, PhD, FACSM
Illinois State University

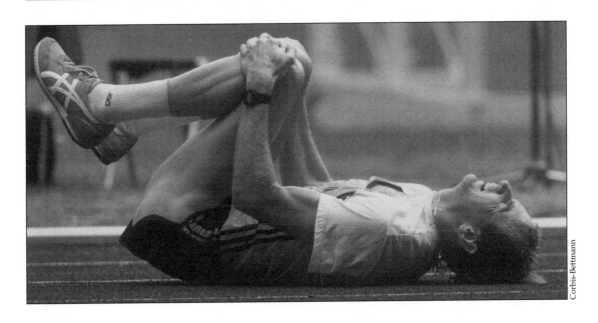

Promising careers in sport, and good intentions in fitness programs, are sometimes cut short by injury. Competitive runners can suffer a variety of ailments in the lower extremity and low back that interrupt training. A baseball pitcher can suffer continuous shoulder or elbow pain that restricts the ability to throw. For many seeking to improve their fitness level, an injury is often cause enough to abandon a cardiovascular training program. Obviously, preventing injury can benefit both competitive and recreational athletes, but injury prevention requires an understanding of why injury occurs. This chapter discusses the biomechanical basis of injury and develops a structured approach to identifying factors related to injury prevention.

OBJECTIVES

When you finish this chapter, you should be able to do the following:

▌ Differentiate between force and pressure or stress

▌ Explain how the stress continuum relates to tissue adaptation and injury

▌ Describe the concept of the stress threshold

▌ Differentiate between intrinsic and extrinsic factors related to injury development

▌ Identify intrinsic factors that predispose an individual to injury

▌ Identify extrinsic factors associated with high stress during skill performance

▌ Explain the concepts of "cross-training" and "within-activity cross-training"

▌ Suggest interventions to decrease the risk of injury when performing a task

Mechanical Stress and Injury

The combination of all forces acting on a body is responsible for the observed motion of the body. Improving performance depends on improving the application of force to the body. That is, by altering the size, direction, line of application, and timing of force production, one can improve performance.

Forces acting on a body are also responsible for the occurrence of injury. An injury is damage to a tissue that inhibits performance. For example, a baseball striking a batter in the face can fracture the zygomatic bone in the cheek; excessive inversion of the ankle can sprain the ligaments on the lateral surface of the ankle joint; muscle strains occur when high levels of tension developed by a muscle disrupt the continuity of the connective tissue within the muscle.

Although the simplistic interpretation is that "high levels of force" cause injury, a simple example demonstrates that more than the size of the force itself is responsible. Consider this offer: for $100, would you lie down on a wooden floor and let my son stand barefoot on your abdomen? Many of you would agree to that (even without knowing how heavy he is). The forces applied to your body would be his body weight pushing down on you, gravity pulling down on you, and the floor pushing back up on you with a magnitude equal to his weight plus yours. The contact between the floor and your back spreads the force over a wide area. The soles of his feet spread his weight over your abdomen. Although you would be uncomfortable, for $100 most of you would put up with the mild discomfort.

But how many of you would lie down on a wooden floor and let him stand on your abdomen wearing golf spikes? Except for the greediest individuals, few would agree to this offer. Why not? The forces are still the same, ignoring the slight increase in his weight because of the shoes. The contact area between your back and the floor is similar. The only change is that his weight is now concentrated over a smaller surface—the narrow tips of the golf spikes. This scenario promises to be so painful that enduring such discomfort would not be worth a paltry $100. It also clearly demonstrates that more than force magnitude and direction are responsible for injury.

The concept of mechanical stress provides the starting point for understanding all injury and clarifies the scenario described above. As stated in chapter 12, mechanical stress (pressure) refers to the distribution of a force over the body it acts on. Graphically, figure 18.1 compares the difference in stress on your abdomen if my son were to stand on

a

b

Figure 18.1   The difference in pressure on your abdomen when my son stands barefoot or in golf spikes. His weight is 356 N (approximately 80 lb), and he stands with his weight spread equally on each of his two feet. (*a*) The surface area of the sole of each foot is about 97 cm² (15 in²). The pressure under the contacting area of each foot is calculated to be P = F/A = 356 N/97 cm² = 3.67 N/cm². (*b*) Each of his golf shoes has 12 spikes, and each spike has a surface area of about 0.03 cm² (0.004 in²). The pressure under the contacting area of his shoes is calculated to be 356 N/(12 × 0.03 cm²). The pressure under the sole of a golf shoe is about 269 times the pressure under a bare foot.

you with bare feet and with golf shoes. His body weight (the downward force) remains the same, but the area changes with the different conditions. Mathematically, stress is defined as force/area (read "force over area"). With force measured in newtons and area measured in meters, the unit of stress is N/m², also known as a pascal. You may be more familiar with the English system unit, which is pounds per square inch (PSI).

Mechanical stress is imposed on the body in different ways. Compressive stress occurs when opposing forces squeeze a body together. For example,

when you land from a jump, a compressive force is applied to your leg by the body weight above it acting downward and the ground acting upward. Each vertebra is continually exposed to compressive stress from the weight of the body positioned above it pushing down and the vertebra below it pushing up. Another form of compressive loading comes from muscle tension applied at a bone. The pull of a muscle can be resolved into two components (see figure 18.2). The rotary component acts perpendicular to the bone and tends to cause rotation of the segment. The parallel component acts along the

Figure 18.2 The force produced by a muscle can be resolved into two components. The rotary component affects the rotation of the segment, and the compressive component affects joint stability by altering the compressive force at the joint.

bone toward the joint crossed by the muscle. This force acts to pull the segment tight against the adjacent bone with which it articulates. Simultaneous with the force acting toward the joint along the adjacent segment, the two forces compress the joint and stabilize it by squeezing the segments together. Compressive force is associated with a variety of injuries. Chondromalacia patella, characterized by deterioration of the cartilage under the kneecap, is caused by the high compressive component of the quadriceps force during knee motion. Compressive force applied down the vertebral column during a head-first collision may collapse the vertebrae, causing severe spinal cord injury.

Tensile stress occurs when a force pulls on a body. Sources of tensile stress include loading of the medial collateral ligament of the knee joint when the leg is struck from the lateral side or the loading imposed on a muscle attachment site such as the tibial tuberosity when tension is present in the patellar ligament. Avulsion fractures, which occur when high tensile stress in a tendon or ligament breaks off bone at the site of attachment, are more likely when muscle activity is high or when a joint is pulled to an extreme end of its range of motion. Common sites of avulsion fracture include the lateral malleolus when the ankle joint is extremely inverted or the medial epicondyle of the humerus during the pitching motion. Incidentally, when calculating tensile stress, area refers to the cross-sectional area of the tissue. For example, the Achilles

tendon in a typical adult is somewhat circular, with a diameter of about 0.7 cm (0.28 in.), or a radius of about 0.35 cm (0.14 in.). Using the equation for the area of a circle, $A = \pi r^2$, the cross-sectional area of the tendon is $A = \pi(0.35 \text{ cm})^2 = 0.385 \text{ cm}^2$. If the force developed by the triceps surae muscle group is 2800 N, the calculated tensile stress in the Achilles tendon is $\sigma = F/A = 2800 \text{ N}/0.385 \text{ cm}^2 = 7273 \text{ N}/\text{cm}^2$, often reported as 72.7 MPa (megapascal, or million pascal).

Shear stress occurs when a force attempts to slide bodies across each other. Examples of shear stress in the human body include the loading of knee joint tissues during foot support in running. The momentum of the body mass above the knee joint tends to cause the femur to slide over the tibial plateau. The hyaline cartilage of the knee joint and the menisci are subjected to shear stress, and the medial collateral ligament and the anterior and posterior cruciate ligaments are under tensile stress. Commonly, when a knee "blows out," the anterior cruciate ligament, the medial collateral ligament, and the menisci are all injured.

In most situations, the human body is subjected to complex loading—a combination of the compressive, tensile, and shear stresses imposed at the same time. For example, consider the forces acting on the calcaneus (heel) when your foot is planted on the ground as you stand in the anatomical position. A compressive force is applied to the calcaneus by the vertical ground reaction force acting upward and your body weight acting downward. Tensile stress is imposed at the insertion point of the Achilles tendon and the plantar fascia, as the soleus is active to maintain upright balance and the plantar fascia supports the longitudinal arch of the foot. Shear stress is imposed where the calcaneus articulates with the navicular bone (a tarsal). Torsion is present in the calcaneus because of its alignment along the longitudinal axis of the foot. Thus, loading of the calcaneus is a combined load, reflecting the combined action of the multiple forces applied. Think of some other activities and identify the nature of the imposed stresses on various tissues.

Tissue Response to Stress

Most tissues in the human body are remarkable in their ability to adapt to imposed stress. In 1892, German anatomist Julius Wolff summarized the nature of the response to mechanical stress. This summary, known as Wolff's law, states that a tissue adapts to the level of stress imposed on it; that is, the level of adaptation in a tissue reflects the level of typical loading. As mechanical stress increases, a tissue gains strength through hypertrophy, or an increase in size. If mechanical stress is removed, a tissue loses strength through atrophy, or a decrease in size. Although Wolff's observations were specific to bone, subsequent study of the body revealed the law to be applicable to other connective tissue such as ligament and tendon.

The study of anatomy offers clear examples of Wolff's law. We learn that the protuberances and landmarks on bones are adaptations to the compressive, tensile, torsional, and shear forces imposed by the tensile forces of tendons and ligaments and to the compressive forces of gravity and muscle. Structurally, bone adapts by increasing mineralization and aligning trabeculae in the direction of the imposed stress. Muscle tissue adapts to training, or the imposition of an overload stress, by increasing cross-sectional area as individual fibers increase in diameter. The arrangement of collagen and elastin in tendon and ligament similarly reflects the exposure to tensile loading and is affected by the level of training. A notable exception to Wolff's law is neural tissue such as sensory receptors and axons, which do not benefit from applied stress. Although learning is characterized by the adaptation and proliferation of synapses, the nerve tissue itself does not respond to the level of stimulus by hypertrophying or atrophying. In some disease or overuse conditions, a nerve or nerves may atrophy, but

this is technically a loss of myelin rather than a change in the diameter of the nerve itself.

The level of stress on a tissue varies with changes in activity level. The magnitude of the imposed stress ranges from a very low level, or no stress, to a very high level. For example, the compressive stress on the femur is less when lying in bed than during aerobic dance. Shear stress on knee cartilage is less when standing upright than during downhill skiing. Tensile stress on the skin increases when an object pushing on the skin stretches it. Controlling the level of imposed stress is important in training various tissues and avoiding injury.

Graphically, the level of imposed stress can be viewed as a stress continuum (see figure 18.3). The continuum of stress ranges from a low level (pathologic underload zone) to a high level (pathologic overload zone). With an active lifestyle, the level of stress is usually kept within the physiologic loading zone, and the tissue maintains its current status. This can be thought of as a maintenance program, where the tissue gets neither stronger nor weaker. When stress is maintained in the physiologic loading zone, muscle maintains the same force-generating capability, bone mineral content stays the same, and tendons and ligaments maintain their ability to withstand tensile stress.

> **Controlling the level of imposed stress is important in training various tissues and avoiding injury.**

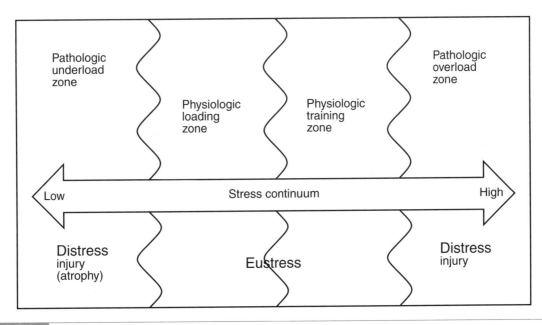

Figure 18.3   The stress continuum.

In the physiologic training zone, a level of stress is imposed above that to which the tissue has adapted. A stress within this zone exceeds the yield strength of the tissue and causes microdamage (microscopic tissue damage) within the tissue. The larger the imposed stress, the greater the extent of this microdamage. The body's response to the microdamage is to initiate remodeling, or rebuilding, of the tissue. The time required for remodeling is related to the extent of damage; the greater the damage, the longer it takes. According to Wolff's law, the healing of the microdamage leads to hypertrophy, or strengthening, of the tissue. Systematic application of loads in the physiologic training zone is the basis of the overload principle, which is responsible for causing the cellular and structural changes within a tissue called the training effect. Systematic loading means that the imposed tissue loading purposely exceeds the tissue's yield threshold to cause microdamage, but adequate rest time is provided to allow for tissue rebuilding prior to the next overload session (figure 18.4). Typically, the mineral content of bone will increase and muscle fibers will hypertrophy. The magnitude of the training response is, of course, dependent on other factors, including genetics, diet, rest, and hormonal status.

At opposite ends of the continuum are areas marked distress (*dis* = bad). These areas reflect stress levels that are too low, below the physiologic loading zone, and too high, above the physiologic training zone. Loading of a tissue at these levels leads to undesirable changes that compromise tissue function.

The pathologic underload zone represents continued low loading applied to a tissue, such as occurs during extended periods of inactivity. An inactive, or sedentary, lifestyle may be selected by choice, or it may be imposed by prolonged bed rest or immobilization with a cast. A low level of mechanical stress is also an unfortunate characteristic of space flight, when the lack of gravity reduces the compressive loading of the skeleton and the amount of muscle tension needed to move body segments and external objects. If the level of imposed stress is not raised out of the underload zone, in accordance with Wolff's law, a detraining effect occurs. An understressed tissue begins to atrophy, or waste away. Muscle decreases in cross-sectional area, bone mineral content decreases, and ligaments and tendons lose their flexibility. You may have observed such tissue changes when a cast is removed and the immobilized limb has noticeably atrophied compared to the uninjured limb. A

Model for Overuse Injuries

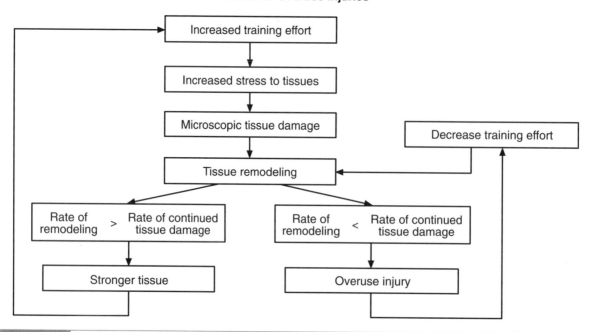

Figure 18.4   Williams' model for the effect of repetitive stress imposed on the body. The imposed stress causes microscopic tissue damage and initiates tissue remodeling. If the rate of remodeling is greater than the rate of tissue damage, a training effect occurs and the tissue gets stronger. If the rate of remodeling is slower than the rate of tissue damage, an overuse injury develops (from Williams, 1993).

chronically underloaded tissue becomes weaker and more prone to injury as the yield threshold of the tissue is reduced.

The pathologic overload zone represents a level of loading that causes substantial damage to the tissue. When a single application of a relatively high level of stress is identified as the injury-causing factor, the injury is referred to as an accidental or traumatic injury. This type of injury occurs during high-impact collisions between two or more objects, such as other players, the ground, or an obstacle. For example, a medially directed blow to the knee can strain the medial collateral ligament, or a fall onto an outstretched arm can fracture the distal radius. An extended period of recuperation and rehabilitation is required to return the tissue to a state where activity can be resumed.

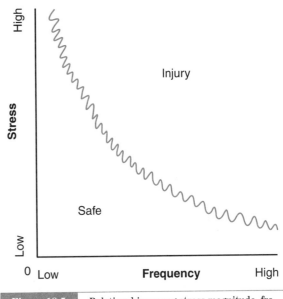

Figure 18.5 Relationship among stress magnitude, frequency of stress application, and injury.

Mechanism of Overuse Injury

As is evident from clinical records, not all injury is caused by a single, identifiable stress. An overuse injury can occur following repeated applications of a stress lower than that required to cause injury in a single application. For example, anterior shin pain may develop due to the repeated footstrikes during a long-distance run. Workers in an office or factory may develop chronic shoulder and neck pain due to prolonged work in an awkward posture involving arm abduction or neck flexion. No single incidence of a running step or incident during the work process can be identified as the cause of the pain. Instead, the injury and associated discomfort reflects a combination of the number of repetitions and the magnitude of the stress.

Figure 18.5 graphically presents the theoretical relationship among stress magnitude, number of repetitions, and development of an injury. The vertical axis is the magnitude, or size, of the stress, and the horizontal axis is the number of repetitions of the stress, or the frequency of application. The curved line represents the injury threshold and reflects the observed interaction of stress magnitude and frequency to the occurrence of injury. The higher the magnitude of stress imposed, the fewer repetitions needed for an injury to develop.

What determines whether imposed stress will cause a training effect or lead to injury appears to be the amount of time provided for remodeling to occur. Williams's model of the effect of imposed stress (figure 18.4) shows that with adequate rest, a training effect follows the tissue damage caused when a load falls within the physiologic training zone. However, if the stress is reimposed without providing adequate time for the tissue to repair, an overuse injury eventually develops. Without adequate recovery time, stress is imposed on tissue that is already damaged, and the repetitive loading increases the extent of damage. Eventually, the pain and discomfort from the repetitively stressed tissue inhibit performance, and an overuse injury is diagnosed.

Because an overuse injury is caused by the frequency and magnitude of loading combined with inadequate rest time for tissue remodeling, eliminating overuse injury has proven to be quite difficult. Symptoms of an overuse injury are often the first indication that an individual is at risk. Frequently, someone with an overuse injury will try to "work through" the pain, hoping that it is temporary and will disappear on its own. This approach, however, violates the need to treat the injury by providing adequate time away from the stress for the tissue damage to heal.

Once the individual is healed and ready to resume the activity, the probable cause of the overuse injury must be eliminated to prevent recurrence. That is, the frequency of application or the magnitude of the stress must be reduced, or the resistance to injury must be increased through a systematic training program. Without such an intervention, the injury is likely to recur.

Individual Differences in Tissue Threshold

Based on the conceptually simple model of injury outlined above, preventing injury would seem to be relatively easy: reduce the magnitude and frequency of loading and provide adequate rest for remodeling (the training effect) to occur. However, injury prevention is complicated by the difficulty in determining the injury threshold and the amount of time to wait before stress is reimposed. Different tissues in the body have different thresholds. For example, the threshold of bone is higher than tendon, which is higher than ligament, which is higher than cartilage. In addition, the injury threshold varies within each tissue according to the direction of stress. For example, bone strength is highest in compression and lowest in shear, with tensile strength falling in between. Establishing a stress threshold that is applicable to all individuals is complicated because the threshold of a tissue such as bone is not the same for everyone. Instead, the injury threshold reflects differences in genetics, training, and tissue adaptation. A level of stress falling within the physiologic loading zone of one individual may fall within the pathologic overload zone of another. Finally, the rate of recovery, or tissue remodeling, also differs among individuals. A recovery period adequate for one subject might not be adequate for another. Individual differences in tissue threshold and recovery rate complicate establishing training guidelines that will be successful in preventing all injury.

For example, consider the guidelines established to reduce the number of elbow injuries among pitchers. In Little League baseball, a rule limits the number of innings a player can pitch in a single game as well as the number of games per week. Obviously, the first part of the rule is intended to limit the number of times the stress of pitching is imposed, and the second part is intended to ensure that a rest period is provided to recuperate from the imposed stress. Although no such formal rule exists for pitchers in professional baseball, a manager typically limits a starting pitcher to between 100 and 120 pitches per appearance and tries to provide four days of rest between starts. Unfortunately, elbow injuries still occur at both the Little League and professional levels. A reduction in the number of elbow injuries in Little League baseball suggests that, for the most part, the guidelines are adequate for reducing the rate of injury. The observed injuries may develop because some players throw many more pitches than expected during the allowed innings, some throw more frequently each week because parents or coaches knowingly violate the rule, and some may throw hard outside of formal games or practices. However, players who do not violate the rules develop injuries too. Obviously, the guidelines are not adequate for these players, but it is not possible to identify those who will be injured even if the guidelines are followed. Individual differences in tissue threshold and level of imposed stress make it impossible to develop foolproof guidelines for injury prevention.

Intrinsic and Extrinsic Factors Affecting Injury

The observed differences in the types of injury recorded suggest that factors related to both the individual and the task are associated with injury development. Characteristics that describe the individual are known as intrinsic factors. Characteristics of the task and the environment in which it is performed are known as extrinsic factors. Generally, intrinsic factors reflect an individual's ability to withstand loading, whereas extrinsic factors reflect the nature of the loading that will be imposed on the individual. Examples of intrinsic and extrinsic factors are listed in table 18.1.

Intrinsic factors related to injury include anthropometrics, skeletal alignment, fitness, and previous history of injury. These factors are related to an individual's ability to cope with imposed mechanical stress—that is, to how an imposed force creates stress within the individual and how well the tissues are adapted to the level of stress. These factors include skeletal structure, such as bone density and joint congruity (alignment); current fitness level, such as muscle strength, endurance, and flexibility; and previous history of injury. If you consider the variety of physical shapes evident among individuals in your community, it will be readily apparent that individual differences in anthropometrics play a potentially large role in protection from or predisposition to injury.

Extrinsic factors related to injury include characteristics of both the task and the environment. Task-related factors include the nature of the task being performed; how a given individual performs the task; the movement patterns involved; and the frequency, speed, and duration of performance. Environmental factors include the type of surface played on, the rules used, the skill level and number

TABLE 18.1   Intrinsic and Extrinsic Factors Related to Injury Development

Intrinsic factors	Influence
Skeletal alignment	Affects the pattern of stress imposed on tissues
Muscle strength	Affects magnitude of loading and shock absorption
Muscle endurance	Affects magnitude of loading and shock absorption
Current level of fatigue	
Joint flexibility	Affects loading pattern of segments
Tissue temperature	
Joint alignment	Affects area of force distribution
Bone mineral density	Affects strength of bone to withstand stress
Diet	
Hormone levels	
Previous injury history (injury status)	Affects tissue threshold
Muscle firing pattern	Affects magnitude of load and pattern of imposed load
Body mass (body weight)	Affects magnitude of imposed load
Body composition	
Psychological factors	Affect pain threshold
Motivation	
Pain tolerance	

Extrinsic factors	Influence
Task	
Nature of the task	
Single (discrete) vs. repetitive (continuous)	Affects magnitude of loading and recovery
Movement pattern	Affects tissues loaded
Intensity of performance	Affects magnitude of loading
Frequency of performance	Affects recovery time
Environment	
Playing surface	
Slope	Affects magnitude and direction of loading
Hardness	Affects magnitude of loading and friction
Material condition	Affects magnitude of loading and friction
Equipment	
Footwear	
Outsole materials	Affect magnitude of friction
Midsole materials	Affect amount of cushioning
Padding	
Level of participation	Affects magnitude of forces imposed
Recreational vs. competitive	
Skill level of opponents	Affect the magnitude and frequency of loading, and the pattern of force distribution
Rules	

of teammates and opponents, the type and condition of protective equipment, the type and condition of implements, and the current weather conditions. These factors primarily affect the magnitude and frequency of the applied stress.

For example, consider how different people walk, run, swim, perform a dance leap, swing a hammer, or throw a ball. Despite there being similar general characteristics for each movement, each person performs a task with unique individual characteristics. For a pitcher, variable characteristics of the task include type of pitch thrown, finger placement on the ball, length of stride, degree of trunk rotation, and the positions and ranges of motion of the joints of the throwing arm during preparation, execution, and follow-through. For an assembly line employee, extrinsic characteristics include joint posture assumed while working, rate of assembly, number of hours at the task, percentage of maximum strength required to complete the task, and use of allotted break time. Obviously, differences among individuals in performance characteristics can increase or decrease the risk of injury when performing a similar task. The risk of injury is increased for assembly workers whose wrists must be in an awkward position when performing their job if similar awkward positions are maintained during scheduled break periods or time away from work. Altering the posture is a critical component of preventing repetitive overuse injuries in the workplace.

When a given task is performed, the various intrinsic and extrinsic factors interact to set the level of risk for a particular individual during a specific performance. The nature of this interaction is indicated in figure 18.6. The intrinsic factors taken together set the threshold value for the stress that may cause an injury. The extrinsic factors taken together reflect the potential for the given performance to impose a stress that exceeds the threshold value determined by the intrinsic factors. For example, the strength, flexibility, and anatomical alignment of the vertebral column each represent an intrinsic factor related to the risk of suffering a back injury. Strength of the abdominal and back muscles affects the support provided to the column as well as the size of the force that will be imposed due to the compressive and tensile forces muscle activity develops. Flexibility of the column reflects the condition of both the annulus fibrosus (fibrous outer covering) and the nucleus pulposus (inner material) of the intervertebral disks and is related to the injury threshold. A person with a weak spot in the fibers of the annulus fibrosus, caused by either a congenital defect or previous loading of the disk, is at in-

creased risk of disk herniation when the back is loaded. Vertebral column alignment (whether scoliotic, kyphotic, or lordotic) affects the pattern of stress distribution over a disk, because the angle between adjacent vertebrae determines the tensile, compressive, and shear loading of the disk. Extrinsic factors for a given lift include the weight and dimensions of the load, the shape of any available grips, and the trunk angle adopted while performing the lift. Greater load weight imposes greater stress on the vertebral column. Dimensions of the load determine whether the load can be kept close to the body, affecting the amount of muscle tension required to perform the lift. Grips affect the potential for the load to slip, which can affect the amount of muscle activity required during the lift. Keeping the back "straight," or upright, with the vertebral column in the aligned anatomical position, creates less stress on an intervertebral disk than does flexing forward to a more horizontal position of the trunk. Together, these extrinsic factors influence the magnitude of the stress imposed on the intervertebral disks of the back and determine whether their stress threshold will be exceeded.

Risk of injury to an individual during a task can be considered an interaction between the intrinsic factors characterizing that individual and the extrinsic factors characterizing the specific task and the environment in which it is performed. Multiple extrinsic and intrinsic factors have been identified

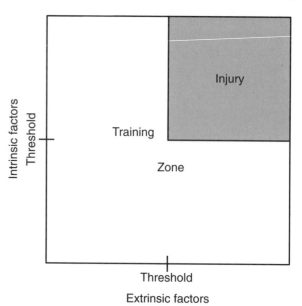

| Figure 18.6 | The interaction between intrinsic factors and extrinsic factors and the development of injury. Figure adapted, with permission, from Adis International. |

for various tasks. Consider the risk of running on a road. Most paved roadways have a raised crown in the middle, tapering off to the sides to improve drainage. The slope of the road, an extrinsic factor, can interact with the leg length of the runner, an intrinsic factor, to create, reduce, or increase the leg length inequality (see figure 18.7). Leg length inequality is an intrinsic factor that increases the risk of hip, knee, or low back injury because it shifts the musculoskeletal alignment of the lower back, hip, knee, and ankle. This shift alters the pattern of stress distribution within these joints. A significant leg length inequality (greater than 0.5 cm or 0.25 in.) poses a risk of premature hip or knee and low back degeneration due to the loads imposed

during walking or running, loads that represent less risk if there is no leg length inequality. However, the slope of the running surface interacts with the length of the runner's legs to cause, reduce, or increase the difference between the lengths of the runner's right and left legs. When feasible, such as when running on paved bike or walking trails, a runner should switch sides of the path to reduce the "environmental" effect of road slope on leg length inequality. If the extrinsic risk factors are minimized by running on a surface that reduces the leg length difference, or at a pace, in shoes, or on a surface that reduces the load across the malaligned joint, the risk of injury may not be increased.

a

b

c

Figure 18.7   The slope of a paved roadway, an extrinsic factor, interacts with the leg length of a runner, an intrinsic factor, to create, reduce, or increase the characteristic of leg length inequality. (*a*) The slope of the road creates a leg length inequality in a runner with equal leg lengths; (*b*) the slope of the road reduces the leg length inequality in a runner with a leg length inequality; (*c*) the slope of the road increases the size of the difference in leg lengths in an individual with a leg length inequality.

Conceptually, considering the development of an injury as an interaction between intrinsic and extrinsic factors provides a basis for evaluating the appropriateness of an activity for a particular individual. It also provides a basis for identifying the steps required to reduce the risk of injury. An activity presents a risk if it exposes a tissue to a high level of stress or to repetitive applications of stress, or if it is performed by an individual with a low stress threshold. An intervention to prevent injury must be based on reducing the magnitude of the stress imposed on a tissue, reducing the frequency of the stress, or increasing the tissue's ability to withstand stress. If a particular intervention does not affect one of these three factors, it will not provide protection from injury. Let's look in more detail at how intrinsic and extrinsic factors relate to the development of injury in running.

Running

Running is a popular pastime for millions of fitness enthusiasts. In addition, running competitively is a sport of choice for many, and running is an important part of other sports. Although considerable cardiovascular training benefit is gained from a running program, the development of an overuse injury is relatively common. Clinical data indicate that the knee is the most common site of running-related injury, followed by the lower leg and the foot.

The development and pattern of overuse injury in running is interesting from a biomechanics perspective. Clinical data reveal that some runners are frequently injured while others are rarely injured. A unilateral pattern of injury development is prevalent in that injury typically develops in only one side of the body despite the cyclical nature of running. These observations raise the question as to whether some individuals are predisposed to or at greater risk of injury. Finally, the type of running program followed and the rate of injury, and the effect of shoes and surfaces on the development of injury, suggest that the nature of loading is important in developing an injury. Patterns of running injury suggest that the risk of injury is very much related to the interaction among the individual, the task, and the environment, or intrinsic and extrinsic factors.

To understand how an overuse injury develops in running, the intrinsic and extrinsic factors that interact to cause injury must be examined. As stated earlier, intrinsic factors are those related to the characteristics of the individual runner, whereas extrinsic factors are related to the movement of running and the nature of the environment in which the runner performs. We'll begin with the general characteristics of the loading imposed when running, taking into account the task and the environment. This will be followed by a discussion of intrinsic factors related to running injury.

General Patterns of Loading When Running

Running is a cyclical activity consisting of alternating periods of single-leg support, stance, and no support—the airborne or flight phase. Most runners exhibit a step rate of between 50 and 70 steps per minute. In other words, each foot contacts the ground 300 to 900 times per mile. Mechanically, each stance phase begins with contact between the downward falling runner and the ground. During stance, the downward motion of the runner must be stopped and then reversed to propel the runner upward and forward into the next airborne phase. With each stance phase, the lower extremity must initially absorb energy as the downward vertical motion of the runner is stopped, briefly support the body during midstance, and finally, generate energy as the runner is pushed back into the air during the propulsion phase. All these actions occur within the 200 to 300 ms that the foot is in contact with the ground.

Most runners exhibit a rearfoot landing with the foot slightly dorsiflexed and inverted at touchdown. Initial contact with the ground occurs along the lateral border of the heel of the shoe. Some runners use a midfoot or forefoot landing, with the foot slightly plantar flexed and inverted at touchdown. Initial ground contact still occurs along the lateral border of the shoe, but closer to the front of the foot.

The pattern of stress distribution on the sole of the foot during the stance phase of running is shown in figure 18.8 for both rearfoot and forefoot strikers. As expected, two main areas of loading are evident in the pattern of rearfoot strikers: both the rearfoot and forefoot portions of the shoe. The rearfoot is loaded during the initial part of stance and the forefoot is loaded during the latter part when the foot is plantar flexed for pushoff. For forefoot strikers, the forefoot region is the main area loaded during stance.

The majority of runners use the rearfoot striking pattern, so most running shoes are designed for this pattern of loading. A cushioned heel is intended to help absorb and distribute the load during initial contact, and a flexible, cushioned forefoot region

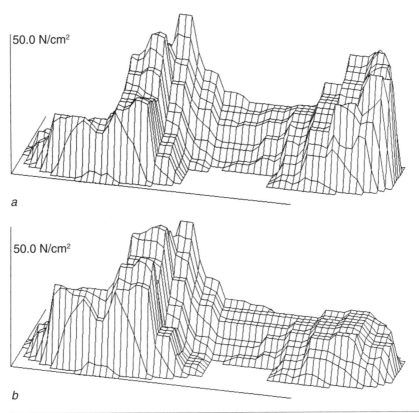

50.0 N/cm²

a

50.0 N/cm²

b

Figure 18.8 Patterns of stress over the sole of the runner's foot. The pattern for a rearfoot striker is shown in (*a*), and the pattern for a forefoot striker is shown in (*b*). Note that both runners exhibit a high level of stress in both the front and rear of the shoe. (Graphics courtesy of Dr. Tom Kernozek, Physical Therapy Department, University of Wisconsin-La Crosse.)

helps absorb the load while allowing the shoe to bend during the latter portion of stance. Cushioning the forefoot region of the shoe helps absorb load during both initial contact and the latter portion of stance for forefoot strikers. Cushioning is provided by a combination of materials and shoe construction.

The pattern of observed injury to the foot closely follows the observed pattern of loading on the sole of the foot. A stress fracture of the head of either the first or second metatarsal is a common injury among runners. The magnitude of the stress applied in this area is relatively high for both forefoot and rearfoot strikers. When the number of steps per mile is considered, it is easy to see that the cumulative effect of this loading may lead to a stress fracture in this area of the foot.

Chapter 4 introduced the concept of reaction force. Ground reaction force is basically the reaction force applied to the body during contact with the ground. The ground reaction force in gait ignores the actual distributed pattern of the ground contact force over the sole of the foot. The point of application of the single vector representing the ground reaction force applied to the foot is called the center

of pressure. Understanding the ground reaction force and the center of pressure simplifies analysis of the effect of the ground contact force on the motion of the runner.

> **The pattern of observed injury to the foot closely follows the observed pattern of loading on the sole of the foot.**

The ground reaction force is applied to the runner's foot beginning with the start of the stance phase, or initial contact with the ground. Figure 18.9 presents patterns of the ground reaction forces typical of rearfoot and forefoot landing styles in running. The ground reaction force consists of three components: the vertical force (acting upward), the anteroposterior force (acting forward or backward on the body), and the mediolateral force (acting from side to side). Measuring the ground reaction force during the support phase of running has long been the focus of research, especially since force platforms became available commercially in the late 1970s. Although the magnitude of the individual curves varies

Center of Pressure

Figure 18.9 (*a*) Typical patterns of the ground reaction force components for a rearfoot (solid line) and a forefoot (dashed line) striker at 3.6 m/s. (*b*) The center of pressure pattern for the left foot (from Williams, 1985).

depending on factors such as running speed, subject weight, shoes, and surface materials, the general pattern and temporal relationship of each curve remain relatively consistent across individuals.

The vertical ground reaction force pushes up on the runner throughout the stance phase. The antero-posterior ground reaction force initially pushes backward as the forward-moving runner contacts the ground and then forward as the runner rotates over the planted foot and pushes back on the ground. The medial-lateral ground reaction force tends to push laterally on the runner, although this pattern is much more variable across runners.

The vertical ground reaction force increases in magnitude until approximately midstance, the point of maximum knee flexion. During this time, the vertical ground reaction force acts upward to decrease the downward motion of the runner. The maximum value of the vertical ground reaction force increases with faster running speeds, ranging in magnitude from two to five times body weight (BW), depending on running speed. With 300 to 900 foot contacts per mile, the cumulative load to a leg can be measured in tons. During the latter half of the stance phase, the vertical ground reaction force continues to act upward on the runner as the leg is extended. Although decreasing in size during the latter half of stance, the vertical force acts to push the runner up into the next flight phase.

Application of the vertical ground reaction force represents what is known as an impulsive load. An impulsive load is an applied force that reaches a relatively high magnitude in a short time. The graph of the vertical force for both forefoot and rearfoot strikers (figure 18.9) shows that within the first 50 ms, the force reaches a magnitude of about 2.0 BW. The force is directed up the body and imposes stress on all tissues of the body. Although the impulsive load is partially absorbed by eccentric muscle action as it passes up the body, traces of it have been recorded at the skull. Tissue responses to this load may be responsible for both the training effect and the injury pattern seen in runners.

Much speculation has focused on the relationship among the magnitude and pattern of ground reaction forces and injury, but research has established no direct relationship. Although the impulsive loading typical of running has been implicated in the development of injury such as osteoarthritis, the observed pattern of running-related injury, the types of injury encountered, and biomechanical models of the runner suggest that factors in addition to ground reaction forces may play a role in overuse injury.

Good running shoes include features for comfort, shock absorption, and rearfoot control. Comfort of a shoe is an obvious selling point, and shock absorption through well-designed cushioning is important for dealing with the impulsive loads applied during stance. Rearfoot control refers to a shoe's capability to allow rearfoot motion, the natural sequential pattern of pronation and supination during stance. Pronation is a combination of dorsiflexion, eversion, and abduction. This motion allows the foot to adapt to the running surface and to absorb shock. Supination is the opposite action, a combination of plantar flexion, inversion, and adduction. This motion stabilizes the foot and lets it serve as a rigid lever during the propulsion that occurs in later stance.

Figure 18.10 is a position-time graph of the typical motion pattern exhibited by rearfoot strikers during the support phase of running. For research and clinical purposes, rearfoot motion is measured in the frontal plane as the angle between the shoe and the lower leg or shank. At contact, the foot is slightly supinated (measured as rearfoot inversion). Following contact, the foot initially pronates (measured as eversion) to reach a maximum everted position at approximately midstance before supinating through to toe-off. Typically, the foot will evert to an angle of about 5-15° past neutral, or straight alignment of the foot and lower leg. Some individuals exhibit excessive pronation, measured as moving more than 18° of inversion past the neutral position. For some individuals, excessive pronation reflects the anatomical structure of the foot;

however, the softer midsole materials intended to provide increased cushioning may actually cause the feet of some runners to pronate beyond the typical 15° of inversion.

The ground reaction force components tend to cause pronation of the foot during the initial part of the stance phase but are resisted by eccentric activity in muscles inserting on the foot. Concentric activity of the same muscles is primarily responsible for supination during the latter portion of stance. Figure 18.11 demonstrates how the ground reaction force acts on the foot of a rearfoot striker to cause the observed motion during the initial stance period. In a rearfoot landing, the ground reaction force is applied behind the ankle joint on the lateral portion of the foot. The vertical force pushes upward, the mediolateral force pushes laterally, and the anteroposterior force pushes posteriorly. All three components have a moment arm around the ankle joint. Torque created by both the vertical and mediolateral components tends to cause the foot to evert. In rearfoot strikers, the torque of the anteroposterior force tends to cause the foot to plantar flex. A shoe made higher by the addition of materials to provide more cushioning will produce longer moment arms for the ground reaction force components. The greater torque created by these components acting with longer moment arms will tend to increase the degree of pronation and plantar flexion. In rearfoot strikers, the anterior and posterior tibialis muscles are active eccentrically to prevent the foot from slapping down and rolling inward too fast. Increased activity in these muscles to resist greater pronation may lead to common injuries reported in runners, such as shin splints, anterior compartment syndrome, and Achilles tendinitis.

Pronation and supination also affect the magnitude of the stress imposed at the knee joint, the most frequent site of overuse injury in running. When running, the knee joint flexes during the initial portion of stance and extends during the latter portion. Knee flexion during running is controlled by the quadriceps muscle group as it is active eccentrically to absorb energy. Knee extension is caused by shortening of the quadriceps muscles as they release the energy while shortening. The tibia, the weight-bearing bone of the lower leg, is rotated internally along its long axis during ankle pronation and externally when the ankle supinates. These actions are caused by forces imposed on the tibia by the motion of the talus bone of the foot as it rotates under the tibia. At its proximal end, the tibia articulates with the femur at the knee joint. Due to the structure of the distal end of the femur, knee flexion and

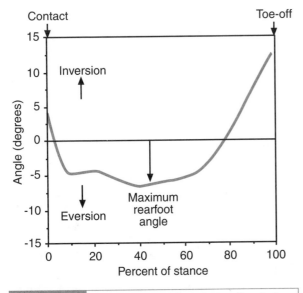

Figure 18.10 | Typical pattern of rearfoot motion during the stance phase of running.

| Figure 18.11 | Influence of the geometry of loading on muscle activity in the lower leg when running. (*a*) Rear view of the foot-leg system. The vertical ground reaction force vector is applied lateral to the axis of the ankle joint, while the mediolateral force is applied below the axis of the ankle joint. These forces create torque that causes the foot to pronate. Eccentric muscle activity in the tibialis anterior and other foot "evertors" is required to control the motion of the foot. (*b*) Side view of the foot-leg system. The vertical ground reaction force vector is applied posterior to the ankle joint and the posteriorly-directed anteroposterior ground reaction force vector is applied below the ankle joint. These forces create torques that cause plantar flexion of the foot. Eccentric muscle activity in the tibialis anterior and other foot "dorsiflexors" is required to control the rate of plantar flexion. |

extension also cause the tibia to rotate along its vertical axis. Flexion causes external rotation of the tibia, and extension causes internal rotation. Ideally, the joint actions of ankle pronation and knee flexion and ankle supination and knee extension should occur simultaneously to avoid placing the tibia in torsion and stressing the knee joint. If the ankle joint overpronates, the synchronization of ankle and knee motion may be disrupted. Knee extension may begin before the ankle reaches maximum pronation. The disruption in synchronization of the coordinated joint action imposes abnormal stress at the knee joint and may alter the muscle activity patterns. The line of pull of the patellar tendon may be altered by misalignment of the femur and tibia, altering the tracking of the patella within the femoral groove. The resulting abnormal stress pattern on the sides and back of the patella may cause injury to the patella or the femur, commonly referred to as patellofemoral pain.

Extrinsic Factors

Extrinsic factors are those characteristics of the environment or task being performed that are related to the potential for injury. For runners, these factors are the physical conditions of the running area, such as surface texture, and the motions involved in running. This section will focus on task- and environment-specific factors that affect running.

Specific Factors Related to the Task of Running

Variable characteristics of the running task include pace, distance of each run, total number of miles run per week, stride length, vertical motion, and joint position and range of motion during each cycle. Task factors affect the size and the number of times the load is applied to the body and how the body is aligned to accept the load.

Running pace, or speed, directly influences the size of the ground reaction force components. Faster speeds are associated with higher loading. The maximum vertical ground reaction force, for example, increases from approximately 2 BW at a slow jog to 6 BW at a fast run. The higher forces associated with faster speeds cause greater torques at the joints. Increased muscle activation is required to control the joint motions incurred by the torques and to create the propulsive ground reaction forces necessary to maintain the faster speed. Thus, a runner must increase running pace slowly over several training sessions to provide time for the neuromusculoskeletal adaptations to occur. Too rapid an increase in training pace is a common training error. Many running overuse injuries are attributable to this "too fast, too soon" syndrome. Setting a more reasonable goal of increasing training pace over a longer period—that is, a smaller increase in pace with each session—will help avoid overuse injury.

A similar error summarized as "too much, too soon" involves either increasing the distance run too quickly or increasing the number of training sessions per week. The increased number of loading cycles involved with increasing distance or number of sessions causes greater microdamage to the tissue. Additional rest time is required for remodeling to occur after an increase in distance. Without adequate rest, the next training session imposes a load on tissues that have not completed remodeling. Typically, 48 hours is recommended between successive training sessions. Extending this period to 72 hours is practical advice when a prior training session has involved an increase in mileage. It may also be beneficial to decrease the pace of the subsequent session to reduce the load on the body.

The total number of miles run per week, a combination of miles per session and number of sessions, reflects the total cumulative load applied to the body. Thus, both errors discussed above—"too fast, too soon" and "too much, too soon"—contribute to the risk of injury related to total number of miles run per week. A systematic approach to increasing both pace and distance can reduce the effect of total number of miles per week on injury risk. As running speed is increased, it is wise to limit the number of miles run at the new pace by restricting the time per session. Similarly, an increase in distance should not be accompanied by an increase in pace.

> **Controlling running pace and the distance run each session, along with the number of sessions per week, is among the most effective injury prevention techniques.**

Understanding the interaction of these three factors relative to the size and repetition of loading provides the best basis for designing a training program. These guidelines apply to runners interested in achieving a greater training effect as well as those preparing for competition.

Extrinsic factors such as stride length, vertical motion, joint position, and the range of motion during each cycle are more closely related to individual running style than the extrinsic factors discussed earlier. These factors reflect an interaction between the individual's body size and running technique. The question of whether changing these factors is effective or useful is more controversial than the guidelines related to controlling increases in pace, distance, and total miles per week. Research on ex-perienced runners indicates that a runner naturally chooses the most efficient stride length for a given pace, one that will minimize energy consumption. The vertical motion of the runner and the observed joint kinematics reflect the efficient gait pattern; however, since the research subjects were experienced runners, whether training with that particular stride length influences energy consumption is unknown. Typically, someone learning to run receives little instruction. Most cues provided in a training program relate to racing strategies such as positioning in a crowd or moving up during a race. Most coaches are reluctant to suggest changes in running style to experienced runners. The effect of providing guidelines on running style to inexperienced runners has not been fully investigated and deserves further research.

Specific Factors Related to the Running Environment

Environmental factors related to injury risk while running include the running surface material (asphalt, concrete, wood chips, grass, sand, tartan), the condition of the surface (wet, dry, icy, loose gravel, sand), the material of the running shoes (midsole material, inserts for shock absorption or rearfoot control [orthotics]), the design characteristics of the shoes (special features for shock absorption or rearfoot control), and temperature (hot, warm, cold). Although the runner has relatively little control over temperature, other than considering how it might affect pace and distance, or even the decision of whether to go for a run, the shoes worn and the terrain run on are selected by the runner.

Running surface directly affects the magnitude of the ground reaction forces and the amount of rearfoot control. Harder surfaces such as asphalt or concrete are associated with higher forces than are surfaces such as grass or sand. The higher coefficient of restitution typical of asphalt and concrete (see chapter 16) means that less force is absorbed by the material and greater forces are imposed on the runner, but these firm surfaces provide good traction or grip. Conversely, the lower coefficient of restitution typical of wood chips, grass, and sand results in lower ground reaction forces, but the tendency for the materials to shift can result in reduced traction. Many communities have created running trails made of such softer materials to provide a safer environment for training. However, some runners, especially highly trained runners, will not train regularly on these surfaces specifically because they feel that the softer surfaces and reduced friction decrease the quality of the workout. In addition, softer materials allow more pronation of the foot, leading to increased muscle pain in the lower extremity due to increased range of motion and muscle activity. Selecting a running surface that provides a balance between ground reaction force magnitude and rearfoot control is an individual preference.

Running shoes are also a matter of personal preference. Shoes marketed today incorporate a variety of features to make them more attractive, comfortable, and safe. The distinction between features is not always clear, though. For example, a highly cushioned shoe may feel extremely comfortable when walking in the salesroom, giving the sensation of "walking on air." However, the materials and shoe construction that provide cushioning may not afford the degree of rearfoot control needed when the ground reaction forces typical of running are imposed on the shoe. In addition, a shoe that provides cushioning and adequate rearfoot control on a harder surface may feel too soft and lacking in rearfoot control on a softer surface.

As a guideline for injury prevention, it makes sense to run on a variety of surfaces with a variety of running shoes. This is a form of "within-activity cross-training." Cross-training refers to a training program that incorporates day-to-day rotation among a variety of activities. For example, an individual training for cardiovascular fitness might rotate training sessions among running, cycling, swimming, and aerobics. The idea is that each activity imposes a different pattern of loading on the body. By alternating activities, the pattern of loading is altered, avoiding the cumulative effect of a consistent loading pattern and reducing the risk of injury. Running on different surfaces and/or with different shoes has a somewhat similar effect. Although the same general pattern of cyclical movement is involved, the slight differences in magnitude and loading pattern represent a form of cross-training within the activity. By varying loads from session to session with different shoe/surface combinations, you avoid placing a cumulative load on a tissue that can lead to overuse injury.

In practice, alter the pace and distance run in accordance with how your body responds to the shoe and surface. You may find that more pain develops during or following a training session on a particular surface. Respond to this by changing the pace and length of your run. A good rule of thumb is that, if an overuse injury develops from using a particular shoe or running on a particular surface,

change the environment. Discard a pair of shoes if it causes more pain than usual. When buying shoes, select a pair similar to ones in which you have been able to run without pain. Use your own experience to select a particular model or style of shoe, and realize that your response to a shoe may differ from advertising hype. Also, shoes wear out with use, and as their condition deteriorates, the shock absorption or rearfoot control they afford decreases. Similar considerations relate to running surface. If you experience increased soreness from running on a particular surface, change surfaces. Some individuals prefer softer surfaces, and others prefer harder surfaces. Recognizing that shoes and surfaces interact to create the injury risk, and that the risk is specific to an individual, leads to individual choice and preferences in shoes and surfaces.

Intrinsic Factors

Intrinsic factors relate to the characteristics of the individual performing a task in a given environment. Researchers have attempted to identify individuals for whom running is an inappropriate activity, believing that some people may be predisposed to injury because of individual characteristics affecting the level of stress imposed and the response to stress. A few of these characteristics will be discussed in this section.

Body Mass

An appropriate starting point is to consider body mass. In mechanics, the mass of a body represents inertia, or resistance to a change in the body's current state of motion. As described earlier, during the stance phase, the runner's downward motion at contact is slowed and stopped by the vertical ground reaction force, whereas the forward motion is slowed by the anteroposterior ground reaction force. The vertical and anteroposterior forces then act simultaneously to propel the runner upward and forward into the next flight phase. In chapter 6, Newton's second law ($\Sigma F = ma$) was used to explain that for a given acceleration, a larger force is required to change the motion of a larger mass. If two runners train together at the same running pace, Newton's second law suggests that the heavier runner will be exposed to higher forces; that is, to provide the same acceleration, higher forces must act on the heavier runner as compared to the lighter runner.

This is why ground reaction force data are presented in multiples of body weight. Generally, when performing a task with similar kinematic patterns, the more massive individual will be exposed to higher forces. Wolff's law indicates that the body of the more massive runner should be adapted to the higher forces. Typically, higher bone density and more muscle strength are associated with greater body mass, as would be expected according to Wolff's law. Presenting ground reaction force in multiples of body weight is an attempt to scale the size of the imposed load to the mass of the individual to account for assumed adaptations to the higher absolute force. The idea is that an imposed load of 2.5 BW will have a similar effect on individuals despite differences in the absolute size of the imposed force. However, as we pointed out earlier, the stress threshold of various tissues is uncertain. The question as to whether absolute or relative loading is more related to tissue response is still unclear.

Lower Extremity Anatomy

The structure and function of the lower extremity have been implicated in the development of overuse injury in runners. As an example, in this section, we will consider how two common anatomical factors relate to overuse injury.

Foot Structure.   Foot structure refers to the type of medial longitudinal arch. The medial longitudinal arch runs along the medial aspect of the foot and is created by the calcaneus, talus, navicular, three cuneiforms, and three medial metatarsals. The arch is supported by muscles and the plantar aponeurosis, the connective tissue band that runs from the calcaneus to the head of the first metatarsal. The arch lengthens during the initial phases of foot contact to help distribute the contact load over the bones of the foot, shortens during midstance to help support the body, then lengthens again to aid in pushoff.

Differences in foot structure are associated with differing types of common overuse injury. A lower, more flexible arch allows for absorption of more energy than does a higher, more rigid arch. As a result, in individuals with rigid high arches, a greater load is passed on to the tibia and up the leg compared to in those with flexible low arches. This difference in stress absorption is reflected in a different pattern of overuse injury. Individuals with high arches are more prone to stress fractures

in the tibia and femur due to reduced absorption associated with less foot motion. Individuals with low arches are more prone to metatarsal stress fractures due to the increased stress imposed on these bones as they contribute to greater load absorption.

The structure of the foot should be considered when selecting running shoes. Individuals with low arches should choose a shoe that provides rearfoot control. Such a shoe will help prevent overpronation (>15°) that may occur due to increased foot mobility. Individuals with high arches should choose a shoe with added shock absorption features to provide increased absorption of the forces imposed during running. Ultimately, shoe choice should be based on comfort and success in preventing overuse injury.

Alignment of the Knee Joint.   The knee joint is the articulation between the femur and the tibia. This joint is exposed to high stress during running because of the imposed ground reaction force and the tension in muscles crossing the joint. Knee joint function is affected by its structure and anatomical factors such as the shape of the femoral condyles and the tibial plateau, and the presence of the medial and lateral menisci. Individual knee joint alignment characteristics are important in assessing the relative risk of injury when running because they affect the magnitude and pattern of stress imposed on the tissues of the joint and the joint's contribution to shock absorption.

The Q angle is a measure of alignment between the femur and the tibia. To measure the Q angle, lines representing the long axis of the tibia and femur are drawn on the front of the leg. As shown in figure 18.12, one line is drawn from the tibial tuberosity to the midpoint of the superior border of the patella to represent the tibia and another line is drawn from the anterior superior iliac spine of the pelvis to the midpoint of the superior border of the patella to represent the femur. The Q angle is measured as the smaller angle between the two lines. A Q angle greater than 20 degrees is considered excessive.

The alignment of the femur and tibia is important because it affects patellar tracking, or the path the patella follows as it rides between the femoral condyles during knee flexion and extension. Because the patella is the distal insertion site for the quadriceps muscles, a higher Q angle (poor alignment of the femur and tibia) means that the patella will not track smoothly between the condyles. This

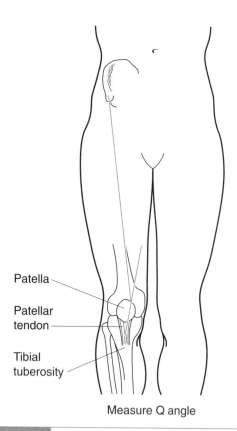

Measure Q angle

Figure 18.12   The Q angle measures the alignment between the femur and the tibia, a factor affecting the pattern and magnitude of stress imposed on the tissues of the joint.

Patella

Patellar tendon

Tibial tuberosity

leads to excessive stress on the posterior and lateral borders of the patella, contributing to the development of pain in this region (commonly referred to as patellofemoral pain).

Despite the obvious link between the Q angle and patellar tracking, factors such as muscle tightness in the soft tissue of the lateral knee and weakness of the vastus medialis also need to be considered in the development of patellofemoral pain. Tightness in the lateral soft tissues will pull the patella laterally, even if the Q angle is within normal range. If the vastus medialis is weak relative to the vastus lateralis, the unequal pull on the patella causes it to glide along the lateral condyle. If a runner reports patellofemoral pain, Q angle, lateral tightness, and muscle strength should all be evaluated to identify the cause of the probable patellar mistracking.

Psychological Considerations

Psychological factors related to overuse injury include the individual's pain tolerance level and mo-

tivation to continue with a training program. Both factors are related to the onset and progression of an injury. Clinical studies of running-related injury frequently report on individuals who continue to run despite medical advice to rest and seek treatment. Surveys of marathon runners, both competitive and recreational, frequently list multiple running-related injuries in a single individual. Common injuries that sufferers try to "run through" include a variety of musculoskeletal ailments such as sprains, joint pain, and even fractures. Although some coaches, sport commentators, and the athletes themselves believe continued participation despite pain to be admirable, this attitude belies the fact that pain is a sign that the body needs time for healing. Continuing to participate through injury exposes the injured tissue to stress when its ability to withstand stress is already compromised. As a result, the injury can worsen, usually to the point of requiring more extensive medical intervention and a longer rest period than if it had been treated promptly.

Continuing through injury requires an adaptation in movement pattern commonly known as favoring the injured body segment. In mechanical terms, a modified movement pattern is used to reduce the stress imposed at the site of injury and eliminate the cause of the pain. However, alterations in the movement pattern necessary to reduce stress at one site means that higher stress will be imposed somewhere else. The change in stress patterns, often made in a short time if the runner continues to perform, frequently leads to a secondary injury. For example, some runners report to a clinic complaining of pain in both limbs, sometimes at the same site and other times at different sites on each leg. The athlete admits to a delay in seeking treatment until the pain became unbearable or constant. The injury history frequently reveals that the pain began unilaterally, limited to a single leg. The development of pain in the other limb results from the altered stress pattern while either consciously or unconsciously favoring the injured limb.

Prompt treatment of the cause of pain is critical to preventing more extensive injury. Conservative intervention such as rest provides time for healing to occur. It is necessary, in fact imperative, to identify the reason for the injury. Typically, it is simply a violation of the "too much, too soon" principle. When returning to running after the injury has healed, the athlete must adhere to a systematic approach of increasing distance or speed and consider using more appropriate shoes or running on a different surface. Individuals with significant skeletal malalignment may need an orthotic to align the musculoskeletal system properly or may need to incorporate specific strength and flexibility exercises to correct the malalignment. Some individuals may have to choose a more appropriate activity, one that reduces loading to a level that does not place excessive stress on damage-prone tissue. These individuals should seek the advice of a practitioner knowledgeable about the biomechanics of common cardiovascular training activities for help in selecting the appropriate training modality.

Summary

Mechanical stress refers to the distribution of force over the body on which it acts. The body responds to chronically low levels of stress by atrophying and to systematically applied high levels of stress by hypertrophying. However, if the stress level imposed is too high or is applied too frequently, a traumatic or overuse injury will occur. Preventing injury can be a matter of reducing either the magnitude or frequency of imposed stress or modifying the tissue's stress threshold to better withstand the stress. Examples of stress imposed during running demonstrate that although the stress-injury relationship is conceptually simple, preventing injury during skill performance is difficult.

The three chapters in part IV provide guidelines for conducting qualitative biomechanical analyses with specific emphases on technique improvement, training, and injury prevention. Although these three types of analysis were presented separately, and indeed are different in their approaches, they are inseparable in any thorough qualitative biomechanical analysis of an activity. The technical improvements suggested by a qualitative biomechanical analysis to improve technique should be examined to ensure that these changes in technique will not impose stresses within the pathological overload zone of the stress continuum and thus lead to injury. Will the technique improvement increase the risk of injury? A qualitative biomechanical analysis to improve training relies on the model of the most effective technique suggested by a qualitative biomechanical analysis to improve technique. What physical parameters must be trained to be able to produce the most effective technique? Practitioners must also be able to identify the anatomical structures that are at risk for injury

as a result of the performance of a skill. This identification results from a qualitative biomechanical analysis to understand injury development. The results of this type of analysis may also be used to improve training to strengthen the tissues that are at risk for injury. What tissues must be strengthened to prevent injury?

Qualitative biomechanical analysis requires a sound knowledge and understanding of mechanical principles. The material presented in the first three parts of this book should have assisted you with gaining this knowledge and understanding. Qualitative biomechanical analysis is also a complex and difficult task. The general guidelines and procedures for conducting qualitative biomechanical analyses have been presented in the last part of this book. These guidelines and procedures are an attempt to simplify the process of conducting a qualitative biomechanical analysis. As with most things in life, learning to do something well requires practice. Qualitative biomechanical analysis also requires much practice and experience to perfect. Opportunities to apply biomechanics and conduct qualitative biomechanical analyses will occur every day in your future as a human movement professional. Take advantage of these opportunities and use your knowledge of biomechanics. The payoff will be more satisfied students, athletes, or clients and an increase in your effectiveness as a teacher, coach, or other human movement professional.

Key Terms

center of pressure (p. 359)

extrinsic factors (p. 354)

impulsive load (p. 360)

intrinsic factors (p. 354)

overuse injury (p. 353)

physiologic loading zone (p. 351)

Q angle (p. 365)

rearfoot motion (p. 361)

remodeling (p. 352)

stress (p. 348)

stress continuum (p. 351)

traumatic injury (p. 353)

Wolff's law (p. 350)

REVIEW QUESTIONS

1. Explain how the magnitude of imposed stress relates to the size of the applied force and the area over which the force is distributed. Give examples of interventions to alter stress by changing the size of the imposed force and examples of interventions to alter stress by changing the area over which the force is distributed.

2. Differentiate between tensile, compressive, and shear stress. Give an example of each type of stress imposed on the human body, identifying the source of the force and the tissue loaded.

3. Draw the stress continuum, identifying the four regions.

4. Describe the concept of the "stress threshold." Relate the concept to the development of traumatic and overuse injury.

5. Differentiate between intrinsic and extrinsic factors related to injury. Identify intrinsic and extrinsic factors related to overuse injury in: (a) swimming, (b) long jumping, (c) pitching, (d) an assembly task in industry.

6. Explain how the concept of within-activity cross-training is similar to and different from that of cross-training. Basing your answer on the mechanical basis for injury development, describe why either mode of training may be effective in reducing injury risk.

A

Trigonometric Functions

The branch of geometry that deals with the study of triangles is called trigonometry. Trigonometry was discussed in more detail when it was applied to resolving forces into components in chapter 4. Trigonometry is primarily concerned with the relationships between or among dimensions of right triangles (see figure A.1). The side opposite the 90° angle in a right triangle is called the hypotenuse of a right triangle. The square of the hypotenuse is equal to the sum of the squares of the other two sides ($c^2 = a^2 + b^2$). This is called the Pythagorean theorem.

The common trigonometric functions are sine, cosine, and tangent. The values of these functions for angles between 0° and 360° are listed in table A.1.

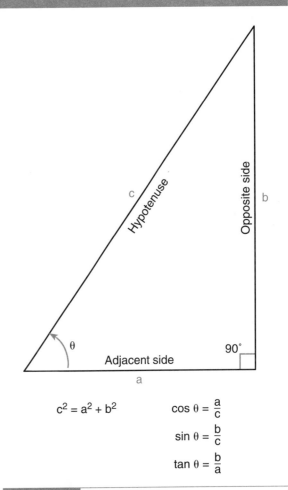

$$c^2 = a^2 + b^2 \qquad \cos\theta = \frac{a}{c}$$

$$\sin\theta = \frac{b}{c}$$

$$\tan\theta = \frac{b}{a}$$

Figure A.1   A right triangle.

TABLE A.1   Table of Trigonometric Functions for Angles Measured in Degrees or Radians

Angle (deg)	Angle (rad)	sine	cosine	tangent	Angle (deg)	Angle (rad)
0	0.0000	0.0000	1.0000	0.0000	−360	−6.2832
1	0.0175	0.0175	0.9998	0.0175	−359	−6.2657
2	0.0349	0.0349	0.9994	0.0349	-358	-6.2483
3	0.0524	0.0523	0.9986	0.0524	-357	-6.2308
4	0.0698	0.0698	0.9976	0.0699	-356	-6.2134
5	0.0873	0.0872	0.9962	0.0875	-355	-6.1959
6	0.1047	0.1045	0.9945	0.1051	-354	-6.1785
7	0.1222	0.1219	0.9925	0.1228	-353	-6.1610
8	0.1396	0.1392	0.9903	0.1405	-352	-6.1436
9	0.1571	0.1564	0.9877	0.1584	-351	-6.1261
10	0.1745	0.1736	0.9848	0.1763	-350	-6.1087
11	0.1920	0.1908	0.9816	0.1944	-349	-6.0912
12	0.2094	0.2079	0.9781	0.2126	-348	-6.0737
13	0.2269	0.2250	0.9744	0.2309	-347	-6.0563
14	0.2443	0.2419	0.9703	0.2493	-346	-6.0388
15	0.2618	0.2588	0.9659	0.2679	-345	-6.0214
16	0.2793	0.2756	0.9613	0.2867	-344	-6.0039
17	0.2967	0.2924	0.9563	0.3057	-343	-5.9865
18	0.3142	0.3090	0.9511	0.3249	-342	-5.9690
19	0.3316	0.3256	0.9455	0.3443	-341	-5.9516
20	0.3491	0.3420	0.9397	0.3640	-340	-5.9341
21	0.3665	0.3584	0.9336	0.3839	-339	-5.9167
22	0.3840	0.3746	0.9272	0.4040	-338	-5.8992
23	0.4014	0.3907	0.9205	0.4245	-337	-5.8818
24	0.4189	0.4067	0.9135	0.4452	-336	-5.8643
25	0.4363	0.4226	0.9063	0.4663	-335	-5.8469
26	0.4538	0.4384	0.8988	0.4877	-334	-5.8294
27	0.4712	0.4540	0.8910	0.5095	-333	-5.8119
28	0.4887	0.4695	0.8829	0.5317	-332	-5.7945
29	0.5061	0.4848	0.8746	0.5543	-331	-5.7770
30	0.5236	0.5000	0.8660	0.5774	-330	-5.7596
31	0.5411	0.5150	0.8572	0.6009	-329	-5.7421
32	0.5585	0.5299	0.8480	0.6249	-328	-5.7247
33	0.5760	0.5446	0.8387	0.6494	-327	-5.7072
34	0.5934	0.5592	0.8290	0.6745	-326	-5.6898
35	0.6109	0.5736	0.8192	0.7002	-325	-5.6723
36	0.6283	0.5878	0.8090	0.7265	-324	-5.6549
37	0.6458	0.6018	0.7986	0.7536	-323	-5.6374
38	0.6632	0.6157	0.7880	0.7813	-322	-5.6200
39	0.6807	0.6293	0.7771	0.8098	-321	-5.6025

Angle (deg)	Angle (rad)	sine	cosine	tangent	Angle (deg)	Angle (rad)
40	0.6981	0.6428	0.7660	0.8391	-320	-5.5851
41	0.7156	0.6561	0.7547	0.8693	-319	-5.5676
42	0.7330	0.6691	0.7431	0.9004	-318	-5.5501
43	0.7505	0.6820	0.7314	0.9325	-317	-5.5327
44	0.7679	0.6947	0.7193	0.9657	-316	-5.5152
45	0.7854	0.7071	0.7071	1.0000	-315	-5.4978
46	0.8029	0.7193	0.6947	1.0355	-314	-5.4803
47	0.8203	0.7314	0.6820	1.0724	-313	-5.4629
48	0.8378	0.7431	0.6691	1.1106	-312	-5.4454
49	0.8552	0.7547	0.6561	1.1504	-311	-5.4280
50	0.8727	0.7660	0.6428	1.1918	-310	-5.4105
51	0.8901	0.7771	0.6293	1.2349	-309	-5.3931
52	0.9076	0.7880	0.6157	1.2799	-308	-5.3756
53	0.9250	0.7986	0.6018	1.3270	-307	-5.3582
54	0.9425	0.8090	0.5878	1.3764	-306	-5.3407
55	0.9599	0.8192	0.5736	1.4281	-305	-5.3233
56	0.9774	0.8290	0.5592	1.4826	-304	-5.3058
57	0.9948	0.8387	0.5446	1.5399	-303	-5.2883
58	1.0123	0.8480	0.5299	1.6003	-302	-5.2709
59	1.0297	0.8572	0.5150	1.6643	-301	-5.2534
60	1.0472	0.8660	0.5000	1.7321	-300	-5.2360
61	1.0647	0.8746	0.4848	1.8040	-299	-5.2185
62	1.0821	0.8829	0.4695	1.8807	-298	-5.2011
63	1.0996	0.8910	0.4540	1.9626	-297	-5.1836
64	1.1170	0.8988	0.4384	2.0503	-296	-5.1662
65	1.1345	0.9063	0.4226	2.1445	-295	-5.1487
66	1.1519	0.9135	0.4067	2.2460	-294	-5.1313
67	1.1694	0.9205	0.3907	2.3559	-293	-5.1138
68	1.1868	0.9272	0.3746	2.4751	-292	-5.0964
69	1.2043	0.9336	0.3584	2.6051	-291	-5.0789
70	1.2217	0.9397	0.3420	2.7475	-290	-5.0615
71	1.2392	0.9455	0.3256	2.9042	-289	-5.0440
72	1.2566	0.9511	0.3090	3.0777	-288	-5.0265
73	1.2741	0.9563	0.2924	3.2709	-287	-5.0091
74	1.2915	0.9613	0.2756	3.4874	-286	-4.9916
75	1.3090	0.9659	0.2588	3.7321	-285	-4.9742
76	1.3265	0.9703	0.2419	4.0108	-284	-4.9567
77	1.3439	0.9744	0.2250	4.3315	-283	-4.9393
78	1.3614	0.9781	0.2079	4.7046	-282	-4.9218
79	1.3788	0.9816	0.1908	5.1146	-281	-4.9044

(continued)

TABLE A.1 *(continued)*

Angle (deg)	Angle (rad)	sine	cosine	tangent	Angle (deg)	Angle (rad)
80	1.3963	0.9848	0.1736	5.6713	-280	-4.8869
81	1.4137	0.9877	0.1564	6.3138	-279	-4.8695
82	1.4312	0.9903	0.1392	7.1154	-278	-4.8520
83	1.4486	0.9925	0.1219	8.1443	-277	-4.8346
84	1.4661	0.9945	0.1045	9.5144	-276	-4.8171
85	1.4835	0.9962	0.0872	11.4301	-275	-4.7997
86	1.5010	0.9976	0.0698	14.3007	-274	-4.7822
87	1.5184	0.9986	0.0523	19.0811	-273	-4.7647
88	1.5359	0.9994	0.0349	28.6363	-272	-4.7473
89	1.5533	0.9998	0.0175	57.2900	-271	-4.7298
90	1.5708	1.0000	0.0000	undefined	-270	-4.7124
91	1.5882	0.9998	-0.0175	-57.2900	-269	-4.6949
92	1.6057	0.9994	-0.0349	-28.6363	-268	-4.6775
93	1.6232	0.9986	-0.0523	-19.0811	-267	-4.6600
94	1.6406	0.9976	-0.0698	-14.3007	-266	-4.6426
95	1.6581	0.9962	-0.0872	-11.4301	-265	-4.6251
96	1.6755	0.9945	-0.1045	-9.5144	-264	-4.6077
97	1.6930	0.9925	-0.1219	-8.1443	-263	-4.5902
98	1.7104	0.9903	-0.1392	-7.1154	-262	-4.5728
99	1.7279	0.9877	-0.1564	-6.3138	-261	-4.5553
100	1.7453	0.9848	-0.1736	-5.6713	-260	-4.5379
101	1.7628	0.9816	-0.1908	-5.1446	-259	-4.5204
102	1.7802	0.9781	-0.2079	-4.7046	-258	-4.5029
103	1.7977	0.9744	-0.2250	-4.3315	-257	-4.4855
104	1.8151	0.9703	-0.2419	-4.0108	-256	-4.4680
105	1.8326	0.9659	-0.2588	-3.7321	-255	-4.4506
106	1.8500	0.9613	-0.2756	-3.4874	-254	-4.4331
107	1.8675	0.9563	-0.2924	-3.2709	-253	-4.4157
108	1.8850	0.9511	-0.3090	-3.0777	-252	-4.3982
109	1.9024	0.9455	-0.3256	-2.9042	-251	-4.3808
110	1.9199	0.9397	-0.3420	-2.7475	-250	-4.3633
111	1.9373	0.9336	-0.3584	-2.6051	-249	-4.3459
112	1.9548	0.9272	-0.3746	-2.4751	-248	-4.3284
113	1.9722	0.9205	-0.3907	-2.3559	-247	-4.3110
114	1.9897	0.9135	-0.4067	-2.2460	-246	-4.2935
115	2.0071	0.9063	-0.4226	-2.1445	-245	-4.2761
116	2.0246	0.8988	-0.4384	-2.0503	-244	-4.2586
117	2.0420	0.8910	-0.4540	-1.9626	-243	-4.2412
118	2.0595	0.8829	-0.4695	-1.8807	-242	-4.2237
119	2.0769	0.8746	-0.4848	-1.8040	-241	-4.2062
120	2.0944	0.8660	-0.5000	-1.7321	-240	-4.1888

Angle (deg)	Angle (rad)	sine	cosine	tangent	Angle (deg)	Angle (rad)
121	2.1118	0.8572	-0.5150	-1.6643	-239	-4.1713
122	2.1293	0.8480	-0.5299	-1.6003	-238	-4.1539
123	2.1468	0.8387	-0.5446	-1.5399	-237	-4.1364
124	2.1642	0.8290	-0.5592	-1.4826	-236	-4.1190
125	2.1817	0.8192	-0.5736	-1.4281	-235	-4.1015
126	2.1991	0.8090	-0.5878	-1.3764	-234	-4.0841
127	2.2166	0.7986	-0.6018	-1.3270	-233	-4.0666
128	2.2340	0.7880	-0.6157	-1.2799	-232	-4.0492
129	2.2515	0.7771	-0.6293	-1.2349	-231	-4.0317
130	2.2689	0.7660	-0.6428	-1.1918	-230	-4.0143
131	2.2864	0.7547	-0.6561	-1.1504	-229	-3.9968
132	2.3038	0.7431	-0.6691	-1.1106	-228	-3.9794
133	2.3213	0.7314	-0.6820	-1.0724	-227	-3.9619
134	2.3387	0.7193	-0.6947	-1.0355	-226	-3.9444
135	2.3562	0.7071	-0.7071	-1.0000	-225	-3.9270
136	2.3736	0.6947	-0.7193	-0.9657	-224	-3.9095
137	2.3911	0.6820	-0.7314	-0.9325	-223	-3.8921
138	2.4086	0.6691	-0.7431	-0.9004	-222	-3.8746
139	2.4260	0.6561	-0.7547	-0.8693	-221	-3.8572
140	2.4435	0.6428	-0.7660	-0.8391	-220	-3.8397
141	2.4609	0.6293	-0.7771	-0.8098	-219	-3.8223
142	2.4784	0.6157	-0.7880	-0.7813	-218	-3.8048
143	2.4958	0.6018	-0.7986	-0.7536	-217	-3.7874
144	2.5133	0.5878	-0.8090	-0.7265	-216	-3.7699
145	2.5307	0.5735	-0.8192	-0.7002	-215	-3.7525
146	2.5482	0.5592	-0.8290	-0.6745	-214	-3.7350
147	2.5656	0.5446	-0.8387	-0.6494	-213	-3.7176
148	2.5831	0.5299	-0.8480	-0.6249	-212	-3.7001
149	2.6005	0.5150	-0.8572	-0.6009	-211	-3.6826
150	2.6180	0.5000	-0.8660	-0.5774	-210	-3.6652
151	2.6354	0.4848	-0.8746	-0.5543	-209	-3.6477
152	2.6529	0.4695	-0.8829	-0.5317	-208	-3.6303
153	2.6704	0.4540	-0.8910	-0.5095	-207	-3.6128
154	2.6878	0.4384	-0.8988	-0.4877	-206	-3.5954
155	2.7053	0.4226	-0.9063	-0.4663	-205	-3.5779
156	2.7227	0.4067	-0.9135	-0.4452	-204	-3.5605
157	2.7402	0.3907	-0.9205	-0.4245	-203	-3.5430
158	2.7576	0.3746	-0.9272	-0.4040	-202	-3.5256
159	2.7751	0.3584	-0.9336	-0.3839	-201	-3.5081
160	2.7925	0.3420	-0.9397	-0.3640	-200	-3.4907

(continued)

TABLE A.1 *(continued)*

Angle (deg)	Angle (rad)	sine	cosine	tangent	Angle (deg)	Angle (rad)
161	2.8100	0.3256	-0.9455	-0.3443	-199	-3.4732
162	2.8274	0.3090	-0.9511	-0.3249	-198	-3.4558
163	2.8449	0.2924	-0.9563	-0.3057	-197	-3.4383
164	2.8623	0.2756	-0.9613	-0.2867	-196	-3.4208
165	2.8798	0.2588	-0.9659	-0.2679	-195	-3.4034
166	2.8972	0.2419	-0.9703	-0.2493	-194	-3.3859
167	2.9147	0.2250	-0.9744	-0.2309	-193	-3.3685
168	2.9322	0.2079	-0.9781	-0.2126	-192	-3.3510
169	2.9496	0.1908	-0.9816	-0.1944	-191	-3.3336
170	2.9671	0.1736	-0.9848	-0.1763	-190	-3.3161
171	2.9845	0.1564	-0.9877	-0.1584	-189	-3.2987
172	3.0020	0.1392	-0.9903	-0.1405	-188	-3.2812
173	3.0194	0.1219	-0.9925	-0.1228	-187	-3.2638
174	3.0369	0.1045	-0.9945	-0.1051	-186	-3.2463
175	3.0543	0.0872	-0.9962	-0.0875	-185	-3.2289
176	3.0718	0.0698	-0.9976	-0.0699	-184	-3.2114
177	3.0982	0.0523	-0.9986	-0.0524	-183	-3.1940
178	3.1067	0.0349	-0.9994	-0.0349	-182	-3.1765
179	3.1241	0.0175	-0.9998	-0.0175	-181	-3.1590
180	3.1416	0.0000	-1.0000	0.0000	-180	-3.1416
181	3.1590	-0.0175	-0.9998	0.0175	-179	-3.1241
182	3.1765	-0.0349	-0.9994	0.0349	-178	-3.1067
183	3.1940	-0.0523	-0.9986	0.0524	-177	-3.0892
184	3.2114	-0.0698	-0.9976	0.0699	-176	-3.0718
185	3.2289	-0.0872	-0.9962	0.0875	-175	-3.0543
186	3.2463	-0.1045	-0.9945	0.1051	-174	-3.0369
187	3.2638	-0.1219	-0.9925	0.1228	-173	-3.0194
188	3.2812	-0.1392	-0.9903	0.1405	-172	-3.0020
189	3.2987	-0.1564	-0.9877	0.1584	-171	-2.9845
190	3.3161	-0.1736	-0.9848	0.1763	-170	-2.9671
191	3.3336	-0.1908	-0.9816	0.1944	-169	-2.9496
192	3.3510	-0.2079	-0.9781	0.2126	-168	-2.9322
193	3.3685	-0.2250	-0.9744	0.2309	-167	-2.9147
194	3.3859	-0.2419	-0.9703	0.2493	-166	-2.8972
195	3.4034	-0.2588	-0.9659	0.2679	-165	-2.8798
196	3.4208	-0.2756	-0.9613	0.2867	-164	-2.8623
197	3.4383	-0.2924	-0.9563	0.3057	-163	-2.8449
198	3.4558	-0.3090	-0.9511	0.3249	-162	-2.8274
199	3.4732	-0.3256	-0.9455	0.3443	-161	-2.8100
200	3.4907	-0.3420	-0.9397	0.3640	-160	-2.7925
201	3.5081	-0.3584	-0.9336	0.3839	-159	-2.7751

Angle (deg)	Angle (rad)	sine	cosine	tangent	Angle (deg)	Angle (rad)
202	3.5256	-0.3746	-0.9272	0.4040	-158	-2.7576
203	3.5430	-0.3907	-0.9205	0.4245	-157	-2.7402
204	3.5605	-0.4067	-0.9135	0.4452	-156	-2.7227
205	3.5779	-0.4226	-0.9063	0.4663	-155	-2.7053
206	3.5954	-0.4384	-0.8988	0.4877	-154	-2.6878
207	3.6128	-0.4540	-0.8910	0.5095	-153	-2.6704
208	3.6303	-0.4695	-0.8829	0.5317	-152	-2.6529
209	3.6477	-0.4848	-0.8746	0.5543	-151	-2.6354
210	3.6652	-0.5000	-0.8660	0.5774	-150	-2.6180
211	3.6826	-0.5150	-0.8572	0.6009	-149	-2.6005
212	3.7001	-0.5299	-0.8480	0.6249	-148	-2.5831
213	3.7176	-0.5446	-0.8387	0.6494	-147	-2.5656
214	3.7350	-0.5592	-0.8290	0.6745	-146	-2.5482
215	3.7525	-0.5736	-0.8192	0.7002	-145	-2.5307
216	3.7699	-0.5878	-0.8090	0.7265	-144	-2.5133
217	3.7874	-0.6018	-0.7986	0.7536	-143	-2.4958
218	3.8048	-0.6157	-0.7880	0.7813	-142	-2.4784
219	3.8223	-0.6293	-0.7771	0.8098	-141	-2.4609
220	3.8397	-0.6428	-0.7660	0.8391	-140	-2.4435
221	3.8572	-0.6561	-0.7547	0.8693	-139	-2.4260
222	3.8746	-0.6691	-0.7431	0.9004	-138	-2.4086
223	3.8921	-0.6820	-0.7314	0.9325	-137	-2.3911
224	3.9095	-0.6947	-0.7193	0.9657	-136	-2.3736
225	3.9270	-0.7071	-0.7071	1.0000	-135	-2.3562
226	3.9444	-0.7193	-0.6947	1.0355	-134	-2.3387
227	3.9619	-0.7314	-0.6820	1.0724	-133	-2.3213
228	3.9794	-0.7431	-0.6691	1.1106	-132	-2.3038
229	3.9968	-0.7547	-0.6561	1.1504	-131	-2.2864
230	4.0143	-0.7660	-0.6428	1.1918	-130	-2.2689
231	4.0317	-0.7771	-0.6293	1.2349	-129	-2.2515
232	4.0492	-0.7880	-0.6157	1.2799	-128	-2.2340
233	4.0666	-0.7986	-0.6018	1.3270	-127	-2.2166
234	4.0841	-0.8090	-0.5878	1.3764	-126	-2.1991
235	4.1015	-0.8192	-0.5736	1.4281	-125	-2.1817
236	4.1190	-0.8290	-0.5592	1.4826	-124	-2.1642
237	4.1364	-0.8387	-0.5446	1.5399	-123	-2.1468
238	4.1539	-0.8480	-0.5299	1.6003	-122	-2.1293
239	4.1713	-0.8572	-0.5150	1.6643	-121	-2.1118
240	4.1888	-0.8660	-0.5000	1.7321	-120	-2.0944
241	4.2062	-0.8746	-0.4848	1.8040	-119	-2.0769

(continued)

TABLE A.1 *(continued)*

Angle (deg)	Angle (rad)	sine	cosine	tangent	Angle (deg)	Angle (rad)
242	4.2237	-0.8829	-0.4695	1.8807	-118	-2.0595
243	4.2412	-0.8910	-0.4540	1.9626	-117	-2.0420
244	4.2586	-0.8988	-0.4384	2.0503	-116	-2.0246
245	4.2761	-0.9063	-0.4226	2.1445	-115	-2.0071
246	4.2935	-0.9135	-0.4067	2.2460	-114	-1.9897
247	4.3110	-0.9205	-0.3907	2.3559	-113	-1.9722
248	4.3284	-0.9272	-0.3746	2.4751	-112	-1.9548
249	4.3459	-0.9336	-0.3584	2.6051	-111	-1.9373
250	4.3633	-0.9397	-0.3420	2.7475	-110	-1.9199
251	4.3808	-0.9455	-0.3256	2.9042	-109	-1.9024
252	4.3982	-0.9511	-0.3090	3.0777	-108	-1.8850
253	4.4157	-0.9563	-0.2924	3.2709	-107	-1.8675
254	4.4331	-0.9613	-0.2756	3.4874	-106	-1.8500
255	4.4506	-0.9659	-0.2588	3.7321	-105	-1.8326
256	4.4680	-0.9703	-0.2419	4.0108	-104	-1.8151
257	4.4855	-0.9744	-0.2250	4.3315	-103	-1.7977
258	4.5029	-0.9781	-0.2079	4.7046	-102	-1.7802
259	4.5204	-0.9816	-0.1908	5.1446	-101	-1.7628
260	4.5379	-0.9848	-0.1736	5.6713	-100	-1.7453
261	4.5553	-0.9877	-0.1564	6.3138	-99	-1.7279
262	4.5728	-0.9903	-0.1392	7.1154	-98	-1.7104
263	4.5902	-0.9925	-0.1219	8.1443	-97	-1.6930
264	4.6077	-0.9945	-0.1045	9.5144	-96	-1.6755
265	4.6251	-0.9962	-0.0872	11.4301	-95	-1.6581
266	4.6426	-0.9976	-0.0698	14.3007	-94	-1.6406
267	4.6600	-0.9986	-0.0523	19.0811	-93	-1.6232
268	4.6775	-0.9994	-0.0349	28.6363	-92	-1.6057
269	4.6949	-0.9998	-0.0175	57.2900	-91	-1.5882
270	4.7124	-1.0000	0.0000	undefined	-90	-1.5708
271	4.7298	-0.9998	0.0175	-57.2900	-89	-1.5533
272	4.7473	-0.9994	0.0349	-28.6363	-88	-1.5359
273	4.7647	-0.9986	0.0523	-19.0811	-87	-1.5184
274	4.7822	-0.9976	0.0698	-14.3007	-86	-1.5010
275	4.7997	-0.9962	0.0872	-11.4301	-85	-1.4835
276	4.8171	-0.9945	0.1045	-9.5144	-84	-1.4661
277	4.8346	-0.9925	0.1219	-8.1443	-83	-1.4486
278	4.8520	-0.9903	0.1392	-7.1154	-82	-1.4312
279	4.8695	-0.9877	0.1564	-6.3138	-81	-1.4137
280	4.8869	-0.9848	0.1736	-5.6713	-80	-1.3963
281	4.9044	-0.9816	0.1908	-5.1446	-79	-1.3788

Angle (deg)	Angle (rad)	sine	cosine	tangent	Angle (deg)	Angle (rad)
282	4.9218	-0.9781	0.2079	-4.7046	-78	-1.3614
283	4.9393	-0.9744	0.2250	-4.3315	-77	-1.3439
284	4.9567	-0.9703	0.2419	-4.0108	-76	-1.3265
285	4.9742	-0.9659	0.2588	-3.7321	-75	-1.3090
286	4.9916	-0.9613	0.2756	-3.4874	-74	-1.2915
287	5.0091	-0.9563	0.2924	-3.2709	-73	-1.2741
288	5.0265	-0.9511	0.3090	-3.0777	-72	-1.2566
289	5.0440	-0.9455	0.3256	-2.9042	-71	-1.2392
290	5.0615	-0.9397	0.3420	-2.7475	-70	-1.2217
291	5.0789	-0.9336	0.3584	-2.6051	-69	-1.2043
292	5.0964	-0.9272	0.3746	-2.4751	-68	-1.1868
293	5.1138	-0.9205	0.3907	-2.3559	-67	-1.1694
294	5.1313	-0.9135	0.4067	-2.2460	-66	-1.1519
295	5.1487	-0.9063	0.4226	-2.1445	-65	-1.1345
296	5.1662	-0.8988	0.4384	-2.0503	-64	-1.1170
297	5.1836	-0.8910	0.4540	-1.9626	-63	-1.0996
298	5.2011	-0.8829	0.4695	-1.8807	-62	-1.0821
299	5.2185	-0.8746	0.4848	-1.8040	-61	-1.0647
300	5.2360	-0.8660	0.5000	-1.7321	-60	-1.0472
301	5.2534	-0.8572	0.5150	-1.6643	-59	-1.0297
302	5.2709	-0.8480	0.5299	-1.6003	-58	-1.0123
303	5.2883	-0.8387	0.5446	-1.5399	-57	-0.9948
304	5.3058	-0.8290	0.5592	-1.4826	-56	-0.9774
305	5.3233	-0.8192	0.5736	-1.4281	-55	-0.9599
306	5.3407	-0.8090	0.5878	-1.3764	-54	-0.9425
307	5.3582	-0.7986	0.6018	-1.3270	-53	-0.9250
308	5.3756	-0.7880	0.6157	-1.2799	-52	-0.9076
309	5.3931	-0.7771	0.6293	-1.2349	-51	-0.8901
310	5.4105	-0.7660	0.6428	-1.1918	-50	-0.8727
311	5.4280	-0.7547	0.6561	-1.1504	-49	-0.8552
312	5.4454	-0.7431	0.6691	-1.1106	-48	-0.8378
313	5.4629	-0.7314	0.6820	-1.0724	-47	-0.8203
314	5.4803	-0.7193	0.6947	-1.0355	-46	-0.8029
315	5.4978	-0.7071	0.7071	-1.0000	-45	-0.7854
316	5.5152	-0.6947	0.7193	-0.9657	-44	-0.7679
317	5.5327	-0.6820	0.7314	-0.9325	-43	-0.7505
318	5.5501	-0.6691	0.7431	-0.9004	-42	-0.7330
319	5.5676	-0.6561	0.7547	-0.8693	-41	-0.7156
320	5.5851	-0.6428	0.7660	-0.8391	-40	-0.6981

(continued)

TABLE A.1 *(continued)*

Angle (deg)	Angle (rad)	sine	cosine	tangent	Angle (deg)	Angle (rad)
321	5.6025	-0.6293	0.7771	-0.8098	-39	-0.6807
322	5.6200	-0.6157	0.7880	-0.7813	-38	-0.6632
323	5.6374	-0.6018	0.7986	-0.7536	-37	-0.6458
324	5.6549	-0.5878	0.8090	-0.7265	-36	-0.6283
325	5.6723	-0.5736	0.8192	-0.7002	-35	-0.6109
326	5.6898	-0.5592	0.8290	-0.6745	-34	-0.5934
327	5.7072	-0.5446	0.8387	-0.6494	-33	-0.5760
328	5.7247	-0.5299	0.8480	-0.6249	-32	-0.5585
329	5.7421	-0.5150	0.8572	-0.6009	-31	-0.5411
330	5.7596	-0.5000	0.8660	-0.5774	-30	-0.5236
331	5.7770	-0.4848	0.8746	-0.5543	-29	-0.5061
332	5.7945	-0.4695	0.8829	-0.5317	-28	-0.4887
333	5.8119	-0.4540	0.8910	-0.5095	-27	-0.4712
334	5.8294	-0.4384	0.8988	-0.4877	-26	-0.4538
335	5.8469	-0.4226	0.9063	-0.4663	-25	-0.4363
336	5.8643	-0.4067	0.9135	-0.4452	-24	-0.4189
337	5.8818	-0.3907	0.9205	-0.4245	-23	-0.4014
338	5.8992	-0.3746	0.9272	-0.4040	-22	-0.3840
339	5.9167	-0.3584	0.9336	-0.3839	-21	-0.3665
340	5.9341	-0.3420	0.9397	-0.3640	-20	-0.3491
341	5.9516	-0.3256	0.9455	-0.3443	-19	-0.3316
342	5.9690	-0.3090	0.9511	-0.3249	-18	-0.3142
343	5.9865	-0.2924	0.9563	-0.3057	-17	-0.2967
344	6.0039	-0.2756	0.9613	-0.2867	-16	-0.2793
345	6.0214	-0.2588	0.9659	-0.2679	-15	-0.2618
346	6.0388	-0.2419	0.9703	-0.2493	-14	-0.2443
347	6.0563	-0.2250	0.9744	-0.2309	-13	-0.2269
348	6.0737	-0.2079	0.9781	-0.2126	-12	-0.2094
349	6.0912	-0.1908	0.9816	-0.1944	-11	-0.1920
350	6.1087	-0.1736	0.9848	-0.1763	-10	-0.1745
351	6.1261	-0.1564	0.9877	-0.1584	-9	-0.1571
352	6.1436	-0.1392	0.9903	-0.1405	-8	-0.1396
353	6.1610	-0.1219	0.9925	-0.1228	-7	-0.1222
354	6.1785	-0.1045	0.9945	-0.1051	-6	-0.1047
355	6.1959	-0.0872	0.9962	-0.0875	-5	-0.0873
356	6.2134	-0.0698	0.9976	-0.0699	-4	-0.0698
357	6.2308	-0.0523	0.9986	-0.0524	-3	-0.0524
358	6.2483	-0.0349	0.9994	-0.0349	-2	-0.0349
359	6.2657	-0.0175	0.9998	-0.0175	-1	-0.0175
360	6.2832	0.0000	1.0000	0.0000	0	0.0000

Answers to Selected Review Questions

This appendix contains the answers to most of the review questions listed at the end of each chapter. However, you should make a genuine effort to answer each question before referring to this appendix—use it only to check your answers.

Chapter 1

1. Kinesiology is the study of human movement. In the broad sense, it includes study of the anatomy, physiology, psychology, sociology, and mechanics of human movement. Biomechanics is the study of forces and their effects on living things. Kinesiology and biomechanics are thus two disciplines of study that overlap in the study of the mechanics of human movement.

8. Performance improvement and injury prevention.

9. Improving techniques, equipment, and training.

10. International Society of Biomechanics, American Society of Biomechanics, International Society of Biomechanics in Sports.

11. Digital computers.

Chapter 2

1. Deltoid.

2. Femur.

3. Thorax and abdomen.

4. Trapezius.

5. Sagittal plane and transverse axis.

6. Horizontal abduction (or horizontal extension) in the transverse plane around a longitudinal axis.

7. Internal rotation, adduction, and extension at the shoulder joint; extension at the elbow joint.

8. Adduction in the frontal plane around the AP axis.

9. In a narrow-grip pull-up, extension and flexion occur at the shoulder joint, while in a wide-grip pull-up, adduction and abduction occur at the shoulder joint.

10. Extension.

Chapter 3

1. Yes, if it was moving to begin with—but it cannot change its state of motion (accelerate). That is, it cannot speed up, slow down, or change direction since an external force is required to cause acceleration.

2. No. A change in direction of motion is an acceleration, and an external force is required to cause acceleration.

3. Yes, but only if those external forces yield a resultant force of zero.

4. The rope exerts 50 pounds on you, according to Newton's third law, the law of action and reaction.

5. Multiply your height in inches by 0.0254 m/in.

6. Divide your weight in pounds by 2.2 lb/kg.

Chapter 4

1. The resultant force is 943.4 N acting downward and backward against the block at an angle of 58° below horizontal.

2. 1000 N.

3. 42.3 N acting in a direction 11.5° north of west or 41.4 N west and 8.6 N north.

4. Yes.

5. Using 10 m/s/s for g: the resultant force acting on the skier is 248.4 N directed down the hill, parallel to the 30° slope.

   Using 9.81 m/s/s for g: the resultant force acting on the skier is 243.5 N directed down the hill, parallel to the 30° slope.

6. The normal reaction force she must exert on the racket is 444.4 N.

Chapter 5

1. 27 yards.

2. The shot from 5 m away will take 0.50 s to reach the goal, while the shot from 10 m away will take 0.25 s to reach the goal. The goalie will have a better chance of blocking the 5-m shot because it will take longer to reach him.

3. The batter has 0.44 s to react to the pitch after the pitcher has released the ball.

4. Generally, the faster the speed of the ball (or puck) the smaller the size of the goal.

5. 14.52 m/s in a direction 23.2° east of north or 11.66 m/s north and 5 m/s east.

6. Yes, the runner must be accelerating since the runner must change directions to move around a curve.

7. a. 1.83 m/s/s in the direction of the finish line.

   b. Zero—her velocity did not change.

   c. Toward the starting line.

8. 4.62 m/s downward.

9. Your acceleration is toward the center of the circle.

Chapter 6

1. 9.81 m/s/s downward.

2. a. 9 m/s forward.

   b. 9.0 m forward.

   c. 5.81 m/s downward.

3. Using 10 m/s/s for g: the dumbbell will accelerate 0.81 m/s/s upward.

   Using 9.81 m/s/s for g: the dumbbell will accelerate 1.0 m/s/s upward.

4. Using 10 m/s/s for g: gravitational force of 10 N pulling downward, contact force of 10 N pushing upward, contact force of 5.0 N pushing horizontally in the direction of the acceleration. The resultant of these forces is 5.0 N acting horizontally in the direction of the acceleration.

   Using 9.81 m/s/s for g: gravitational force of 9.81 N pulling downward, contact force of 9.81 N pushing upward, contact force of 5.0 N pushing horizontally in the direction of the acceleration. The resultant of these forces is 5.0 N acting horizontally in the direction of the acceleration.

5. a. 1000 N in the opposite direction of the force Tonya exerts on Nancy.

   b. Tonya will accelerate 16.67 m/s/s in the direction of the 1000 N force that Nancy exerts on her. Nancy will accelerate 20 m/s/s in the opposite direction.

6. Using 10 m/s/s for g: an upward force just greater than 150.0 N.

   Using 9.81 m/s/s for g: an upward force just greater than 147.15 N.

7. Using 10 m/s/s for g: 1.02 m/s/s backward (south).

   Using 9.81 m/s/s for g: 1.00 m/s/s backward (south).

8. 20.0 N in the opposite direction of the ball's velocity.

9. Using 10 m/s/s for g: 2800 N upward.
   Using 9.81 m/s/s for g: 2781 N upward

10. Using 10 m/s/s for g: 4.73 m/s/s down the
    30° slope.
    Using 9.81 m/s/s for g: 4.64 m/s/s down the
    30° slope.

follow-through, the smaller the force required by the muscles to slow down the limb or implement.

11. The longer limbs allow the performer to exert forces through a greater displacement, thus increasing the work done and consequently increasing the change in energy of the limb or implement.

Chapter 7

1. 400 J.

2. Using 10 m/s/s for g: 1775 J.
   Using 9.81 m/s/s for g: 1742 J.

3. a. Yes, you are doing positive mechanical work against your opponent since the force you are exerting and the displacement are in the same direction.

   b. Yes, your opponent is doing negative work against you since the force your opponent is exerting and the displacement are in opposite directions.

4. a. 120 J.

   b. Positive.

   c. 40 m/s horizontally in the direction of the force.

5. a. 120 J.

   b. –120 J.

   c. Negative.

   d. 1500 N in the opposite direction of the ball's motion.

6. The 5-kg ball rolling at 4 m/s has more energy.

7. Using 10 m/s/s for g: 1000 W.
   Using 9.81 m/s/s for g: 981 W.

8. The 45-cm jump by the 90-kg student was more powerful.

9. The sand does negative work on the jumper during impact. It reduces the jumper's energy by displacing while exerting a force on the jumper. The displacement of the sand reduces the impact force on the jumper.

10. During the follow-through, muscles do negative work on the swinging limb or implement to reduce their kinetic energy. The greater the displacement of the limb or implement during the

Chapter 8

1. 15000 N · cm or 150 Nm.

2. a. 3000 N · cm or 30 Nm flexor torque.

   b. 750 N.

3. Yes. The flexor muscles create a flexor torque of 40 Nm, which is sufficient to overcome the extensor torque of 37.5 Nm created by the weight.

4. a. Using 10 m/s/s for g: 1.03 kg or 10.3 N.
      Using 9.81 m/s/s for g: 1.01 kg or 10.1 N.

   b. Using 10 m/s/s for g: 10.3 Nm
      Using 9.81 m/s/s for g: 10.1 Nm

   c. Using 10 m/s/s for g: 205.9 N.
      Using 9.81 m/s/s for g: 202.0 N.

5. When leaning over to lift something, a large moment arm is created between the object being lifted and the lumbar spine. When the lift begins, this moment arm will create a large torque about the lumbosacral joint and require large forces in the back extensor muscles and create large stresses on the intervertebral disks in the lower back. The muscles and structures of the lower back are thus susceptible to injury when lifting heavy objects using this technique.

6. Using 10 m/s/s for g: the left hand lifts up on the pole with a force of 62.5 N and the right hand pushes down on the pole with a force of 40 N.
   Using 9.81 m/s/s for g: the left hand lifts up on the pole with a force of 61.3 N and the right hand pushes down on the pole with a force of 39.2 N.

Chapter 9

1. 96° or 1.68 rad of flexion.

2. 3.75 rev/s; or 1350°/s; or 23.6 rad/s.

3. 20 rad/s/s in the direction of his spin.

4. 20 in.

5. a. 35.6 m/s tangent to the circular path of the hammer.

   b. 1056 m/s/s toward the axis of rotation.

6. If both clubs can be swung with the same angular velocity, then the longer swing radius of the driver will cause a faster linear velocity of the club head according to the relationship $v = \omega r$.

7. 1500 cm/s or 15 m/s.

8. Besides the ability to do greater work due to greater displacement (see question 11 in chapter 7), a longer-limbed individual also has a longer swing radius, which may cause a faster linear velocity at the end of the limb according to the relationship $v = \omega r$.

9. The angular displacement of the leg during a step is related to the linear distance traveled by the foot during a step (which is related to the step length) as shown in the equation arc length = θr.

10. a. Yes, at an average angular velocity of 12 rad/s, the bat will rotate through an angular displacement of 1.65 rad to reach a hitting position above the plate by the time the ball reaches the plate.

    b. 37.5 m/s.

Chapter 10

1. a. 0.33 rad/s.
   b. 1.33 rad/s.

2. Carol.

3. Carol, Ann, Barbara, Donna.

4. Carol.

5. Donna.

6. Flexing the recovery leg more during the swing-through is more effective since it reduces the moment of inertia of the leg about the transverse axis through the hip joint, thus speeding up the swing-through.

7. Her legs rotate in the opposite angular direction.

8. He should rotate his arms clockwise, in the same direction as the fall.

9. a. Dive 1.
   b. Dive 3.

10. 5 rad/s.

11. 694 rad/s or 110.5 rev/s.

Chapter 11

1. Form drag.

2. The prone position would create the least amount of drag force.

3. As objects increase in size (volume), the surface area to mass ratio decreases (larger objects have less surface area relative to their mass). A bug's surface area to mass ratio is much larger than a human's. The effect of air resistance on a bug is therefore much larger as well since the drag force acting on the bug is larger relative to its mass (due to the relatively larger surface area), and the effect of this drag force is greater due to the smaller mass of the bug.

4. By inducing areas of turbulent flow around the ball.

5. By inducing the Magnus effect.

6. Air resistance has the largest effect on 500-m speedskating due to the greater speeds in this event.

7. Air resistance may provide a lift force on the discus, thus prolonging its flight. The vertical component of the initial velocity could be reduced because of this. So the optimal projection angle would be smaller than if air resistance had no effect.

8. From largest to smallest: Frisbee, beach ball, Nerf ball, tennis ball, baseball, discus, basketball, shot.

9. The sail acts as a wing and maximizes lift force while minimizing drag force due to the wind. The keel or centerboard of the boat also provides a lift force that prevents the boat from being pushed sideways. The sum of these forces, along with the drag forces from the water and the wind, is a resultant force that may have a component into the wind, thus providing the boat with an acceleration component into the wind.

11. To reduce drag forces.

Chapter 12

1. 0.001 cm/cm or 0.1%.

2. 200 N/mm^2; or 20,000 N/cm^2; or 200,000,000 Pa; or 200 MPa.

3. The lateral side of your knee undergoes compressive stress, while the medial side undergoes tensile stress.

4. 0.005 cm/cm or 0.5%.

5. The tendon will elongate 0.6 cm, i.e., stretch to a length of 10.6 cm.

6. The anterior cruciate ligament is mechanically tougher.

7. The shear stress will be greatest at the outermost surfaces of the bone.

8. Bone.

9. Ligament.

10. a. Compressive stress.
    b. Shear stress.

Chapter 13

1. Support, movement, and protection.

2. Mineral storage and hemopoeisis.

3. a. Long bone.
    b. Irregular bone.
    c. Short bone.
    d. Long bone.
    e. Sesamoid bone.

4. The mineral component.

5. Puberty.

6. Gliding, pivot, hinge, ellipsoidal, saddle, and ball and socket.

7. Ball and socket.

8. Synovial fluid lubricates a synovial joint, provides nourishment to the articular cartilage, cleanses the joint cavity, and provides some shock absorption.

9. The synovial membrane and articular cartilage keep the synovial fluid in the joint cavity.

10. Articular cartilage improves bone-to-bone fit, increases joint stability, reduces bearing pressure, reduces friction, and provides some shock absorption.

11. Stability of a synovial joint is affected by the degree of fit between the articulating surfaces, the articular capsule and ligaments that cross the joint, the muscles and tendons that cross the joint, and the negative pressure gradient between the atmosphere and the joint cavity.

Chapter 14

1. Movement and posture maintenance, heat production, protection, and pressure alteration.

2. The sarcomere.

5. Concentric activity occurs during the delivery phase of the throw, while eccentric activity occurs during the follow-through.

6. Eccentric activity occurs during the down (coiling or preparatory) phase followed by concentric activity during the up or propulsive phase while the jumper is on the ground.

7. Biceps brachii and brachialis.

8. Rectus femoris, pectineus, iliopsoas, tensor fascia latae, and sartorius.

9. Pronator teres and the pronator quadratus.

10. Muscles with longitudinal fiber arrangements usually have greater range of motion, while muscles with pennate fiber arrangements are usually stronger than similarly sized muscles with longitudinal fiber arrangements.

11. Your hip extensor muscles produce greater forces during landing because their action is eccentric during this phase.

12. Your shoulder muscles produce greater forces to put a shot because their contraction velocity is slower during this activity than during a baseball throw.

Chapter 15

1. Peripheral nervous system.

2. Away from the cell body.

3. The central nervous system can increase the active tension in a muscle by recruiting more motor units or by increasing the rate of stimulation.

4. Muscle spindles.

5. Golgi tendon organs.

6. Muscle spindles.

7. Golgi tendon organs.

9. Flexing the neck and turning the head to the opposite side of the arm doing the curl exercise.

10. Pressure.

Chapter 16

1. Propose a model, examine the skill, test or evaluate the performance, and educate the performer.

2. A quantitative biomechanical analysis involves measuring performance parameters and representing them numerically, while in a qualitative biomechanical analysis they are only observed and described subjectively.

4. a. To the right or left side of the runway for a sagittal view of the takeoff, flight, and landing of the jumper.

   b. In front of or behind the hurdler for a frontal view.

   c. To the right or left side of the swimmer for a sagittal view.

   d. To the rear to view the flight of the ball toward the goal or to the side for a sagittal view of the kicker.

   e. To the side (perpendicular to the arrow's path) for a frontal view of the archer.

Chapter 17

1. Palpation of superficial muscles during activity; identification of sore muscles following activity; EMG; inverse dynamic analysis.

2. Elbow extensors and shoulder horizontal flexors.

3. Knee extensors; hip extensors; ankle plantar flexors.

4. Hip extensors; knee extensors; ankle plantar flexors.

Chapter 18

1. Stress is defined as force/area. Stress may be reduced by either decreasing the magnitude of the applied force or increasing the area over which the force is distributed. Using a softer material to construct a striking implement, such as a bat or hockey stick, reduces the size of the imposed force at contact because of the deformation of the softer material. The hard outer shell of a modern protective helmet is supported with a suspension system that rests on the head. At contact, the hard shell and suspension system distribute the force over the surface of the head, reducing the stress at any single point.

4. The stress threshold refers to the level of stress that exceeds the level of stress to which a tissue has adapted. Loads above the stress threshold cause microdamage and stimulate the remodeling process that leads to a tissue getting stronger, or hypertrophying. However, if a load above the stress threshold is imposed before the tissue has time to remodel, the cumulative effect of stressing a tissue that has not repaired itself is an overuse injury. Traumatic injury occurs when a single level of force far exceeding the stress threshold is applied, causing damage to the tissue and impairing performance.

6. Cross-training and within-activity cross-training are both intended to alter the pattern and magnitude of stress imposed on the body. Cross-training refers to performing two different activities, such as walking and swimming, while within-activity cross-training refers to performing the same task with slight modifications, such as jogging on a track compared to jogging on a wood-chip trail. Either may be successful in preventing injury because the change in movement pattern imposes different levels of stress on tissues of the body, reducing the cumulative effect that may lead to an overuse injury.

Glossary

abduction—Starting from anatomical position, the joint action that occurs around an anteroposterior (AP) axis through a joint and causes limb movements in a frontal plane through the largest range of motion; the opposite of adduction; when referring to shoulder girdle movement, the movement of the scapula away from the midline of the body.

absolute angular position—Orientation of a line relative to another line or plane that is fixed relative to the earth; expressed in degrees or radians.

Achilles tendon—The large tendon on the posterior side of the distal leg; attaches the triceps surae (soleus and gastrocnemius) to the calcaneus.

adduction—The joint action that occurs around an AP axis through a joint and causes limb movement in a frontal plane back toward anatomical position; the opposite of abduction; when referring to shoulder girdle movement, the movement of the scapula toward the midline of the body.

agonist—Role of a muscle whose torque aids the action referred to; prime mover; opposite in meaning to antagonist.

anatomical analysis—An examination of movement to determine joint actions and active muscles.

anatomical position—The position assumed by the body when it is standing erect, facing forward, both feet aligned parallel to each other, toes forward, arms and hands hanging straight below the shoulders at the sides, fingers extended, and palms facing forward.

angular acceleration—Rate of change of angular velocity; measured in units of angular displacement divided by time squared or expressed as radians per second per second, degrees per second per second, or revolutions per second per second; a vector quantity.

angular displacement—Change in absolute angular position experienced by a rotating line with the direction of the change indicated; the angle between a line segment in its initial position and in its final position with the direction of rotation noted; measured as units of angular position or expressed as radians, degrees, or revolutions.

angular impulse—Average torque times the duration of application of the torque; causes and thus is equal to change in angular momentum; measured in units of torque times units of time or expressed as newton-meter-seconds in SI; a vector quantity.

angular inertia—Property of an object to resist changes in its angular motion; also referred to as rotary inertia.

angular momentum—Moment of inertia times angular velocity; measured in units of mass times units of velocity or expressed as kilogram-meters squared per second in SI; a vector quantity.

angular motion—Change in position that occurs when all points on a body or object move in circular paths about the same fixed axis; also referred to as rotary motion or rotation.

angular position—Orientation of a line relative to some other line or plane; expressed in degrees or radians.

angular speed—Rate of change of orientation with no regard for direction or changes in direction of rotation; measured in units of angular position divided by units of time or expressed as radians per second, degrees per second, or revolutions per second; a scalar quantity.

angular velocity—Rate of change of angular displacement with direction of rotation indicated; measured in units of angular displacement divided by units of time or expressed as radians per second, degrees per second, or revolutions per second; a vector quantity.

anisotropic—Material properties are dependent on the direction; that is, an anisotropic material may have greater yield strength and a stiffer elastic modulus when pulled in one direction compared to another.

annulus fibrosus—The fibrous cartilage that forms the outer ring of an intervertebral disk.

antagonist—Role of a muscle whose torque opposes the action referred to or the muscle referred to; opposite in meaning to agonist.

anterior—Toward the front of the body; ventral.

anterior cruciate ligament—One of two internal ligaments of the knee joint; attaches anteriorly on the tibia and posteriorly on the medial surface of the lateral condyle of the femur.

anteroposterior axis—Any one of the imaginary lines running from anterior to posterior and perpendicular to the frontal planes; abbreviated as AP axis; sagittal axis; sagittal-transverse axis.

AP axis—See anteroposterior axis.

aponeurosis—A flat sheet of tendon that connects a broad muscle to a line of attachment on a bone.

appendicular skeleton—That part of the skeleton composed of the bones of the appendages, including the bones of the shoulder girdle (scapula and clavicle) and the bones of the pelvic girdle (ilium, ischium, and pubis) with the exception of the sacrum.

applied biomechanics research—Study directed toward answering specific, practical, useful questions.

articular capsule—Sleeve of ligamentous connective tissue surrounding a synovial joint and attached to the bones on either side of the joint; also referred to as joint capsule.

articular cartilage—Hyaline cartilage covering the articular surfaces of bones in a synovial joint.

autonomic nervous system—Part of the nervous system involved with unconscious sensations and actions; also referred to as involuntary nervous system.

average acceleration—Change in velocity divided by the time it took for the change to occur; expressed as units of length per unit of time squared or as meters per second per second in SI; a vector quantity.

average angular velocity—Angular displacement divided by the time it took for that angular displacement to occur with direction of rotation indicated; measured in units of angular displacement divided by units of time or expressed as radians per second, degrees per second, or revolutions per second; a vector quantity.

average speed—Distance traveled divided by the time it took to travel that distance; expressed as units of length per unit of time or as meters per second in SI; a scalar quantity.

average velocity—Displacement divided by the time it took for the displacement to occur; expressed as units of length per unit of time or as meters per second in SI; a vector quantity.

axial skeleton—That part of the skeleton composed of the bones of the vertebral column, ribs, and skull.

base of support—Area beneath and between the points of contact an object has with the ground.

bending load—Combination of forces that produce tensile stresses near one surface of an object, compressive stresses near the opposite surface, and shear stress throughout the object so that the object becomes a beam; as a result of a bending load, an object will deform by deflecting in a curved shape.

Bernoulli's principle—Lateral pressure in a moving fluid decreases as the velocity of the fluid increases.

biomechanics—The study of forces and their effects on living systems.

buoyant force—Upward force acting on an object in a fluid that is equal to the weight of the fluid displaced by the object.

cancellous bone—Porous, less dense bone tissue found deep to cortical bone near the ends of long bones; also referred to as spongy bone or trabecular bone.

cardinal plane—A plane that passes through the midpoint or center of gravity of the body.

carpal—One of eight short bones of the wrist.

Cartesian coordinate system—A system for locating the position of a point in two (or three) dimensions; the two (or three) coordinates represent displacements from a fixed point or origin in specific directions at right angles to each other.

cartilaginous joint—Joint in which fibrous cartilage or epiphyseal cartilage joins bones together, usually allowing slight movement; examples include the pubic symphysis and the epiphyseal growth plates in immature long bones.

center of gravity—Imaginary point through which the resultant force of gravity acts on an object; the point at which the entire weight of the body may be assumed to be concentrated; the point about which the torques created by the weights of each of the body parts balance; the point of balance of the body.

center of pressure—The theoretical point of application of the dynamic fluid force to an object.

central nervous system—All nervous tissue lying within the skull and vertebral column; the brain and the spinal cord.

centric force—Force whose line of action passes through the center of gravity of an object.

centripetal acceleration—Linear acceleration of a point on a rotating object measured in the direction perpendicular to the circular path of the object (along a line through the axis of rotation or along a radial line); measured in units of length divided by units of time squared or expressed as meters per second per second in SI; a vector quantity.

centripetal force—An external force directed toward the axis of rotation of an object moving in a circular path.

circumduction—Flexion combined with abduction and then adduction or extension and hyperextension combined with abduction and then adduction; the trajectory of a limb being circumducted forms a cone-shaped surface, and the end of the circumducted limb traces a circle.

colinear forces—Two or more forces that have the same line of action (but not necessarily the same direction along this common line of action).

combined load—Combination of forces that produce axial compression, axial tension, bending, torsion, shear loads, or any combination of these loads on an object.

compact bone—See cortical bone.

compression—State of an object as a result of forces pushing on it that are producing compressive stress; compressive stress is axial stress that tends to push molecules together and squash the object.

compressive force—Pushing force whose direction and point of application would tend to shorten or squeeze an object along the dimension coinciding with the line of action of the force.

concentric contraction—Muscular activity that occurs when the muscle develops tension and its points of attachment move closer together; muscular activity that occurs when the muscle does positive work; also referred to as concentric action or concentric activity.

concurrent forces—Two or more forces whose lines of action intersect at a single point.

contact force—A force that results when two objects touch each other.

cortical bone—Solid, dense bone tissue found in the outer layer of bones; also referred to as compact bone.

creep—Material behavior characterized by a continued increase in strain under a constant stress.

curvilinear translation—Linear motion that occurs when an object maintains its orientation during a movement so that all points on the object move the same distance, in the same direction, in the same time, but not in straight lines; also referred to as curvilinear motion.

deep—Describes the relative position of an anatomical structure and indicates that it is farther from the exterior surface of the body.

density—Mass of an object divided by its volume; measured as units of mass divided by units of length cubed or expressed as kilograms per cubic meter in SI.

depression—Inferior movement of the scapula in a frontal plane; the opposite of elevation.

diaphysis—Central part or shaft of a long bone, which is separated from the ends or epiphyses by the epiphyseal cartilage prior to ossification of the epiphyseal cartilage.

displacement—Change in location of a point expressed as the length and direction of the vector from the starting position to the ending position; expressed as units of length or as meters in SI.

distal—A location or direction toward or closer to the end of the limb farthest from its attachment to the body when referring to direction or position along a limb.

distance traveled—Length of the path followed by an object when moving from a starting position to an ending position; expressed as units of length or as meters in SI.

dorsiflexion—Starting from anatomical position, the ankle joint action that occurs around a transverse axis and causes the foot to move in a sagittal plane such that it moves forward and upward toward the leg; the opposite of plantar flexion.

downward rotation—Rotation of the scapula in the frontal plane such that its medial border moves superiorly and the shoulder joint moves inferiorly; the opposite of upward rotation.

drag force—Component of resultant dynamic fluid force that acts on an object in opposition to the relative motion of the object through the fluid.

dynamic fluid force—The resultant of the lift and drag forces that act on an object; the result of pressures exerted on the surfaces of the object.

dynamic friction—Frictional force that develops between two surfaces in contact that are moving or sliding relative to each other; sliding friction; kinetic friction.

dynamics—Branch of rigid-body mechanics concerned with the accelerated motion of objects.

eccentric contraction—Muscular activity that occurs when a muscle develops tension and its points of attachment move farther apart; muscular activity that occurs when the muscle does negative work; also referred to as eccentric action or eccentric activity.

eccentric force—Force whose line of action does not pass through the center of gravity of an object.

elastic—Able to return to its original dimensions from a deformed shape when the stress causing the deformation is removed.

elastic cartilage—Cartilage with more elastin than fibrous cartilage; found in the ears, the epiglottis, and part of the larynx; also referred to as yellow fibrocartilage.

elastic limit—Point on the stress-strain curve beyond which plastic deformation will occur; also referred to as proportional limit.

elastic modulus—Ratio of stress to strain; slope of the elastic region of the stress-strain curve for a material; also referred to as modulus of elasticity and Young's modulus.

elevation—Superior movement of the scapula in a frontal plane; the opposite of depression.

endolymph—Thick fluid that fills the inner ear.

endomysium—Connective tissue sheath or envelope encasing a muscle fiber.

energy—The capacity to do work; expressed as units of force times units of length or as joules in SI; a scalar quantity.

epimysium—Connective tissue sheath or envelope encasing a whole muscle.

epiphyseal cartilage—Cartilage that separates the diaphysis from the epiphyses of long bones; responsible for the growth in length of long bones prior to its ossification, usually during or shortly after puberty; also referred to as epiphyseal plate, epiphyseal disk, or growth plate.

epiphysis—Part of a long bone that is separated from the diaphysis or shaft of the long bone by the epiphyseal cartilage prior to ossification of the epiphyseal cartilage.

eversion—Starting from anatomical position, the joint action that occurs around an AP axis through the ankle when the lateral side of the sole of the foot is lifted; the opposite of inversion.

exercise and sport biomechanics—The study of forces and their effects on humans in exercise and sport.

extension—The joint action that occurs around a transverse axis through a joint and causes limb movement in a sagittal plane back toward anatomical position; the opposite of flexion.

external force—A force that acts on an object as a result of its interaction with the environment surrounding it.

external rotation—Shoulder or hip joint actions that occur around longitudinal axes through these joints and cause limb movements in the transverse plane such that the knees turn away from each other or the palms of the hands turn away from the body; the opposite of internal rotation; lateral rotation; outward rotation.

exteroceptor—Sensory receptor that responds to stimuli from sources external to the body.

extrinsic factors—Factors related to injury and to characteristics of the task and the environment; they include the nature of the task being performed, the intensity and frequency of the performance, the playing surface, equipment, and level of participation.

failure strain—Strain at which a material breaks or fails.

failure strength—Stress at which a material breaks or fails.

fascicle—Bundle of muscle fibers; also referred to as a fasciculus.

femur—The long bone of the thigh.

fibrous cartilage—Cartilage with heavier collagen fibers than hyaline cartilage; the intervertebral disks and articular disks such as the menisci are examples of fibrous cartilage; also referred to as fibrocartilage or white fibrocartilage.

fibrous joint—Joint in which fibrous connective tissue joins bones together, usually in a rigid manner; sutures and syndesmoses are types of fibrous joints.

flexion—Starting from anatomical position, the joint action that occurs around a transverse axis through a joint and causes limb movement in a sagittal plane away from anatomical position through the largest range of motion; the opposite of extension.

force—A push or pull.

force couple—Torque created by a pair of oppositely directed forces about an axis; expressed as units of force times units of length or as newton-meters in SI; a vector quantity.

form drag—Drag force acting on an object within a fluid and caused by the impact forces of the fluid molecules with the object; also referred to as shape drag, profile drag, or pressure drag.

free-body diagram—A tool for analyzing forces and torques; a drawing of the analysis object with all external forces acting on the object represented as arrows showing their points of application and directions.

friction—The component of a contact force that acts parallel to the surfaces in contact; the magnitude of friction is the product of the coefficient of friction and the normal contact force (the component of the contact force acting perpendicular to the surfaces in contact).

frontal plane—An imaginary plane running side to side and superior to inferior that divides the body into anterior and posterior parts; coronal plane; lateral plane.

general motion—A change in position that results from a combination of linear and angular motion.

Golgi tendon organ—Proprioceptor that responds to increases in muscle tension.

gravitational acceleration—The rate of change in velocity caused by the force of gravity; approximately 9.81 m/s/s downward or 32 ft/s/s downward; also referred to as acceleration due to gravity.

gravitational potential energy—Energy due to the vertical position of an object; weight times height above some reference or mass times acceleration due to gravity times height above some reference; expressed as units of force times units of length or as joules in SI; a scalar quantity.

hamstrings—The group of three two-joint muscles of the posterior thigh: biceps femoris, semitendinosus, and semimembranosus.

horizontal abduction—Starting from a position of hip or shoulder flexion, the shoulder or hip joint action that causes movement of the arm or thigh in the transverse plane around a longitudinal axis such that the arm or leg moves away from the midline of the body; the opposite of horizontal adduction; horizontal extension.

horizontal adduction—Starting from a position of hip or shoulder abduction, the shoulder or hip joint action that causes movement of the arm or leg in the transverse plane around a longitudinal axis such that the arm or thigh moves back toward the midline of the body; the opposite of horizontal abduction; horizontal flexion.

humerus—The long bone of the arm.

hyaline cartilage—Shiny, white, smooth cartilage that makes up the articular cartilage covering the articular surfaces of bones in synovial joints.

hyperextension—The joint action that occurs around a transverse axis and is a continuation of extension past anatomical position.

iliopsoas—The iliacus and psoas muscles; deep muscles that cross anterior to the axis of the hip joint and thus are major flexor muscles of the hip.

impulse—Average force times the duration of application of the force; causes and thus is equal to change in linear momentum; measured in units of force times units of time or expressed as newton-seconds in SI; a vector quantity.

impulsive load—An applied force that reaches a relatively high magnitude in a short time.

inertia—The property of an object to resist changes in its motion.

inferior—Toward or closer to the feet; caudal.

instantaneous acceleration—Rate of change in velocity measured at an instant in time rather than over a duration of time; expressed as units of length per unit of time squared or as meters per second per second in SI; a vector quantity.

instantaneous angular velocity—The rate of change of angular displacement measured at an instant in time rather than over a duration of time with direction of rotation indicated; measured in units of angular displacement divided by units of time or expressed as radians per second, degrees per second, or revolutions per second; a vector quantity.

instantaneous speed—Rate of change in distance traveled measured at an instant in time rather than over a duration of time; expressed as units of length per unit of time or as meters per second in SI; a scalar quantity; a car's speedometer measures instantaneous speed.

instantaneous velocity—The rate of change of displacement measured at an instant in time rather than over a duration of time; expressed as units of length per unit of time or as meters per second in SI; a vector quantity.

internal force—A force that acts within the object or system whose motion is being investigated; forces between the molecules of an object that hold the object together.

internal rotation—Starting from anatomical position, shoulder or hip joint actions that occur around longitudinal axes through these joints and cause limb movements in the transverse plane such that the knees turn inward toward each other or the palms of the hands turn toward the body; the opposite of external rotation; also referred to as medial rotation or inward rotation.

interneuron—Nerve cell that transmits impulses between neurons; also referred to as connector neuron.

interoceptor—Sensory receptor that responds to stimuli from sources within the body.

intervertebral disk—The cartilaginous disk found between the bodies of adjacent vertebrae.

intrinsic factors—Factors related to injury and to an individual's ability to cope with imposed mechanical stress; they include anthropometrics, skeletal alignment, fitness, and previous history of injury.

inversion—Starting from anatomical position, the joint action that occurs around an AP axis through the ankle when the medial side of the sole of the foot is lifted; the opposite of eversion.

isometric contraction—Muscular activity that occurs when a muscle develops tension and its points of attachment do not move relative to each other; muscular activity that occurs when a muscle develops tension and does zero work; also referred to as isometric action or isometric activity.

isotropic—Material properties are the same in all directions for an isotropic material.

joint—Where two bones join or meet and connect to one another.

joule—SI unit of measure for work and energy; equal to one newton-meter. 1 J = 1 Nm.

kinematics—The branch of dynamics concerned with the description of motion.

kinesiology—The study of human movement.

kinetic energy—Energy due to the motion of an object; half of the mass times the square of the velocity of the object; expressed as joules in SI; a scalar quantity.

kinetics—The branch of dynamics concerned with the forces that cause or tend to cause motion.

laminar flow—Movement of fluid molecules such that adjacent layers of fluid flow parallel to one another and closely follow the shape of an object in the fluid.

lateral—Away from or farther from the midline of the body.

lateral flexion—Movement of the trunk or neck to the left or right in a frontal plane around an AP axis.

lift force—Component of resultant dynamic fluid force that acts on an object in a direction perpendicular to the relative motion of the object through the fluid.

linear momentum—Mass of an object times the linear velocity of the object; measured in units of mass times units of length divided by units of time or as kilogram-meters per second in SI; a vector quantity.

linear motion—Change in position that occurs when all points on an object move the same distance, in the same direction, and at the same time; also referred to as translation.

longitudinal axis—Any of the imaginary lines running from superior to inferior and perpendicular to the transverse planes; vertical axis; frontal-sagittal axis; twist axis.

Magnus effect—An object's deviation from its normal flight path due to the effect of air resistance and the spin of the object.

mass—Measure of inertia; the quantity of matter in an object.

mechanics—The study of forces and their effects on objects.

medial—Toward or closer to the midline of the body.

meniscus—One of two half-moon–shaped cartilaginous disks found within the knee joint.

metacarpal—One of the five long bones of the hand.

metatarsal—One of the five long bones of the foot.

modulus of elasticity—See elastic modulus.

moment arm—Perpendicular distance between the line of action of a force and the axis about which a moment of force or torque is being measured; determined by measuring the shortest distance between a line drawn along the line of action of the force and another line drawn parallel to this line but through the axis about which the torque is being measured.

moment of force—Torque created by a force about an axis; force times moment arm; expressed as units of force times units of length or as newton-meters in SI; a vector quantity.

moment of inertia—Measure of angular inertia; sum of the product of the mass of each part of an object and the square of the distance each part is from the center of gravity of the object; mass times radius of gyration squared; measured in units of mass times units of length squared or expressed as kilogram-meters squared in SI; a scalar quantity.

motor neuron—Nerve cell that transmits impulses away from the central nervous system; also referred to as efferent neuron.

motor unit—A single motor neuron and all the muscle fibers with which it synapses; the fundamental unit of the neuromuscular system.

muscle fiber—A single muscle cell.

muscle spindle—Proprioceptor that responds to increases in muscle length.

net force—The vector sum of all the external forces that act on an object; the resultant of all the external forces that act on an object; the resultant force.

neutralizer—Role of a muscle whose torque cancels or eliminates the undesired effect of the torque produced by another muscle at that joint to allow a desired movement only.

nucleus pulposus—The spongy gel-like structure found at the center of an intervertebral disk.

overuse injury—An injury caused by repeated applications of a stress lower than that required to cause injury in a single application; also known as repetitive motion injury, cumulative trauma injury, or predictable injury.

pacinian corpuscle—Sensory receptor for pressure.

patella—The sesamoid bone found anterior to the knee joint; otherwise known as the kneecap.

pelvic girdle—The bony structure formed by the joining of the three hip bones from each side (ilium, ischium, and pubis) anteriorly at the pubis symphis and posteriorly with the sacrum.

perimysium—Connective tissue sheath or envelope encasing a bundle of muscle fibers (fascicle).

peripheral nervous system—All nervous tissue lying outside the skull and vertebral column.

phalange—The long bones of the fingers or toes, otherwise known as phalanx.

physical training—Training directed at altering performance limitations due to the physical condition of the performer; involves improving the components of physical fitness, including muscular strength, flexibility, and cardiovascular fitness.

physiologic loading zone—Level of imposed stress within which tissue maintains its current status; muscle maintains the same force-generating capability, bone mineral content stays the same, and tendons and ligaments maintain their ability to withstand tensile stress.

plantar flexion—The ankle joint action that occurs around a transverse axis and causes the foot to move in a sagittal plane downward and away from the leg.

plastic—Able to maintain its deformed shape when the stress causing the deformation is removed.

Poisson's ratio—Ratio between transverse strain and strain in the axial direction for an object loaded axially.

position—Location of a point in space relative to some fixed point.

posterior—Toward the rear of the body; dorsal.

potential energy—Energy stored within an object due to its vertical position or deformation; expressed as units of force times units of length or as joules in SI; a scalar quantity.

power—Rate of doing work; work done divided by time; measured in units of work divided by units of time or expressed as watts in SI; a scalar quantity.

pressure—External force divided by the area over which this force acts; measured in units of force divided by units of length squared or expressed as newtons per square meter in SI.

principal axis—Axis about which an object's moment of inertia is largest; axis about which an object's moment of inertia is smallest; axis perpendicular to the two previously defined principal axes.

projectile—An object that has no external forces acting on it other than the force of gravity.

projectile motion—Motion that results when the only external force acting on an object is the force of gravity; the path a projectile follows is described as a parabolic curve.

pronation—Starting from anatomical position, the radioulnar joint action that occurs around the longitudinal axis of the forearm and causes the palm to turn toward the body; the opposite of supination.

proportional limit—See elastic limit.

proprioceptor—Interoceptor that monitors the status of the musculoskeletal system, including joint position, change in joint position, muscle length, change in muscle length, and muscle tension.

proximal—A location or direction toward or closer to the attachment of a limb to the trunk when referring to direction or position along a limb.

quadriceps femoris—The group of four anterior thigh muscles: rectus femoris, vastus lateralis, vastus medialis, and vastus intermedius.

qualitative anatomical analysis—An analysis in which the predominant muscular activity and the active muscle groups during each phase of a movement are identified.

qualitative biomechanical analysis—A biomechanical analysis in which the mechanical characteristics of the performance are not quantified but are observed and subjectively evaluated.

quantitative biomechanical analysis—A biomechanical analysis in which the mechanical characteristics of the performance are measured and quantified.

radial deviation—The joint action that occurs around an AP axis through the wrist joint and causes movement of the hand in a frontal plane in the direction of the thumb; the opposite of ulnar deviation; also referred to as abduction or radial flexion.

radian—Ratio of the arc length the end of a radial line travels through to the length of the radial line when the radial line rotates about one end; arc length divided by radius; 1 rad equals approximately 57.3°.

radius—The long bone that lies on the lateral side of the forearm; its distal end is larger than its proximal end.

radius of gyration—The distance from an axis of rotation to where the mass of an object would have to be concentrated to create the same moment of inertia of the entire object about that axis; measured in units of length or expressed as meters in SI; a scalar quantity.

reaction force—An external contact force resulting from one object touching another.

rearfoot motion—The natural sequential pattern of pronation and supination during the stance phase of running; measured for research and clinical purposes in the frontal plane as the angle between the shoe and the lower leg.

rectilinear translation—Linear motion that occurs when an object maintains its orientation during a movement so that all points on the object move the same distance, in the same direction, in the same time, in straight lines; also referred to as rectilinear motion.

rectus femoris—One of the four quadriceps muscles of the anterior thigh; the only one that crosses the hip as well as the knee joint.

relative angular position—Orientation of a line relative to another line or plane that may not be fixed; expressed in degrees or radians.

remodeling—The rebuilding of tissue, or the healing of microdamage caused by an imposed stress; the healing of the tissue leads to hypertrophy, or strengthening, of the tissue.

resultant displacement—Change in location of a point expressed as the length and direction of the vector from the starting position to the ending position.

resultant force—The vector sum of two or more forces; the force that results from the vector addition of two or more forces.

rigid-body mechanics—The branch of mechanics concerned with the effects of forces on objects that are assumed to be perfectly rigid.

rotary inertia—See angular inertia.

sagittal plane—An imaginary plane running anterior to posterior and superior to inferior that divides the body into right and left parts; anteroposterior plane.

sarcolemma—Thin membrane covering a muscle cell.

sarcomere—Fundamental contractile unit of muscle, found between two adjacent Z lines in a myofibril.

scapula—The shoulder blade, an irregular bone of the superior back that articulates distally with the humerus and clavicle.

sensory neuron—Nerve cell that transmits sensory impulses to the central nervous system; also referred to as afferent neuron.

sesamoid bone—Bone completely encased in connective tissue such as tendon or ligament; for example, the patella (kneecap).

shear—Force or stress acting parallel to the analysis plane or perpendicular to the long axis of the object; shear stress tends to slide molecules past each other and skew the object.

shoulder girdle—The structure formed by the left and right scapulas and clavicles and their attachments to the sternum.

soleus—The deep muscle of the posterior leg that attaches to the Achilles tendon.

somatic nervous system—Part of the nervous system involved in conscious sensations and actions; also referred to as voluntary nervous system.

specific gravity—Density of an object divided by the density of water.

spongy bone—See cancellous bone.

stability— The resistance of an object to being toppled; the likelihood of an object returning to its original position after it is displaced.

stabilizer—A muscle whose torque prevents movement at a joint.

static equilibrium—The state or condition of an object that results when the object is not moving and the net force and net torque acting on the object are zero.

static friction—Frictional force that develops between two surfaces in contact that are not moving relative to each other; limiting friction.

statics—The branch of rigid-body mechanics concerned with the mechanics of objects at rest or moving at constant velocity.

strain energy—Energy due to the deformation of an object; for stretching or compressing, it is equal to half the stiffness constant of the material times the square of change in length of the object; expressed as units of force times units of length or as joules in SI; a scalar quantity.

stress—Internal force divided by the cross-sectional area of the surface upon which the internal force acts; measured in units of force divided by units of length squared; expressed as newtons per square meter or pascals in SI.

stress continuum—A graphical representation of the level of imposed stress, including zones in both pathologically high and low levels (distress) and physiological levels (eustress).

superficial—Relative position closer to the exterior surface of the body.

superior—Toward or closer to the head; cranial; cephalo.

supination—The radioulnar joint action that occurs around the longitudinal axis of the forearm and causes limb movement in a transverse plane and returns the forearm and hand to anatomical position after being pronated or moves them beyond anatomical position; the opposite of pronation.

surface drag—Drag force acting on an object within a fluid and caused by friction between the fluid and the surface of the object; also referred to as skin friction or viscous drag.

synergy—When two or more muscles crossing the same joint act together to produce a desired result.

synovial joint—A freely movable joint characterized by a joint cavity filled with synovial fluid contained within a synovial membrane that lines a joint capsule.

synovial membrane—Thin membrane that is the interior lining of the joint capsule; it produces synovial fluid.

tangential acceleration—Linear acceleration of a point on a rotating object measured in the direction tangent to the circular path of the object; measured in units of length divided by units of time squared or expressed as meters per second per second in SI; a vector quantity.

tarsals—The short bones of the ankle and foot.

technical training—Training devoted to improving technique; may involve performing the actual skill or performing drills that mimic specific aspects of the skill.

tendon—Cord or sheet of collagenous connective tissue that attaches muscle to bone.

tensile force—Pulling force whose direction and point of application would tend to lengthen or stretch an object along the dimension coinciding with the line of action of the force.

tension—State of an object as a result of forces pulling on it and producing tensile stress; tensile stress is axial stress that tends to pull molecules apart and stretch the object.

tibia—The large, long bone of the leg.

tibialis anterior—The superficial muscle of the anterior leg.

tonic neck reflex—Rudimentary reflex associated with the position of the neck and affecting the muscles of the upper extremity.

torque—The turning effect created by a force about an axis; force times moment arm; expressed as units of force times units of length or as newton-meters in SI; a vector quantity.

torsion load—Torques in opposite directions acting on either end and around the longitudinal axis of an object produce shear stresses in the analysis plane that become larger with increased distance from the longitudinal axis; as a result of a torsion load, an object will deform by twisting.

toughness—Ability of a material to absorb energy before failing.

trabecular bone—See cancellous bone.

transverse axis—Any of the imaginary lines running from left to right and perpendicular to the sagittal planes; horizontal axis; frontal axis; mediolateral axis; frontal-transverse axis.

transverse plane—An imaginary plane running from side to side and anterior to posterior that divides the body into superior and inferior parts; horizontal plane.

trapezius—The broad superficial muscle of the upper back that originates from the vertebral column (cervical and thoracic) and attaches to the spine of the scapula.

traumatic injury—An injury caused by a single application of a relatively high level of stress; an accidental injury.

turbulent flow—Movement of fluid molecules such that adjacent layers of fluid do not flow parallel to one another and they separate from the surface of the object in a fluid.

type I muscle fiber—Smaller-diameter muscle fiber characterized by aerobic metabolism, slow development of maximum tension, smaller maximum tension, and longer duration of tension development (endurance); first to be recruited; also referred to as slow-twitch oxidative (SO) fiber.

type IIA muscle fiber—Larger-diameter muscle fiber characterized by aerobic and anaerobic metabolism, faster development of maximum tension, larger maximum tension, and long duration of tension development (endurance); second to be recruited; also referred to as fast-twitch oxidative-glycolytic (FOG) fiber.

type IIB muscle fiber—Largest-diameter muscle fiber characterized by anaerobic metabolism, fastest development of maximum tension, largest maximum tension, and shortest duration of tension development (low endurance); last to be recruited; also referred to as fast-twitch glycolytic (FG) fiber.

ulna—The long bone that lies on the medial side of the forearm; its proximal end is larger than its distal end.

ulnar deviation—Starting from anatomical position, the joint action that occurs around an AP axis through the wrist joint and causes hand movement in a frontal plane toward the little finger; the opposite of radial deviation; also referred to as adduction or ulnar flexion.

ultimate strength—Maximum stress a material can withstand.

upward rotation—Rotation of the scapula in a frontal plane such that its medial border moves inferiorly and the shoulder joint moves superiorly; the opposite of downward rotation.

vector—Mathematical representation of any quantity that is defined by its size or magnitude (a number) and its direction (its orientation); vector quantities may be represented graphically by arrows, with the length of the arrow scaled to represent the magnitude of the vector and the shaft and head of the arrow oriented to represent the direction and sense of the vector along that direction.

vector components—Vectors that can be added together to form another vector are referred to as components of the vector sum.

viscosity—Measure of resistance of a fluid to shear forces; measure of internal friction between layers of molecules of fluid.

watt—SI unit of measure for power; equal to 1 J/s.

weight—Measure of the force of gravity acting on an object; mass times the acceleration due to gravity; expressed as newtons in SI.

Wolff's law—A tissue adapts to the level of stress imposed on it; the level of adaptation in a tissue reflects the level of typical loading.

work—The product of force exerted on an object times the displacement of the object at the point of application of the force along the line of action of the force; expressed as units of force times units of length or as joules in SI; a scalar quantity.

yield strength—Stress above which plastic deformation will occur.

Young's modulus—See elastic modulus.

References
and Suggested Readings

Abott, A.V., and Wilson, D.G. (Eds.). (1996). *Human-powered vehicles*. Champaign, IL: Human Kinetics.

Adrian, M.J. (1980). The true meaning of biomechanics. In J.M. Cooper and B. Haven (Eds.), *Proceedings of the Biomechanics Symposium* (pp. 14-21). Indianapolis: Indiana State Board of Health.

Alexander, R.M. (1992). *The human machine*. New York: Columbia University Press.

Arend, S., and Higgins, J.R. (1976). A strategy for the classification, subjective analysis and observation of human movement. *Journal of Human Movement Studies*, 2:36-52.

Aristotle. (1912). *De motu animalium* (A.S.L. Farquharson, Trans.) In J.A. Smith and W.D. Ross (Eds.), *The works of Aristotle* (Vol. V, pp. 698-704). Oxford: Clarendon Press.

Atwater, A.E. (1980). Kinesiology/biomechanics: Perspectives and trends. *Research Quarterly for Exercise and Sport*, 51:193-218.

Bennell, K.L., Malcolm, S.A., Wark, J.D., and Brukner, P.D. (1996). Models for the pathogenesis of stress fractures in athletes. *British Journal of Sports Medicine*, 30(3):200-204.

Brancazio, P.J. (1984). *Sports science: Physical laws and optimum performance*. New York: Simon and Schuster.

Braun, G.L. (1941). Kinesiology: From Aristotle to the twentieth century. *Research Quarterly*, 12:163-173.

Brody, D.M. Running injuries: Prevention and management. *Clinical Symposia*, 39(3). New Jersey: Ciba-Geigy Corporation.

Brown, E.W. (1982). Visual evaluation techniques for skill analysis. *Journal of Physical Education, Recreation and Dance*, 53(1):21-26, 29.

Brown, R.M., and Councilman, J.E. (1971). The role of lift in propelling swimmers. In J.M. Cooper (Ed.), *Selected topics on biomechanics—Proceedings of the C.I.C. Symposium on Biomechanics* (pp. 179-188). Chicago: Athletic Institute.

Brüggemann, G.P., and Glad, B. (1989). Time analysis of sprint events. In G.P. Brüggemann and B. Glad (Eds.), *International Amateur Athletic Federation scientific research project at the Games of the XXIVth Olympiad, Seoul 1988, final report* (pp. 11-89). Monaco: International Athletic Foundation.

Bunn, J. (1955). *Scientific principles of coaching*. Englewood Cliffs, NJ: Prentice-Hall.

Cavanagh, P.R. (1990). The mechanics of distance running: A historical perspective. In P.R. Cavanagh (Ed.), *Biomechanics of distance running* (pp. 1-34). Champaign, IL: Human Kinetics.

Cureton, T.K., Jr. (1930). Mechanics and kinesiology of the crawl flutter kick. *Research Quarterly*, 1(4):93-96.

Cureton, T.K., Jr. (1939). Elementary principles and techniques of cinematographic analysis. *Research Quarterly*, 10(2):3-24.

Damask, A.C., and Damask, J.N. *Injury causation analyses: Case studies and data sources*. Charlottesville, VA: Michie Co.

Fenn, W.O. (1930). Frictional and kinetic factors in the work of sprint running. *American Journal of Physiology*, 92:583-611.

Fenn, W.O. (1931a). Work against gravity and work due to velocity changes in running. *American Journal of Physiology*, 93:433-462.

Fenn, W.O. (1931b). A cinematographic study of sprinters. *The Scientific Monthly*, 32:346-354.

Frey, C. (1997). Footwear and stress fractures. *Clinics in Sports Medicine*, 16(2):249-257.

Grimston, S.K., Engsberg, J.R., Kloiber, R., and Hanley, D.A. (1991). Bone mass, external loads, and stress fractures in female runners. *International Journal of Sport Biomechanics*, 7:293-302.

Grimston, S.K., Willows, N.D., and Hanley, D.A. (1993). Mechanical loading regime and its relationship to bone mineral density in children. *Medicine and Science in Sports and Exercise*, 25(11):1203-1210.

Haapasalo, H., Sievanen, H., Kannus, P., Heinonen, A., Oja, P., and Vuori, I. (1996). Dimensions and estimated mechanical characteristics of the humerus after long-term tennis loading. *Journal of Bone and Mineral Research*, 11(6):864-872.

Hall, S.J. (1995). *Basic biomechanics* (2nd ed.). St. Louis: Mosby.

Hamill, J., and Knutzen, K.M. (1995). *Biomechanical basis of human movement*. Baltimore: Williams and Wilkins.

Hatze, H. (1974). The meaning of the term 'Biomechanics.' *Journal of Biomechanics*, 7:189-190.

Hay, J.G. (Spring 1983, No. 10). Biomechanics of sport—exploring or explaining (Part I). *International Society of Biomechanics Newsletter*, pp. 9-12.

Hay, J.G. (Summer 1983, No. 11). Biomechanics of sport—exploring or explaining (Part II). *International Society of Biomechanics Newsletter*, pp. 5-9.

Hay, J.G. (1984). The development of deterministic models for qualitative analysis. In R. Shapiro and J.R. Marett (Eds.). *Proceedings: Second National Symposium on Teaching Kinesiology and Biomechanics in Sports* (pp. 71-83). Colorado Springs: NASPE.

Hay, J.G., and Reid, J.G. (1982). *The anatomical and mechanical bases of human movement*. Englewood Cliffs, NJ: Prentice-Hall.

Hill, A.V. (1928). The air resistance to a runner. *Proceedings of the Royal Society*, B, 102:43-50.

James, S.L., Bates, B.T., and Osternig, L.R (1978). Injuries to runners. *American Journal of Sports Medicine*, 6(2):40-50.

Jenkins, D.B. (1991). *Hollinshead's functional anatomy of the limbs and back* (6th ed.). Philadelphia: W.B. Saunders.

Knudson, D., and Morrison, C. (1997). *Qualitative analysis of human movement*. Champaign, IL: Human Kinetics.

Kreighbaum, E.F., and Smith, M.A. (Eds.). (1995). *Sports and fitness equipment design*. Champaign, IL: Human Kinetics.

Lane, F.C. (1912). One hundred and twenty-two feet a second. *The Baseball Magazine*, 10(2):22-25, 104, 106, 110.

LeVeau, B.F. (1992). *Williams & Lissner's biomechanics of human motion* (3rd ed.). Philadelphia: W.B. Saunders.

Maffulli, N., and King, J.B. (1992). Effects of physical activity on some components of the skeletal system. *Sports Medicine*, 13(6):393-407.

Marey, E.J. (1972). *Movement* (E. Pritchard, Trans.). New York: Amo. (Reprint edition; original translation published 1895 by D. Appleton Co., New York.)

McCaw, S.T. (1992) Leg length inequality: implications for running injury prevention. *Sports Medicine*, 14(2):422-429.

McClay, I., and Manal, K. (1997). Coupling parameters in runners with normal and excessive pronation. *Journal of Applied Biomechanics*, 13:109-124.

McNitt-Gray, J. (1991). Kinematics and impulse characteristics of drop landings from three heights. *International Journal of Sport Biomechanics*, 7:201-224.

McNitt-Gray, J., Yokoi, T., and Millward, C. (1993). Landing strategy adjustments made by female gymnasts in response to drop height and mat composition. *Journal of Applied Biomechanics*, 9:173-190.

McPherson, M.N. (1988). The development, implementation, and evaluation of a program designed to promote competency in skill analysis. *Dissertation Abstracts International*, 48:3071A.

Messier, S.P., and Pittala, S.P. (1988). Etiologic factors associated with selected running injuries. *Medicine and Science in Sports and Exercise*, 20(5):501-505.

Nelson, R.C. (1970). Biomechanics of sport: An overview. In J.M. Cooper (Ed.), *Selected topics on biomechanics—Proceedings of the C.I.C. Symposium on Biomechanics* (pp. 31-37). Chicago: Athletic Institute.

Nelson, R.C. (1980). Biomechanics: Past and present. In J.M. Cooper and B. Haven (Eds.), *Proceedings of the Biomechanics Symposium* (pp. 4-13). Indianapolis: Indiana State Board of Health.

Newton, I. (1934). *Principia* (Vol. I-II, Andrew Motte's translation revised by Florian Cajoari). Berkeley: University of California Press. (Original work published 1686, Motte's English translation, 1729.)

Nigg, B.M. (Ed.) (1986). *Biomechanics of running shoes*. Champaign, IL: Human Kinetics.

Nordin, M., and Frankel, V.H. (1989). *Basic biomechanics of the musculoskeletal system* (2nd ed.). Philadelphia: Lea and Febiger.

Norman, R.W. (1977). An approach to teaching the mechanics of human motion at the undergraduate level. In C.J. Dillman and R.G. Sears (Eds.), *Proceedings: Kinesiology, A National Conference on Teaching* (pp. 113-123). Champaign, IL: University of Illinois.

Rodgers, M.M. (1993). Biomechanics of the foot during locomotion. In Grabiner, M.D. (Ed.). *Current issues in biomechanics*. Champaign, IL: Human Kinetics.

Scott, S.H., and Winter, D.M. (1990). Internal forces at chronic running injury sites. *Medicine and Science in Sports and Exercise*, 22(3):357-369.

Steindler, A. (1935). *Mechanics of normal and pathological locomotion in man*. Springfield, IL: Charles C Thomas.

Viano, D.C., King, A.I., Melvin, J.W., and Weber, K. (1989). Injury biomechanics research: an essential element in the prevention of trauma. *Journal of Biomechanics*, 22(5):403-417.

Webster's New World Dictionary (3rd Coll. ed.). (1988). New York: Simon and Schuster.

Westfall, R. (1993). *The life of Isaac Newton*. Cambridge: Cambridge University Press.

Whitt, F.R., and Wilson, D.G. (1982). *Bicycling science* (2nd ed.). Cambridge, MA: MIT Press.

Williams, K.R. (1993). Biomechanics of distance running. In Grabiner, M.D. (Ed.). *Current issues in biomechanics*. Champaign, IL: Human Kinetics.

Williams, K.R. (1985). Biomechanics of running. In Terjung, R.L. (Ed.) *Exercise and Sport Science Reviews*, 13.

Woodburne, R.T. (1978). *Essentials of human anatomy* (6th ed.). New York: Oxford University Press.

Yamada, H. (1970). *Strength of biological materials*. Baltimore: Williams and Wilkins.

Zatsiorsky, V.M. (1978). The present and future of the biomechanics of sports. In F. Landry and W.A.R. Orban (Eds.), *Biomechanics of Sports and Kinanthropometry* (pp. 11-17). Miami: Symposia Specialists, Inc.

Zebas, C., and Chapman, M. *Prevention of sports injuries: A biomechanical approach*. Dubuque, IA: Eddie Bowers.

Zernicke, R.F., Garhammer, J., and Jobe, F.W. (1977). Human patellar-tendon rupture. *The Journal of Bone and Joint Surgery (American)*, 59-A(2):179-183.

Index

A

abdomen 18-19
abduction 30, 32-35
absolute angular position 159, 167
acceleration 44, 90-93, 100-105, 166-167, 207
activity, effect of 230-231, 352-353
adduction 30, 32-35
adductor longus muscle 22
age, effect of 230-231
agonist muscles 262
airfoils 204-205
anatomical analysis. *See* qualitative anatomical analysis
anatomical axes 26-29
anatomical planes 25-29
anatomical position 24
angles 158-159, 161
angular acceleration 166-167, 328-329, 334, 337, 342
angular displacement 158-163, 169
angular impulse 184-185, 188
angular inertia 172-178, 186
angular kinetics 171-188
angular momentum 178-185
    angular impulse and 184-185
    human body and 179-180
    Newton's laws and 180-187
    overview 186-188
    rigid body 178-179
angular motion 77
angular position 159, 167
angular velocity 163-166, 181-182, 187
anisotropic materials 230
ankle 19-20, 30-32, 250, 330-339
antagonist muscles 263
anterior, definition of 24
anteroposterior (AP) axis 26-27, 30-33, 35
aponeuroses 257
Archimedes 194
arm curls 134-135, 138-139
arms 18-19. *See also* elbow
articular capsule 247
articular cartilage 232-233, 247-249, 251
autonomic nervous system 281
average acceleration 90-91
average angular velocity 163
average speed 85-89

average velocity 88-89
avulsion fracture 231
axes, anatomical 26-29, 174-175
axis of rotation 165, 174-175

B

balance 286-287
ball-and-socket joints 246-247
balls
    action-reaction and 110
    as projectiles 97-98, 100-101, 208
    spin on 206, 315-316
baseball 13, 306-310
base of support 148-154
bench presses 116, 326-329
bending loads 220-223, 225, 235
Bernoulli's principle 204-206
biceps brachii muscle 19, 22, 134-135, 262-264
biceps femoris muscle 23-24
biomechanical analysis 293-321. *See also* qualitative
      anatomical analysis
    evaluation of performance 304-305
    fastball pitch 306-310
    football punt 331-335
    forehand drive in tennis 310-316
    ideal technique 295-300
    identifying goals 296-297
    instructing performer 305-306
    observation of performance 300-304
    qualitative 294-295
    quantitative 294-295
    sprint running 316-321, 334-339
    technique improvement 295-306
    training and 324-325
    vertical jump 330-331
biomechanics 3-14
    definition 5-6
    friction and 56-57
    future of 14
    goals of 6-12
    history of 12-14
body fat 195
bones
    anatomy and classification 241-243
    elastic modulus 231
    growth and development 243-245

bones (continued)
    loads on 216-218, 220, 222-223, 230-232, 235-236
    long 242-245, 260
    marrow 242
    overview 240
    surface irregularities 242-243
    tendon attachment 259
    terminology 18-19
brittleness 229-230
buoyant force 192-196, 207
bursa 259

C
cancellous (trabecular or spongy) bone 231, 241
cardinal plane 25-26
carpals 18-20
Cartesian coordinate system 79-81, 82
cartilage 232-233, 242-249, 251
cartilaginous joints 245-246
cartwheel axis (anteroposterior axis) 26-27, 30-33, 35
catching technique 108
center of gravity
    definition 140
    in free-body diagrams 67-68
    in human body 143-145
    mathematical determination 140-143
    movement and 145-147, 151-153, 299
    stability and 147-154
center of pressure 206, 359-360
central nervous system 280-282. See also nervous system
centric force 128-129
centripetal force 166-169
children, bones in 243-244
circumduction 33
clavicle 19-21
clockwise 160
colinear forces 57-59
collagen 230, 232-234, 236, 259
combined loads 224-225, 235
compact bone 231
compression 51, 217-218, 220, 223, 235, 349-350
concentric contractions 260-262, 273-274, 276, 327
concurrent forces 59-62
condyle 242
connective tissue 230-231
conservation of angular momentum 182
conservation of mechanical energy 122-123
contact forces 52
cortical bone 231-232, 241
cosine function 65-66
counterclockwise 160
creep 232-234
cross-country skiing 8, 10-11, 56-57
cross-training 364
Cureton, Thomas 13
curvilinear translation 76-77

D
dancing, angular momentum 185
deceleration 91
deep, definition of 24
deformation 40, 216-217, 220, 223, 225-226, 233
deltoid muscle 19, 22

density 194-196, 200
depression 33, 35
Descartes, René 79
diagonal planes 29, 33
diaphysis 243
discus throwing 115, 119-120, 185
displacement 81-85, 93, 97, 115-116, 158-163, 169
distal, definition of 24
distance traveled 80-81, 83, 85. See also displacement
diving, moments of inertia 175-176
dorsiflexion 30-31, 35
downward rotation 33, 35
drafting 202, 296
drag force 197-202, 208
ductility 229-230
dynamic fluid force 196-208. See also fluid mechanics
dynamic friction 53-56
dynamics 41, 76, 96, 326

E
eccentric contraction 261-263, 273-274, 277, 286, 327
eccentric force 128-129
effective radius 165
elastic behavior 227-228, 231, 233
elastic cartilage 232
elastin 230, 233
elbow 19-21, 30-31, 250-251, 338, 343
elevation 33, 35
ellipsoidal joints 246
endolymph 286
endomysium 257, 259
epicondyle 242
epimysium 257, 259
epiphyseal cartilage 242-244
epiphysis 243
equipment 8-9, 12, 272. See also running, shoes
Erasquin, Felix 7
erector spinae muscle group 24
eversion 30, 32, 35
exercise machines 272
extension 30-31, 35
external forces 52. See also forces
external oblique muscle 22, 24
external rotation 33-35
exteroceptors 284, 287-288

F
failure strength 229, 235-236
fascicle 257, 259
fastball pitch 306-310
fatigue, muscle 275
femur 19-21, 222-225, 244
Fenn, Wallace 13
fibrous cartilage 232
fibrous joints 245-246
fibula 19-21
figure skating 10-11, 67-68, 182
fingers 19, 22-23
flexibility exercises 286
flexion 30-31, 35
fluid friction 52-53
fluid mechanics 191-208
    buoyancy 192-196

center of pressure 206
drag force 197-202
lift force 202-206
overview 40
relative velocity 196-197, 202
foot 18-19, 223, 358-362, 365
football punt 331-335
force couples 128
forces. *See also* Newton's laws of motion
  addition of 57-62
  buoyant 192-196, 207
  centric 128-129
  centripetal 166-169
  colinear 57-59
  compressive 51, 349-350
  concurrent 59-62
  contact 52
  definition of 43, 50
  drag 197-202, 208
  dynamic fluid 196-208
  eccentric 128-129
  graphical determination of 63-64
  gravitational 45-46, 52, 59
  impact 108-109, 328-329, 334, 337
  injury and 348, 350
  lift 202-206, 208
  moment of. *See* torque
  in muscles 138-139, 273-274
  net 57, 66-67, 106
  normal 53-54
  reaction 68-71, 98, 103-104, 109-110, 359-361
  resultant 57-62
  static equilibrium 67-72
  tensile 51, 266-274, 350
  torque and 136-140
  trigonometric determination of 64-67
  types of 51
  zero net 98-100
forearm 18-19
forehand drive in tennis 310-316
form drag 199-202
Fosbury, Dick 7, 9
fossa 242
fovea 242
free-body diagrams 67-71
friction 52-57
frontal plane 25-27, 35
frontal-sagittal (longitudinal) axis 26-27, 29, 33, 35
frontal (transverse) axis 26-27, 30, 34-35

G
gastrocnemius muscle 23-24, 263
general motion 77-78
gliding joints 245
gluteus maximus muscle 19, 23
gluteus medius muscle 19, 23
golf 164-165, 178-179, 206
Golgi tendon organ 286, 289
gracilis muscle 22
gravitation 45-46, 52, 59, 101, 118
groin muscle 24
gymnastics 6, 12, 59-61, 182, 185

H
hammer throwing 167-168
hamstring muscle 270-272
hand 18-19
head 18-19
heel 350
Held, Dick 8-9
Held, Frank "Bud" 8-9
high jumping 3, 7-9, 144, 298-300, 302-304
Hill, Archibald V. 13
hinge joints 245, 248
hip
  in football punt 331-333, 335
  in javelin throwing 338, 341, 344-345
  joint movement 30-34, 246
  overview 19-21
  in sprint running 338-339
  in vertical jump 330
horizontal motion 96-98, 101-102. *See also* displacement
hormones, bones and 244
human body. *See also* qualitative anatomical analysis
  angular momentum 179-181
  axes 26-29, 174-175
  buoyancy 195-196
  center of gravity 143-145
  mechanical properties 230-235
  moments of inertia 175-176
  response to stress 350-353
humerus 18, 20-21, 216-218
hurdling 182-183
hyaline cartilage 232
hyperextension 30-31, 35

I
ice hockey 296-297
ice skating 10-11, 67-68, 182
impact force 108-109, 328-329, 334, 337
impulse 105-109, 184-185, 187-188
impulse-momentum equation 105-108, 110
impulsive load 360
inertia 42-44, 96-100, 172-179, 186, 188
inferior, definition of 24
injury 347-367
  biomechanics and 11-12, 235, 304
  cross-training and 364
  factors affecting 354-358
  individual differences and 354, 356
  loading patterns and 359-360
  mechanical stress and 348-351
  overuse 352-353, 358-367
  psychological considerations 366-367
  traumatic 353
instantaneous acceleration 91
instantaneous angular velocity 163
instantaneous speed 88
instantaneous velocity 89, 318
internal forces 51
internal rotation 33-35
International Amateur Athletic Federation (IAAF) 7-9
interneurons 282
interoceptors 284
inverse dynamic analysis 326

inversion 30, 32, 35
iron cross 10-11
isometric activity 273-274, 276-277, 327
isotropic materials 230

J

javelin throwing 7-8, 206-207, 338-342
Johnson, Ben 86-87
Johnson, Walter 13
joints. *See also specific joint*
    analysis of motions 29-35, 326-329, 330-332
    angular acceleration 328-329, 334, 337, 342
    cavities 248-249, 294
    classification 244-247
    flexibility 249-251
    javelin throw and 338
    range of motion 249-250, 252, 329, 334-335, 337-339, 342
    stability 248-252
    terminology 18-19
joules 115-116
jumping
    anatomical analysis 330-331
    center of gravity 145-147
    high jump 3, 7-9, 144, 298-300, 302-304
    landing techniques 108-109
    long jump 301-302
    momentum 106-108
    prestretch and 274

K

kinesiology 12-14
kinetic energy 117-118
kinetics 41, 96
knee
    in football punt 331-333, 335
    injuries 350
    joint movement 30-31, 249
    overview 19-21
    running and 361-362, 365-366
    in sprint running 334-335, 337, 339
    tensile loading 250
    in vertical jump 330
Koch, Bill 8

L

laminar flow 199-200, 202
lateral, definition of 24
lateral flexion 30, 32-33, 35
lateral (transverse) axis 26-27, 34-35
latissimus dorsi muscle 23-24
legs 18-19, 136
length 42-43
Lewis, Carl 86-87
lift force 202-206, 208
ligaments 231-234, 249-251
linear momentum 105, 187
linear strain 225-226
load
    bending 220-223, 225, 235
    on bones 216-218, 220, 222-223, 230-236
    combined 224-225, 235
    impulsive 360
    injury and 359-360
    mechanical 219-224
    during running 358-362
    shear 219
    strength increase from 230-231
    torsion 223-224
    uniaxial 219, 235
loading rate 231-232
longitudinal axis 26-27, 29, 33, 35
longitudinal fibers 261, 266-270, 276
long jump 301-302

M

Magnus effect 205-207
Marey, Etienne Jules 13
mass 42-43, 364-365. *See also* weight
mechanics 39-46. *See also* biomechanical analysis;
        biomechanics
medial, definition of 24
mediolateral (transverse) axis 26-27, 30, 34-35
men, center of gravity 143
metacarpals 18, 20
metatarsals 19-20
modulus of elasticity 228, 233
moment arm 131, 134-135, 150
moment of force. *See* torque
moment of inertia 172-179, 188
momentum 105-110, 187. *See also* angular momentum
motion 76-78
motor neurons 282-286
movement, center of gravity and 151-153
muscles 255-277
    contractions
        concentric 260-263
        eccentric 261-263
        in football punt 335
        in javelin throwing 338-345
        mechanics of 264
        muscle length and 268-271
        qualitative anatomical analysis of 327-328, 333-334
        recruitment of 284
        in sprint running 334-337, 339
        types of 260-262
        velocity of 271-274
    effect of inactivity 352-353
    fibers
        contractions and 265
        fatigue and 275
        longitudinal 261, 266-270, 276
        nervous system and 282-283
        pennate 261, 266-270, 276
        structure of 257-261
        Type I/II 275
    force and 138-139, 265-276, 283
    length of 268-271
    line of pull 260
    maximum tension in 270, 284
    mechanical properties of 234-236
    moment arms in 162-163
    nervous system and 282-283
    overview 19-24
    power output and velocity 271-274, 276-277

qualitative analysis of 325-329
range of motion and 252
roles of 262-264
single- *vs.* multiple-joint 270-271
spindles 285-286
structure of 256-260, 265
torque in 132-134, 186, 262-263, 276
types of 256
myofibrils 257-259, 265
myofilaments 257-258, 265

N

neck 18-19, 31-34, 242
nervous system 279-289
exteroceptors 287-288
motor unit 282-284, 288
muscle contractions and 264
neurons 280-282, 288
proprioceptors 285-287, 289
receptors and reflexes 284-285
net force 57, 66-67, 106
neurons 281-282, 288
neutralizer muscles 263-264
newton meters 131
newtons 50
Newton, Sir Isaac 44
Newton's laws of motion
action-reaction 109-110
angular momentum 180-187
overview 44-46
projectile motion 98-100, 102-105
static equilibrium and 67
Nike Sport Research Laboratory 12
Nordic ski racing 8, 10-11
normal contact force 53-54

O

O'Brien, Parry 122
otoliths 286

P

pacinian corpuscles 287, 289
pain receptors 287-288
pascals 215, 349
patellofemoral pain 362
pectineus muscle 22
pectoralis major muscle 22, 24
pelvis 19-21
pennate fibers 261, 266-270, 276
performance
biomechanics and 6-11, 298-300
center of gravity and 145-147
evaluation of 304-305
observation of 300-304
perimysium 257, 259
peripheral nervous system 280-281
phalanges 18-20
physiologic loading zone 351-352
pitching 306-310
pivot joints 245-246
planes, anatomical 25-29
plantar flexion 30-31, 35, 361-362
plastic behavior 228-229

Poisson's ratio 226-227
pole vaulting 10-11, 139-140, 144, 175-177
position 24, 78-81, 158-161, 167
posterior, definition of 24
potential energy 117-119, 149-152
power 123-125, 271-274, 276-277
pressure 193-194, 206, 294, 349, 359-360
prestretch 274
principal axes 174-175
projectile motion 96-98, 100-102, 207-208
pronation 33-35, 361-362
proprioceptors 284-287
proximal, definition of 24
proximal radioulnar joint 19-20
pubic symphysis 245-246
Pythagorean theorem 62, 64

Q

Q angle 365-366
quadriceps femoris 263
qualitative anatomical analysis
football punt 331-335
javelin throwing 338-345
overview 325-329, 342
sprint running 334-339
vertical jump 330-331
quantum mechanics 41

R

radial deviation 30, 32, 35
radians 160, 162
radioulnar joint 19-20, 34
radius 18, 20-21, 165
radius of gyration 174, 186
range of motion 249-250, 252, 329, 334-335, 337-339, 342
reaction forces 68-71, 98, 103-104, 109-110, 359-361
rearfoot motion 361
reciprocal inhibition 286
rectilinear translation 76
rectus abdominis muscle 22, 24
rectus femoris muscle 22, 24
reference frames 78-80
reflexes 284-289
relativistic mechanics 40-41
remodeling 352-353
resultant displacement 81, 83-85
resultant force 57-62
ribs 19-21
right-hand thumb rule 160
righting reflex 287, 289
rigid-body mechanics 40-41, 46
rotary (angular) inertia 172-178, 186
running
anatomical analysis 334-338
angular momentum 180
average speed 86-89
biomechanical analysis 316-321
body mass 364-365
effect of road slope 357
foot structure and 365
knees and 365-366
loading patterns 358-362
overuse injuries 358-367

running (continued)
    position-time graph 361
    Q angle 365-366
    shoes 12, 56, 358-359, 361, 364-365
    speed and injury 362-363
    surfaces 363-364
Ryan, Nolan 90

S
saddle joints 246
sagittal (anteroposterior) axis 26-27, 30-33, 35
sarcolemma 257, 259
sarcomeres 257-258, 265, 268-269, 272-273, 276
sartorius muscle 22, 24
scapula 18, 21
segments of the body 18-19
self-experiments
    acceleration 100, 103
    angular motion 162, 165, 173, 181, 187
    bending loads 221
    buoyant force 193
    center of gravity 141-142, 145
    deformation 233
    elastic behavior 228
    energy 117-119
    friction 53-56
    impulse and momentum 106-108
    joint motion 251
    lift force 203
    linear motion 77-78
    linear strain 225
    moment of inertia 174
    muscle contraction force 265-267, 272, 274
    net force 104
    projectile motion 97, 102
    reaction force 98-99, 104, 110
    speed, velocity, and acceleration 89, 91-92
    stress 215, 218
    stretch reflex 285
    torque 129-130, 132, 135, 137-138
    work 114
semimembranosus muscle 23-24
semitendinosus muscle 23-24
sensory neurons 281
sesamoid bones 242
shear strain 226
shear stress 217-220, 224, 235, 350
shock absorbance 122
shoes 12, 56, 358-359, 361, 364-365
shotput 106-107, 121-122
shoulder
    abduction 27-28
    in javelin throwing 338, 341, 343
    joint movement 30-34, 246, 330-331
    overview 19-20
    in sprint running 337, 339
sine function 65-66
skating 10-11, 67-68, 182
skeletal system 239-252. See also bones; joints
skiing 8, 10-11, 56-57
skull 19-20, 246
soleus muscle 23-24

somatic nervous system 281
space travel 352
specific gravity 194-195
speed 85-90, 164
spin, ball 205-206
spinning actions 185
sport biomechanics. See biomechanical analysis;
        biomechanics
sprinting 175-176, 316-321, 334-339. See also running
stability 147-154, 151-153, 248-252
stabilizer muscles 263
stances 148-149, 152-154
static equilibrium 67-72, 136-140
static friction 53-56
statics 41
Steindler, Arthur 13
stiffness 234-235
stimulus duration 274-275, 284
strain 118-119, 225-227, 235
streamlined shapes 200-202, 204
strength 229-231, 235-236, 249-250
strength-training 134-136
stress 214-224
    compression 217
    continuum 351
    definition of 214-215, 235, 349
    in feet of runners 358-359
    injury and 348-354
    relaxation of 232-234
    shear 217-220, 224, 235, 350
    stress-strain relationship 227-230, 232, 234-235
    tension 215-216, 220, 230, 235
    tissue response to 350-353
    Williams's model of 352-353
stretching exercises 234-235
stretch reflex 285-286, 289
superficial, definition of 24
superior, definition of 24
supination 33-35, 361-362
surface area 54-55, 196
surface drag 198-199, 202
surface texture 200
swimming 7, 195, 203
swinging 71
synergy 264
synovial joints 245-251
Systeme-Internationale d'Unités (SI units) 42-43, 46

T
tangent function 65
tangential acceleration 166, 169
tarsals 19-20
technique improvement 6-8, 12, 295-306. See also
        performance
tendon reflex 286
tendons 231, 233-235, 249, 257-259
tennis 310-316
tensile forces 51, 266-274, 350
tensile strength 249-250
tensile stress 215-216, 220, 230, 235
tension 215-216, 223
thigh 18-19

thorax 18-19
throwing sports 106-108, 121-122, 306-310
tibia 19-21
tibialis anterior muscle 22, 24
tightrope walking 186
time 42-43, 326
tonic neck reflex 287-289
torque
    angular momentum 183-188
    examples of 131-133
    force and 136
    muscular 132-134, 186, 262-263, 276
    net 137-140
    overview of 128-131, 153
torsion load 223-224
toughness 229, 235
training 323-345. *See also* qualitative anatomical
        analysis
    cross-training 364
    improvement of 9-11
    physical 325
    technical 324-325
training effect 352-353
trajectories 96-98, 100-102, 207-208
transverse axis 26-27, 30, 34-35
trapezius muscle 23-24
triceps brachii muscle 19, 23
trigonometry 64-67, 84-85, 369-378
trunk 31-34, 340-341, 344
turbulent flow 199-200, 202
Type IIA muscle fibers (FOG) 275
Type IIB muscle fibers (FG) 275
Type I muscle fibers (SO) 275

U
ulna 18, 20-21
ulnar deviation 30, 32, 35
ultimate strength 229, 235-236
uniaxial loads 219, 235
uniaxial tension 224
units 42-43, 46
upward rotation 33, 35

V
vastus intermedius muscle 24
vastus lateralis muscle 22, 24

vastus medialis muscle 22, 24
vectors
    acceleration 91-93
    addition of 57-64
    angular velocity 57-64, 163
    components 96-98
    definition of 50
    displacement and 81, 83-85
    torque 131, 139-140
    velocity 88-89
velocity
    angular 163-166, 181-182, 187
    average 88-89
    of fastball pitch 307
    injury and 362
    instantaneous 89, 318
    linear 163-166
    moment of inertia and 177-178
    muscle contraction and 271-274
    overview 85, 88-90
    during performance 303
    relative 196-197, 202, 208
vertebral column 19-21, 227
vertical (longitudinal) axis 26-27, 29, 33, 35
vertical motion 100-102
vestibular system 286-287, 289
viscosity 198-199, 202

W
water pressure 193-194
watts 123-124
weight 43, 59, 69-70, 364-365
weight-lifting 69-70, 98-99, 135-136, 273-274. *See also*
        arm curls; bench presses
withdrawal reflex 288
Wolff's law 350-352, 365
women, center of gravity 143
work 113-117, 119-125
wrestling 132, 134, 154
wrist 19-23, 31-32

Y
yield strength 229, 235
Young's modulus 228, 233

Z
zero net force 98-100

About the Author

Andrew Alindogan

Peter M. McGinnis is an associate professor in the Physical Education Department at the State University of New York, College at Cortland, where he has taught since 1990. The students at Cortland named McGinnis the "Outstanding Faculty Member" for the 1996-97 academic year.

From 1987 to 1990 McGinnis was an assistant professor in the Department of Kinesiology at the University of Northern Colorado. During that time he served as a sport biomechanist in the Sports Science Division of the U.S. Olympic Committee in Colorado Springs, where he conducted applied sports biomechanics research, tested athletes, taught biomechanics courses to coaches, and developed educational materials for coaches.

McGinnis is the author of numerous articles and technical reports about the biomechanics of pole vaulting as well as other specific athletic events, including running, pole vaulting, skiing, rowing, and swimming.

The author is a sport scientist for USA Track and Field. In addition he is the biomechanist on the Pole Vault Development Committee of USA Track and Field, which in January of 1996 recognized him for making "outstanding contributions to the development of the pole vault."

McGinnis was an instructor and assistant professor in the Department of Physical Education and Human Movement Studies at the University of Oregon from 1983 to 1987.

McGinnis is a member of a number of professional organizations including the American Alliance for Health, Physical Education, Recreation and Dance; the American Association of Physics Teachers; and the International Society of Biomechanics in Sports.

McGinnis received a PhD in physical education from the University of Illinois in 1984. He and his wife Boodie live in Cortland, New York.